CLIMATE CHANGE POLICY

CATRINUS J. JEPMA received an M.A. in law and a Ph.D. in economics from the University of Groningen. Presently, he is a Professor of Economics at the Netherlands Open University and at the University of Groningen, and holds a special chair in international environmental economics at the University of Amsterdam. Dr. Jepma has served as an adviser to the Netherlands government, the Organization for Economic Cooperation and Development, the UN Food and Agriculture Organization, the Secretariat of the UN Framework Convention on Climate Change, and the Intergovernmental Panel on Climate Change. He has (co)authored more than thirty books and monographs, and published more than a hundred articles and papers. He is editor in chief of *Joint Implementation Quarterly* and has lectured on international economics for twenty years at the University of Groningen and other institutions.

MOHAN MUNASINGHE received a B.A. and an M.A. in engineering from Cambridge University, an S.M. and an E.E in electrical engineering from the Massachusetts Institute of Technology, a Ph.D. in physics from McGill University, and an M.A. in economics from Concordia University. Presently, he is Distinguished Visiting Professor of Environmental Management at the University of Colombo; Senior Adviser to the Minister of Environment, Sri Lanka; and President, Lanka International Forum on Environment and Sustainable Development. He has been Division Chief for Environmental Policy and for Energy at the World Bank. He has served as Senior Adviser to the President of Sri Lanka and Adviser to the U.S. President's Council on Environmental Quality. He has won several international research awards and authored more than sixty books and many technical papers. He is a fellow of several leading science academies and serves on the editorial boards of many international journals.

CLIMATE CHANGE POLICY

FACTS, ISSUES, AND ANALYSES

CATRINUS J. JEPMA

University of Groningen,
University of Amsterdam, and
Netherlands Open University

MOHAN MUNASINGHE

University of Colombo, Sri Lanka, and
Beijer Institute of Ecological Economics,
Royal Swedish Academy of Sciences, Stockholm

Sponsored by the government of the Netherlands
Cosponsored by the Stockholm Environment Institute
and the Beijer Institute of Ecological Economics

PUBLISHED BY THE PRESS SYNDICATE OF THE UNIVERSITY OF CAMBRIDGE
The Pitt Building, Trumpington Street, Cambridge CB2 1RP, United Kingdom

CAMBRIDGE UNIVERSITY PRESS
The Edinburgh Building, Cambridge CB2 2RU, United Kingdom
40 West 20th Street, New York, NY 10011-4211, USA
10 Stamford Road, Oakleigh, Melbourne 3166, Australia

First published 1998

Printed in the United States of America

Typeset in New Baskerville

Library of Congress Cataloging-in-Publication Data
Jepma, C. J.
Climate change policy : facts, issues, and analyses / Catrinus J.
Jepma, Mohan Munasinghe.
p.   cm.
Includes index.
ISBN 0-521-59314-X. – ISBN 0-521-59688-2 (pbk.)
1. Climatic changes.   2. Climatic changes – Decision making.
3. Greenhouse effect, Atmospheric.   I. Munasinghe, Mohan.
II. Title.
QC981.8.C5J47   1998
551.6 – dc21   97-7559
                                                  CIP

*A catalog record for this book is available from the
British Library.*

ISBN 0 521 59314 X hardback
ISBN 0 521 59688 2 paperback

To our children,
Anusha, Jitske, Ranjiva, and Sietske,
and their progeny,
who may face more serious challenges
in the world they inherit

CONTENTS

FOREWORD

BERT BOLIN AND ROBERT WATSON
Chair and Chair-Elect, IPCC

The work of the Intergovernmental Panel on Climate Change (IPCC) has helped to draw the attention of policy makers and the general public alike to the unprecedented challenge posed by global climate change. The IPCC was created in 1988 by the World Meteorological Organization (WMO) and the United Nations Environment Program (UNEP) to (a) assess scientific information on climate change; (b) evaluate the environmental and socioeconomic effects of climate change; and (c) determine response strategies. The First Assessment Report published in 1990 and the Second Assessment Report released in 1996 are the most authoritative and comprehensive works on climate change, which include contributions from several thousand of the world's leading scientists and experts as both authors and reviewers.

Climate change is a complex and difficult topic. At the same time, the work of the IPCC, as well as the writings of other researchers, have resulted in a flood of information that is rather complicated even for experts to keep abreast of. Therefore, this well-written synthesis volume by Professor Jepma and Professor Munasinghe, which pulls together the most important policy issues, is most welcome. Both men are eminently qualified to write on the subject, having served as convening lead authors of various chapters of the Second Assessment Report (SAR). The book, which draws on the SAR as well as other recent research, is lucid, stimulating, and up-to-date.

This volume reviews the key scientific data on climate change, the impact on ecological and socioeconomic systems, the range of decision-making tools and processes, potential response options, and areas for further work. While acknowledging the complexity of the problem and the formidable analytical difficulties involved, the authors respond to questions such as when, where, and how to initiate action, what information is required, and, most important, how humanity might organize itself to make and implement rational decisions to deal with climate change. The book will therefore be a very useful and convenient resource for decision makers, policy analysts, climate change researchers, students, and the concerned public. The authors have confirmed their reputations as leading authorities on climate change policy by producing a concise, balanced, and highly readable volume.

FOREWORD

JAMES P. BRUCE AND HOESUNG LEE

Co-chairs, Working Group III,
IPCC Second Assessment Report

The conclusion of the Second Assessment Report of the Intergovernmental Panel on Climate Change (IPCC), that "the balance of evidence suggests a discernible human impact on the global climate," has heightened public interest in issues surrounding human-induced climate change. The amount of information now available on climate change is formidably extensive and growing rapidly. The four 1995 reports of the IPCC are a compendium and assessment of climate science, the effects of climate change, potential response strategies, and the economic and social dimensions of the issue. The IPCC reports were written or reviewed by several thousand of the world's leading climate scientists, economists, engineers, and other specialists from many countries. There are some two thousand pages of text, which include both technical summaries and summaries for policy makers. The IPCC volumes exist in addition, of course, to the rapidly growing scientific and popular literature on the alteration of the climate system by human-induced emission of greenhouse gases and aerosols.

For interested lay people, or even specialists on specific aspects of the issue, there is simply too much to absorb. Thus, this volume, *Climate Change Policy: Facts, Issues, and Analyses,* was born. The two authors, Catrinus Jepma (of the Netherlands) and Mohan Munasinghe (of Sri Lanka), were convening lead authors who directed the writing of major chapters of the IPCC volume entitled *Climate Change 1995: Economic and Social Dimensions of Climate Change* (Cambridge University Press, 1996). While the emphasis in this present book reveals their backgrounds as prominent economists, they have also provided a concise overview of the scientific basis for concern about rapidly increasing global emissions of greenhouse gases and the potential impact of resulting climate change on ecosystems and socioeconomic systems.

Jepma and Munasinghe have not hesitated to point out where our knowledge is limited, requiring more research and analysis. Nevertheless, they argue convincingly that we have sufficient knowledge to take actions now to reduce greenhouse gas emissions and to modify those actions as new information becomes available.

In the last chapter, they venture some conclusions, placing them in the context of the United Nations Framework Convention on Climate Change, originally signed in 1992 and now ratified by almost all countries. Equity issues, among countries and among generations, are given special attention.

The authors make it clear that addressing the climate change issue poses unprecedented challenges for the international community. However, they also show how meeting these challenges will represent a major step toward sustainable development worldwide.

For serious students of specific aspects of the climate change problem, this volume does not replace the detailed analyses and assessments available in the IPCC volumes. It should, however, serve as an invaluable companion volume that provides guidance for understanding particular aspects of the issue. Policy makers as well as natural science and social science communities throughout the world owe a debt of gratitude to the authors for writing this essential overview.

For interested general readers, this volume will be a valuable and authoritative place to start exploring the many complexities of the issues surrounding the impact of humans on the global climate system.

ACKNOWLEDGMENTS

The authors are sincerely grateful to all their fellow lead authors and colleagues who contributed to the IPCC Working Group III report, from which much of the information in this volume is drawn. We would also like to thank the lead authors of the Working Group I and II reports, who provided useful information. Dr. N. Sundararaman kindly arranged permission from the IPCC to reproduce key figures from the Second Assessment Report in our book.

We are especially indebted to Erik Haites and Henk Merkus for their unfailing encouragement and advice throughout the preparation of the book, and to Bert Bolin, James Bruce, Hoesung Lee, and Robert Watson for their generous contributions. Valuable assistance was forthcoming from Wytze van der Gaast, Elise Kamphuis, Thijs Knaap, Pradeep Kurukulasuriya, Sria Munasinghe, and Sanath Ranawana during the preparation of the manuscript. We also thank Elise, Wytze, and Darco Janssen for their contributions to the workbook accompanying this volume. Others who provided helpful comments and input include Michael Chadwick, Charles Feinstein, Brian Fisher, Jose Goldemberg, John Houghton, Kenneth King, Rajendra Pachauri, David Pearce, Janaka Ratnasiri, and Arno Rosemarin.

We would like to express our deep appreciation to the Netherlands Ministry of Housing, Spatial Planning and the Environment for providing the main resources to support our writing effort and to promote the dissemination of this volume in developing countries, as well as to the Stockholm Environment Institute and the Beijer Institute of Ecological Economics for their additional support.

Finally, we wish to thank Catherine Flack, Mary Racine, and others at Cambridge University Press who helped to produce this book.

1

SCOPE AND IMPLICATIONS OF CLIMATE CHANGE

The potential threat of global climate change is a very serious problem collectively faced by humanity as a result of its own activities. This book sets out, in a simple but accurate manner, the main issues and risks involved, possible solutions, and – most important – how we might organize ourselves to make, and implement, rational decisions concerning this unprecedented problem. While the focus is on the main policy issues and remedial options, the book starts from a solid foundation based on the best scientific evidence available on global climate change. Not surprisingly, the discussion touches on many inadequacies of society in dealing with generic issues that arise from the increasing globalization of actions and their consequences. Thus, the remedial processes and measures presented here will also have broad relevance to international environmental problems other than climate change.

The gravity of the climate change problem is underlined by the great degree of international attention recently paid to the subject (see Box 3.2, Chapter 3). Among the most important responses made through the United Nations system was the establishment of the Intergovernmental Panel on Climate Change (IPCC) in 1988, based on an initiative led by the World Meteorological Organization (WMO). The IPCC is a body consisting of the world's foremost experts on climate science and related fields, whose task is to provide a sound scientific basis for making decisions on the problem. Another key step was the drafting of the UN Framework Convention on Climate Change (UNFCCC), which was approved by the heads of state of more than 150 nations at the UN Earth Summit in Rio de Janeiro in June 1992.

1.1 OVERVIEW OF THE BOOK

This book deals with the ultimate objective of the UNFCCC, as stated in Article 2: ". . . stabilization of greenhouse gas concentrations in the atmosphere at a level that would prevent dangerous anthropogenic interference with the climate system. Such a level should be achieved within a time-frame sufficient to allow ecosystems to adapt naturally to climate change, to ensure that food production is not threatened, and to enable economic development to proceed in a sustainable manner" (UNFCCC 1992). In this context, the climate system includes "the totality of the atmosphere, hydrosphere, biosphere and geosphere, and their interactions" – in simple terms, the entirety of the planetary surface. Furthermore, climate change is defined as "a change of climate which is attributed directly or indirectly to human activity that alters the composition of the global atmosphere and which is in addition to natural climate variability observed over comparable time periods."

1.1.1 Broad Conclusions

The findings of the IPCC Second Assessment Report (accepted in December 1995 by member governments of the IPCC – basically the countries of the United Nations), as well as the results of other recent studies, constitute the basis for drawing the following important conclusions, which are explored in this book (for a more detailed account, see Houghton et al. 1996). The key finding is that there are serious scientific grounds for concern about the effects of human actions on the global climate since the industrial revolution. At the same time, there are many scientific, economic, social, and technological uncertainties that make it difficult to set forth detailed predictions about the future effects of global climate change. Nevertheless, climate models indicate that the worldwide impact is likely to be significantly negative overall, although some regions might actually benefit.

Given the serious risks posed by climate change, it is necessary (but equally problematic) to determine the desirable target conditions required to stabilize the global climate, as well as the most effective human actions necessary to achieve such an outcome. However, preliminary studies indicate that the ultimate stabilization of atmospheric greenhouse gas (GHG) concentrations, which are the major cause of global climate change, will require reductions in future GHG emissions below present levels. Such reductions could entail significant adjustments and costs, since the primary causes of GHG emissions – energy and land use – lie at the heart of modern economic activities.

Nevertheless, there are sound economic and scientific grounds for undertaking at least "no-regrets" measures – that is, activities such as energy conservation that would be undertaken anyway because they are relatively costless. One such reason is that if and when the effects of climate change materialize, it will take a further 50 to 100 years to stabilize GHG concentrations at levels higher than the prevailing ones, and much longer to actually reduce concentrations. More costly response measures, involving even more extensive climate change mitigation and adaptation policies, also appear to be justified.

However, the scope of such measures cannot be defined in detail until their implementation costs can be compared with the value of averted global warming damage. Under these circumstances, it would be prudent to rely on a flexible overall climate change response strategy that could be adjusted systematically in the light of new information, especially to deal with the high levels of uncertainty. Such a strategy would include a portfolio of mitigation, adaptation, and other options (such as improving knowledge) based on the coordinated application of a variety of market-based, regulatory, and other instruments.

It is clear that international cooperation is essential to achieve a globally coherent response strategy, and could both significantly reduce the costs and increase the effectiveness of GHG mitigation measures. Implementation costs might be further reduced by introducing efficient policies that provide correct economic signals through such instruments as carbon taxes, subsidies, quota schemes, and tradable emission permits, as well as international incentive programs like joint implementation schemes.

It is especially important to address equity issues, because those who are responsible for the problem and those who will suffer the consequences are different, and are separated both geographically and over time. In particular, both the collective decision-making process and the choices made should incorporate equity considerations – not only for ethical and legal reasons, but also to improve the chances of reaching international agreement and implementing decisions. Finally, research and data gathering appear to have an especially high payoff in terms of reducing uncertainty and improving the effectiveness of the response strategy.

1.1.2 Chapter Outline

The chapters in this book are organized according to a circular causal chain that begins with net emissions and atmospheric concentrations of GHGs, continues on to biogeophysical effects, as well as social, economic, and environmental damage, and then arrives at possible response strategies, which in turn will influence future net GHG emissions (Figure 1.1).

In the remainder of Chapter 1, we explain why the subject of global climate change is so extraordinarily important for human society – basically covering modules A, B, and C and submodule D1 in Figure 1.1. First, the basic scientific aspects of the "greenhouse" phenomenon and the present status of the global climate are described. Evidence is presented for the discernible effects that human activities are beginning to have on the global climate, primarily due to the increasing accumulation of GHGs since the industrial revolution. Next, likely climatic changes over the next century are explained in relation to several scenarios of unconstrained GHG emissions. The extent of global warm-

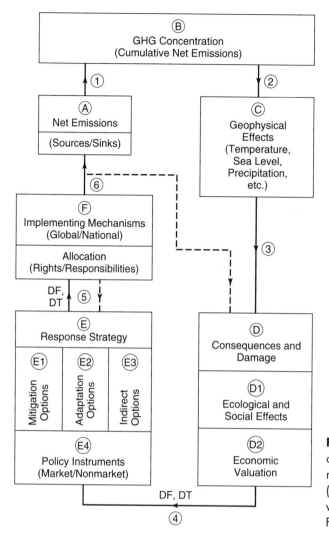

FIGURE 1.1. Causal links that characterize the global climate change phenomenon (DF denotes decision framework; DT, decision tools). From Munasinghe (1996).

ing, sea level rise, and related effects, like extreme weather events, are of particular concern. Some additional scenarios that require policy intervention to stabilize future GHG concentrations are outlined. The possible socioeconomic and environmental consequences of global climate changes are summarized, corresponding to the benchmark atmospheric carbon dioxide concentration equivalent to a doubling of the preindustrial level. The focus is on the vulnerability of ecosystems and natural habitats, hydrology and water resources, food and agriculture, human health, and human infrastructure and habitats. Finally, the case is made for a more proactive human intervention in the form of a broad-based international climate change mitigation program that would reduce the risks of global warming.

Chapter 2 focuses on linkages 1 to 6 in Figure 1.1 and attempts to convey the flavor of the formidable complications and uncertainties surrounding our understanding of the problem. Sustainable development, which includes economic, social, and environmental dimensions, is presented as an important goal for society that has emerged at the end of the twentieth century. The threats to it that climate change poses are discussed. The conceptual and methodological problems associated with predicting climate change are explained in relation to generic complications, which include the global and centuries-long scale of events, complex interactions, and the potential for irreversible, nonlinear, and catastrophic effects. Uncertainties are caused by large gaps in scientific, economic, social, and technological knowledge. The chapter also deals with a host of sociopolitical issues, especially the difficulties of ensuring equity and fairness to all (living as well as future generations), in both the climate change decision-making process and its outcome.

In the third chapter, we address the difficult problem of how decisions might be made – dealing with some aspects of linkages 4 and 5, as well as module F. The fundamental challenges are to determine what might constitute "dangerous anthropogenic interference with the climate system" and how to achieve "stabilization of greenhouse gas concentrations" at the appropriate level. Accordingly, the main objectives of climate change decision making are formulated as a series of issues. Then a framework for making decisions is set out that seeks especially to disentangle issues concerning efficiency (i.e., relating to overall global well-being) and equity (i.e., fairness and distribution). The framework consists of three key elements. The first is global optimization – that is, the provision of efficient benchmark responses (i.e., without considering equity) to questions such as what the total level of future GHG emissions should be, as well as a cost-effective portfolio of remedial measures. The second is collective decision-making principles that will help to ensure fairness in the allocation of both the rights to emit and the responsibilities for undertaking future GHG abatement measures based on a variety of considerations, including responsibility for past emissions, the incidence of future climate change effects, vulnerability, and distribution of resources. The third key element includes procedures and mechanisms by which a global consensus might be reached and implemented, including practical incentives to encourage international cooperation. A stepwise process that emphasizes flexibility and incorporates lessons learned along the way is desirable. A sustainable energy development framework is set out for implementing internationally agreed-upon GHG abatement measures within individual countries.

The fourth chapter completes the discussion of linkages 4 and 5 with a description of the criteria and analytical tools by which rational decisions can be reached. The ways in which conventional approaches must be adapted are explained to suit the unique features of global climate change. A number of decision criteria are outlined on the basis of ethical, economic, social, and legal principles, and the circumstances under which they might be applied are described. The limitations of traditional cost–benefit analysis, as well as the relative advantages of a family of decision methods that have been developed to overcome these shortcomings, are reviewed. Particular attention is paid to the role of valuation in economic cost–benefit analysis, the application of cost-effectiveness analysis in selecting least-cost strategies to achieve predetermined targets, the use of multicriteria analysis when economic valuation is difficult, the need for decision analysis to deal with uncertainty, and the growing importance of integrated assessment models.

Chapter 5 returns to submodule D2 by reviewing the likely social and economic costs of global warming if present trends continue into the future. It not only assesses the extent of such damage in the absence of GHG mitigation efforts, but also identifies what we know about the value of benefits arising from the avoidance of potential damage costs if a climate change mitigation strategy were adopted. The advantages and disadvantages of adaptation options are discussed here (submodule E2) and compared with those of mitigation options (submodule E1). In addition, some information is provided at the end of the chapter concerning the secondary benefits that can be expected to arise from abatement action. This is important, because the combination of damage avoided and secondary benefits constitutes the total benefits of abatement. Throughout this chapter the recurrent problems of valuing damage continue to underline the fact that it is impossible to fully separate efficiency from equity issues.

Chapters 6 and 7 focus on the costs of GHG abatement. In particular, Chapter 6 examines the most promising generic technical options available for dealing with climate change (including GHG emission reduction technologies and carbon sequestration measures), reviews the corresponding costs of implementing them, and assesses their feasibility and applicability. Particular attention is paid to the set of options that may be applicable in developing countries and economies in transition.

The seventh chapter deals with climate change mitigation costs, especially in relation to submodules E1 and E3. It focuses on the various factors that might affect such costs, like scale, location, and speed of technological progress. The sensitivity of implementation costs to these factors makes it quite difficult to provide firm conclusions about the likely pattern of most economically efficient, effective, and politically acceptable climate change response options. The chapter goes on to examine the elements of submodule E4 by assessing the efficiency, effectiveness, and political feasibility of various policy instruments that might facilitate the adoption of desirable climate change response options. This analysis is complicated by the fact that the international framework for decision making and implementation (module F) is still evolving.

Finally, in Chapter 8, three possible scenarios for global climate change during the next century are explored, based on the most likely (or expected), best, and worst cases. The main conclusions of the book are brought together, and several specific recommendations are made. In particular, we reiterate (a) why global climate change is a topic of very great concern for humanity, (b) how we might organize ourselves to make rational decisions in addressing the problem, and (c) what menu of technological and policy options might be available to support climate change abatement and adaptation responses. We try to capture the main insights derived from the earlier chapters, in particular by comparing the benefits and costs of abatement action on the basis of the efficiency criterion. Such an exercise is useful even if actual quantitative calculations are rather imprecise at present. The chapter also focuses on which economic instruments policy makers might use to implement a response strategy – based on a review of the pros and cons of the various instruments – and the implications of their use for allocation, on the one hand, and distribution, on the other. We conclude by highlighting some interesting issues that will continue to be debated in the decades to come.

1.2 CAUSES OF CLIMATE CHANGE AND PRESENT STATUS

1.2.1 Natural and Enhanced Greenhouse Effects

The earth's climate has remained quite stable during the present interglacial period of about 10,000 years, with mean temperature changes not exceeding 1°C per century (Nicholls et al. 1996). Normally, the average global climate is determined by the balance between the inflow of solar radiation into the atmosphere, the trapping of some of this heat (especially the outgoing longer-wavelength terrestrial radiation) due to a natural "greenhouse effect," and the reradiation of heat back into space (see Box 1.1). In the context of this broader heat balance, the pattern of solar radiation falling on the earth sets up a temperature gradient between the hotter equator and the frozen poles. The poleward transport of heat and the earth's rotation give rise to tropospheric westerly winds in each hemisphere, as well as other dramatic weather events like cyclones. The contrasting thermal characteristics of land and oceans, as well as obstacles such as mountain ranges, also affect the climate on a large scale (Dickinson et al. 1996).

Increasing energy use since the industrial revolution has led to the rapid accumulation of greenhouse gases – primarily carbon dioxide (CO_2), methane (CH_4), and nitrous oxide (N_2O) – well above their naturally occurring or historical levels. Figure 1.2 shows typical data for the growth of observed CO_2 concentrations in the atmosphere for the past millennium. Of particular concern is the still-growing present-day concentration of about 360 parts per million by volume (ppmv), compared with preindustrial levels of around 280 ppmv (Schimel et al. 1996). Continuing emissions of GHGs from anthropogenic (and natural) sources will enhance the natural greenhouse effect and

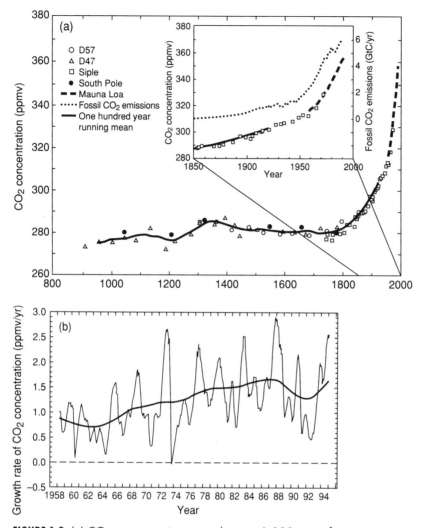

FIGURE 1.2. (a) CO_2 concentrations over the past 1,000 years from ice-core records (D47, D57, Siple, and South Pole) and (since 1958) from the Mauna Loa, Hawaii, measurement site. The smooth curve is based on a 100-year running mean. The rapid increase in CO_2 concentration since the onset of industrialization is evident and has followed closely the increase in CO_2 emissions from fossil fuels (see inset of period from 1850 onward). (b) Annual growth rate of CO_2 concentration since 1958 in parts per million by volume at the Mauna Loa station, showing the high growth rates of the late 1980s, the decrease in the early 1990s, and the recent increase. The smooth curve shows the same data but filtered to suppress any variations on timescales of less than approximately 10 years. From Houghton et al. (1996).

increase the risk of global warming above normal levels as more heat becomes trapped in the atmosphere. However, the extent of such warming critically depends on the relationship between GHG emissions and concentrations, which is determined by the global carbon cycle.

BOX 1.1 What Causes Climate Change?

Weather and Climate

The UNFCCC defines climate change as "a change of climate which is attributed directly or indirectly to human activity that alters the composition of the global atmosphere and which is in addition to natural climate variability observed over comparable time periods." To help the reader comprehend the climate change problem more fully, this box briefly describes what constitutes climate, what variability arises due to natural causes, and how anthropogenic alteration of the atmosphere might have an additional impact on the global climate.

Phenomena associated with the atmosphere are broadly divided into *weather* and *climate* events. Hourly or daily fluctuations in the atmosphere (e.g., precipitation and storms) determine the weather. Such changes occur as weather systems develop, evolve, move, and dissipate. They are characterized by nonlinear and chaotic behavior, and are therefore not generally predictable beyond a period of 1 or 2 weeks into the future.

Climate can be described broadly as the "average" of a series of weather events, or more precisely defined by the statistical measurement of weather variability over a period of time. In the longer timescale (typically, weather statistics averaged over a 30-year period), individual weather events disappear from the mean climate statistics but could influence measures of climate variability (i.e., extreme events or deviations from the mean). Climate too can vary due to alterations in the internal exchanges of energy or in the internal dynamics of the climate system over months or years. The earth's climate is affected by factors that cause a change in the redistribution of energy within the atmosphere or between the atmosphere, land, and ocean. For example, the El Niño–Southern Oscillation (ENSO) events arise from natural coupled interactions between the atmosphere and the ocean centered in the tropical Pacific.

The primary source of energy that drives the climate is radiation from the sun. Incident solar energy is absorbed by the earth's surface and redistributed by atmospheric and oceanic circulation. In turn, infrared radiation is reemitted from the earth into space to maintain a zero average net energy balance between the top of the atmosphere and outer space.

The Global Energy Balance and Natural Greenhouse Effect

The average amount of solar energy incident on a flat surface at the top of the atmosphere is about 342 watts per square meter ($W\,m^{-2}$), as shown in Figure B1.1. About 77 $W\,m^{-2}$ of this energy is scattered or reflected back directly into space by molecules, microscopic airborne particles

(continued)

BOX 1.1 What Causes Climate Change? *(continued)*

(known as aerosols), and clouds in the atmosphere, and another 30 W m^{-2} is reflected back by the earth's surface itself. Of the remaining 235 W m^{-2} of solar radiation, about 168 W m^{-2} is absorbed by the earth's surface and the rest (approximately 67 W m^{-2}) by the atmosphere. Some of the energy intercepted by the surface is released to the atmosphere in the form of evapotranspiration (78 W m^{-2}), and sensible heating or thermals (24 W m^{-2}).

To maintain its long-term thermal equilibrium, the earth must reradiate back to space, on average, the same amount of energy that is absorbed. It does so by emitting thermal "long-wave" radiation in the infrared part of the spectrum. The amount of thermal radiation emitted by a warm surface depends on its temperature and how absorbing it is. If the earth had a perfectly absorbing surface, it would reemit the required 235 W m^{-2} of thermal radiation at a rather low temperature of about $-19°C$. This is much colder than the conditions that actually exist near the earth's surface, where the annual average global mean temperature is about 15°C. This apparent discrepancy arises from the natural green-

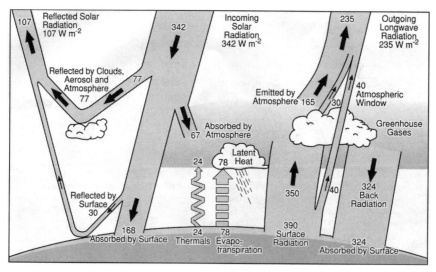

FIGURE B1.1. The earth's radiation and energy balance. The net incoming solar radiation of 342 W m^{-2} is partially reflected by clouds and the atmosphere, or at the surface, but 49% is absorbed by the surface. Some of that heat is returned to the atmosphere as sensible heating and most as evapotranspiration that is realized as latent heat in precipitation. The rest is radiated as thermal infrared radiation, and most of that is absorbed by the atmosphere, which in turn emits radiation both up and down, producing a greenhouse effect, as the radiation lost to space comes from cloud tops and parts of the atmosphere much older than the surface. From Trenberth, Houghton, and Meira Filho (1996).

house effect, which effectively keeps the surface about 34°C warmer than it would otherwise be. Thus, most of the present life forms on the earth depend on the natural greenhouse effect for their existence.

As indicated in the figure, the earth's surface reemits about 390 W m^{-2} of infrared energy. Most of the atmosphere consists of nitrogen and oxygen (99% of dry air), which are transparent to infrared radiation. Thus, a small portion of radiation leaving the surface (40 W m^{-2}) is transmitted relatively unimpeded back through the atmosphere (i.e., the radiation spectrum falls within the atmospheric "window"). However, the bulk of the infrared radiation (350 W m^{-2}) is intercepted and absorbed by atmospheric GHGs like water vapor (which varies in amount from 0 to about 2%), CO_2, and other minor gases such as ozone, CH_4, and N_2O (present in the atmosphere in much smaller quantities). The GHGs in turn reemit this absorbed energy in all directions. Ultimately, a further infrared emission of 195 W m^{-2} to space occurs from the atmosphere (165 W m^{-2}), and the tops of clouds (30 W m^{-2}).

In brief, atmospheric GHGs act as a partial blanket that traps some of the thermal radiation from the surface and makes the earth substantially warmer than it would otherwise be. By analogy with the common garden greenhouse, such atmospheric blanketing of infrared radiation is said to give rise to the natural greenhouse effect – because infrared radiation is reemitted into space from regions of the atmosphere that are generally much colder than the surface.

The Enhanced Greenhouse Effect

Any change in the average net radiative balance at the top of the atmosphere (due to variations in the natural intensity of incoming solar radiation or outgoing infrared energy) is said to cause *radiative forcing* of the atmosphere. For example, increases in the concentrations of GHGs will reduce the efficiency of the radiative cooling of the earth, resulting in a positive radiative forcing. This causes the enhanced greenhouse effect, which tends to further warm the lower atmosphere and surface – the anthropogenic enhancement of a phenomenon that has operated in the earth's atmosphere for billions of years due to naturally occuring GHGs. The amount of warming depends on the size of the increase in concentration of each GHG, the radiative properties of the gases involved, and the concentrations of other GHGs already present in the atmosphere (see Table 1.1 and main text).

Any radiative forcing (e.g., due to changes in atmospheric GHG or aerosol concentrations from either natural or anthropogenic causes) will tend to alter atmospheric and oceanic temperatures, weather patterns, and the entire hydrological cycle. Human-induced changes in climate will be superimposed on a background of natural climactic variations

(continued)

BOX 1.1 What Causes Climate Change? *(continued)*

that occur on a whole range of space- and timescales. Natural climate variability can occur as a result of changes in the forcing of the climate system, for example due to aerosol derived from volcanic eruptions. Climate variations can also occur without any external forcing, as a result of complex interactions between components of the climate system such as the atmosphere and ocean. The ENSO phenomenon is an example of such natural "internal" variability. The sun's output of energy also changes by small amounts (0.1%) over an 11-year cycle. The intensity of solar radiation can vary over longer periods as well. On timescales of tens to thousands of years, slow variations in the earth's orbit (which are well understood) have led to changes in the seasonal and latitudinal distribution of solar radiation. Such changes have played an important part in variations of climate in the distant past, such as during glacial cycles. To distinguish anthropogenic climate changes from natural variations, it is necessary to identify the anthropogenic "signal" against the background "noise" of natural variability.

As explained in the main text, the amount of GHG in the atmosphere has increased significantly since the industrial revolution, worsening the risk of future climate change. For example, if the amount of CO_2 in the atmosphere suddenly doubled with respect to preindustrial levels (but with other things remaining the same), the outgoing long-wave radiation would be reduced by about $4\ W\ m^{-2}$. To restore the radiative balance, the atmosphere must warm up and, in the absence of other changes, the warming at the surface and throughout the troposphere would be about 1.2°C. However, given the many other factors and various forms of feedback that come into play, the best estimate of the global mean temperature increase for doubled CO_2 concentration is approximately 2.5°C. Such a change is very large by historical standards and would be associated with major climate changes around the world.

The Carbon Cycle

Since CO_2 is by far the most important GHG, the basic carbon cycle is reviewed briefly here. Exchanges of carbon take place among the atmosphere, oceans, and terrestrial biosphere and, less rapidly, with sediments and sedimentary rocks. The average carbon fluxes associated with the more rapid components of this cycle are illustrated in Figure B1.2. While the overwhelming bulk of carbon stocks (almost 40,000 gigatons of carbon, or GtC) reside in the intermediate and deep oceans, the crucial factor in climate change is the exchange of carbon between the atmosphere and land and ocean surfaces, due to biological and physical processes as well as human activities. Net anthropogenic carbon emissions of about 7 GtC (in 1990) due to fossil fuel use, cement production, and land-use changes now dominate the net natural uptake of about 3 to

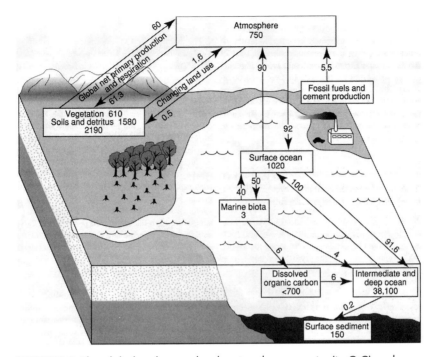

FIGURE B1.2. The global carbon cycle, showing the reservoirs (in GtC) and fluxes (GtC/yr) relevant to the anthropogenic perturbation as annual averages over the period 1980–9 (Eswaran, Van den Berg, and Reich, 1993; Potter et al. 1993; Siegenthaler and Sarmiento 1993). The component cycles are simplified and subject to considerable uncertainty. In addition, this figure presents average values. The riverine flux, particularly the anthropogenic portion, is currently very poorly quantified and so is not shown here. Evidence is accumulating that many of the key fluxes can change significantly from year to year. In contrast to the static view conveyed by figures such as this one, the carbon system is clearly dynamic and coupled to the climate system on seasonal, interannual, and decadal timescales. From Schimel et al. (1996).

4 GtC by the terrestrial biosphere and surface ocean. There is no doubt that this is the main cause of the increase in atmospheric CO_2 concentrations from about 280 ppmv in preindustrial times to 358 ppmv in 1994, which in turn has increased the radiative forcing by about 1.6 W m^{-2}. For comparison, CO_2 concentrations over the past 1,000 years (when the global climate was relatively stable) fluctuated only by about ±10 ppmv around the mean of 280 ppmv.

The net release of carbon from tropical land-use change (mainly forest clearing without regrowth) is roughly balanced by carbon accumulation in other land ecosystems due to forest regrowth outside the tropics, by transfer to other reservoirs stimulated by CO_2 and nitrogen fertilization, and by decadal timescale climatic effects. The fastest process for CO_2 removal is uptake into vegetation and the surface layer of the

(continued)

BOX 1.1 What Causes Climate Change? *(continued)*

oceans, which occurs over a few years. Various other sinks operate on the century timescale (e.g., transfer to soils and to the deep ocean) and so have a less immediate, but no less important effect on the atmospheric concentration. Within 30 years, about 40 to 60% of the CO_2 currently released to the atmosphere is removed. However, if emissions were reduced, the CO_2 in vegetation and ocean surface water would soon equilibrate with that in the atmosphere, and the rate of removal would then be determined by the slower response of woody vegetation, soils, and transfer into the deeper layers of the ocean. Consequently, most of the excess atmospheric CO_2 would be removed over about a century, although a portion would remain airborne for thousands of years because transfer to the ultimate sink – ocean sediments – is very slow.

There is a large degree of uncertainty associated with the future role of the terrestial biosphere in the global carbon budget, for several reasons. First, future rates of deforestation and regrowth in the tropics and midlattitudes are difficult to predict. Second, mechanisms such as CO_2 fertilization remain poorly quantified at the ecosystem level. Over decades to centuries, anthropogenic changes in atmospheric CO_2 content and climate may also alter the global distribution of ecosystem types. Carbon could be released rapidly from areas where forests die, although regrowth could eventually sequester much of this carbon. Estimates of this loss range from nearly zero to as much as 200 GtC over the next one to two centuries, depending on the rate of climate change.

Sources: Adapted from Houghton et al. (1995, 1996).

Carbon dioxide in the atmosphere provides the key link in the global carbon cycle between human activities and naturally occurring biological and physical processes. Thus, carbon is exchanged among the oceans, the terrestrial biosphere, and the atmosphere, and less quickly with sediments and rocks (see Box 1.1). This cycle had reached an equilibrium over past millennia, until the dawn of the industrial revolution and consequent increases in anthropogenic emissions of GHGs. Figure 1.3 shows the clear correlation between changes in the main reservoirs of carbon and rising carbon emissions from fossil fuel use. The atmosphere has been the main absorber of carbon, while the terrestrial biosphere shifted from being a net source to being a net sink for carbon after about 1940.

Another important piece of evidence concerning climate change is provided by the relationship between past concentrations of atmospheric carbon and variations in global mean temperature. Figure 1.4 clearly indicates that such temperature changes have been closely synchronized with levels of CO_2

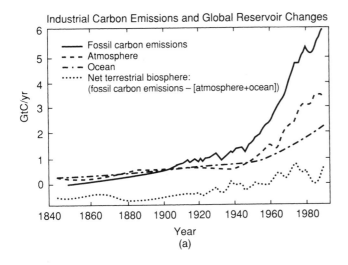

(a)

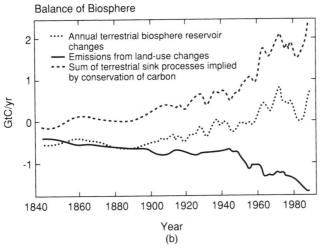

(b)

FIGURE 1.3. (a) Fossil carbon emissions (based on statistics of fossil fuel and cement production) and representative calculations of global reservoir changes: atmosphere (deduced from direct observations and ice-core measurements), ocean (calculated with the ocean carbon model of the Geophysical Fluid Dynamics Lab), and net terrestrial biosphere (calculated as remaining imbalance). The calculation implies that the terrestrial biosphere represented a net source to the atmosphere before 1940 (negative values) and a net sink after about 1960.
(b) The carbon balance of the terrestrial biosphere. Annual terrestrial biosphere reservoir changes, land-use flux (plotted negative because it represents a loss of biospheric carbon), and the sum of the terrestrial sink processes (e.g., Northern Hemisphere regrowth, CO_2 and nitrogen fertilization, climate effects) as implied by conservation of carbon mass. From Schimel et al. (1995).

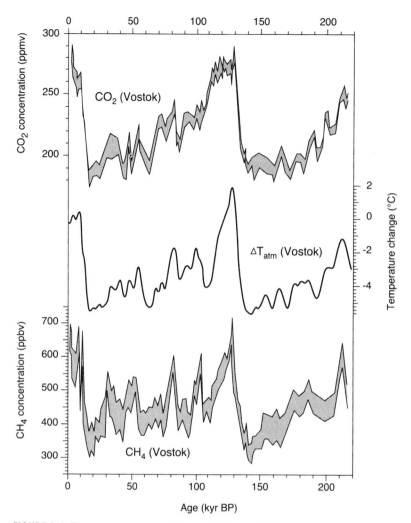

FIGURE 1.4. Temperature anomalies and CH_4 and CO_2 concentrations over the past 220,000 years as derived from the ice-core record at Vostok, Antarctica. From Schimel et al. (1995).

and CH_4 in the air, derived from Antarctic ice-core samples dating back over 200,000 years. Furthermore, current CO_2 concentrations already exceed the highest levels recorded in the past. More generally, there is some evidence that atmospheric concentrations of CO_2 have both influenced and been influenced by climate change in the past. In summary, irrespective of whether atmospheric carbon has been the cause or the effect of past global warming, the correlation between these two indicators provides a sobering perspective on the potential risks posed by the enhanced greenhouse effect.

Table 1.1 summarizes the increases in GHG levels since 1750; for example, atmospheric concentrations of CO_2, CH_4, and N_2O have risen by about 30%, 100%, and 15%, respectively, and are continuing to increase steadily. Figure

TABLE 1.1 Main Greenhouse Gases and Key Characteristics

GHG	Concentration		Rate of Increase in the 1980s		Main Sources/ Causes	(Relative) Global Warming Potential			Atmospheric lifetime (Years)
	Pre-industrial Level	Level in 1992	Average % per Year	Incremental Conc. per Year		20 Years	100 Years	500 Years	
CO_2	280 ppmv	355 ppmv	0.40	1.5 ppmv/yr	Natural and anthropogenic	1	1	1	50–200[a]
CH_4	700 ppbv	1,714 ppbv	0.80	13 ppbv/yr	Natural and anthropogenic	62	24.5	7.5	12–17[b]
N_2O	275 ppbv	311 ppbv	0.25	0.75 ppbv/yr	Natural and anthropogenic	290	320	180	120
CFC-12	0	503 pptv	4.00	18–20 pptv/yr	Entirely anthropogenic	7,900	8,500	4,200	102
HCFC-22c (FC substitute)	0	105 pptv	7.00	7–8 pptv/yr	Anthropogenic, low concentration now, but rising	4,300	1,700	520	13.3
CF_4 (perfluorocarbon)	0	70 pptv	2.00	1.1–1.3 pptv/yr	Anthropogenic, very long life	4,100	6,300	9,800	50,000

Abbreviations: ppmv, parts per million by volume; ppbv, parts per billion by volume; pptv, parts per trillion by volume; CFC, chlorofluorocarbon; FC, fluorocarbon; HCFC, hydrochlorofluorocarbon.

[a]No single lifetime for CO_2 can be defined because of the different rates of uptake by the various sink processes.

[b]Defined as an adjustment time that takes into account the indirect effects of CH_4 on its own lifetime.

Source: Houghton et al. (1995, Tables 3 and 5).

1.5 shows how the main GHGs have contributed cumulatively to total global warming from 1850 to 1990, measured in units of global mean radiative forcing.[1] Clearly, CO_2 is the most important GHG, having contributed about 60% of the total. At the same time, the total radiative forcing effect of a GHG depends on several factors, including its concentration, global warming potential (GWP),[2] and atmospheric lifetime. As Table 1.1 indicates, while the concentration of CO_2 is much higher than that of the other GHGs, smaller increases in the concentrations of the latter gases could be as dangerous, because their GWPs are tens to thousands of times greater relative to CO_2. For example, CH_4, which is far less abundant in the atmosphere than CO_2 (Table 1.1), still contributes about one-fourth as much as CO_2 does to global warming (Figure 1.5). Note that the relative GWPs of gases that are more long-lived than CO_2 tend to increase over time, since these gases are more persistent, and vice versa.

Table 1.1 also indicates the principal human activities that give rise to each GHG. Because these gases have long lifetimes in the atmosphere (ranging from decades to centuries), their concentrations respond rather slowly to changes in emission rates. Furthermore, the exchange of both GHGs and heat between the oceans and atmosphere takes place over long periods, resulting in significant time lags before the final climatic equilibrium is reached.

Small airborne particles (10^{-3} to 10^{-6} m in diameter) or aerosols that are trapped in the troposphere (or lower atmosphere)[3] can give rise to a cooling effect, which opposes global warming. The primary human causes of aerosols are the combustion of fossil fuels (like coal), biomass burning, and smelting of ores. Unlike GHGs, such aerosols are more localized and have a limited lifetime, so that their concentrations are closely connected to emission rates.

---

[1] Radiative forcing measures the additional effect of a GHG on the normal energy equilibrium of the earth–atmosphere system (see Box 1.1 for details).

[2] The GWP indicates the cumulative warming effect or radiative forcing caused by a unit mass of gas from the moment of emission up to some distant time in the future. It is expressed as an index relative to the reference gas, CO_2 (i.e., the GWP of CO_2 is unity). GWPs must take into account not only the direct radiative forcing effects of each gas, but also the indirect effects (e.g., the formation of new GHGs due to chemical changes). Except for CH_4, the indirect components of the GWPs of other GHGs have not yet been estimated, owing to inadequate knowledge of atmospheric processes. The GWPs in Table 1.1 are given relative to the absolute radiative forcing value of CO_2 at its current concentration. As this concentration increases in the future, the absolute radiative forcing value will decrease, and the relative GWPs of other GHGs will tend to rise.

[3] The troposphere extends from the earth's surface up to heights of about 9 km (in high latitudes) to 16 km (in low latitudes). The temperature generally falls with increasing height in the troposphere, and most of the cloud- and weather-related phenomena occur in this region.

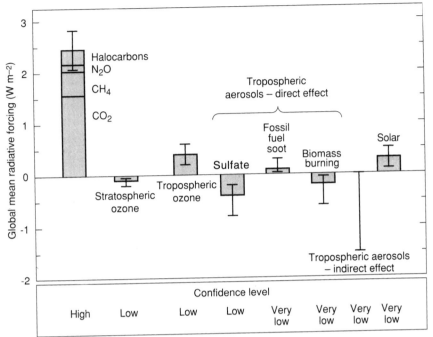

FIGURE 1.5. Estimates of the globally averaged radiative forcing due to changes in GHG and aerosol concentrations from preindustrial times to the present and changes in solar variability from 1850 to the present. The height of the bar indicates a midrange estimate of the forcing, while the lines show the possible range of values. An indication of relative confidence in the estimates is given below each bar. The contributions of individual GHGs are indicated on the first bar for direct GHG forcing. The major indirect effects are a depletion of stratospheric ozone (caused by chlorofluorocarbons and other halocarbons) and an increase in the concentration of tropospheric ozone. The negative values for aerosols should not necessarily be regarded as an offset against the GHG forcing, because of doubts over the applicability of global mean radiative forcing to the case of nonhomogeneously distributed species such as aerosols and ozone. From Houghton et al. (1996).

Aerosols also exist in the stratosphere (or middle atmosphere),[4] and are injected into these higher altitudes primarily by volcanic eruptions and other violent explosions. They usually persist for several years and could spread widely across the globe. The negative radiative forcing effect of such aerosols can be strong enough on a continental or hemispherical scale to offset the warming impact of GHGs over relatively short periods of time. For example, the slight cooling of the earth's surface during 1992–3 was caused in part by the eruption of Mount Pinatubo in mid-1991 (Dutton and Christy 1992; Graf, Kirchner, and Schult 1993).

---

[4] The stratosphere lies immediately above the troposphere and extends up to a height of about 50 km. It is a stable and highly stratified region.

1.2.2 Current Evidence

A broad consensus has emerged among the several thousand world's leading experts who have reviewed current information for the IPCC that the impact of human activities on the global climate is just becoming detectable. The concern is that these changes might continue to intensify because of the interplay between past actions and the normal time lags within the global climate response system. In order to establish a link between climate change and past human interventions, it has been necessary to (a) detect whether a particular change in weather patterns is exceptional in a statistical sense and (b) attribute such an effect to some specific human cause.

On the issue of detection, much of the scientific effort has been devoted to statistically separating out the effects of human intervention (mainly emissions of GHGs) from the background "noise" consisting of the natural variability in the climate due to both internal fluctuations and external influences such as changes in solar radiation and volcanic eruptions. Nevertheless, Figure 1.5 confirms that anthropogenic contributions to radiative forcing (through past GHG emissions up to 1990) are already quite significant compared with natural effects (e.g., variations in solar output). At the same time, the complexity of the various forces at work is evident from Figure 1.5. For example, in addition to the complicated effects of GHGs, tropospheric ozone enhances global warming, whereas stratospheric ozone has an opposite (but lesser) effect. Meanwhile, sulfate and biomass aerosols give rise to cooling effects, whereas soot aerosols tend to increase radiative forcing. There is a significant range of uncertainty in the estimates, especially with respect to the potential indirect cooling effects of tropospheric aerosols.

The evidence that a distinct component of climate change can be attributed to human actions is based on several results that have emerged from recent studies. First, a range of climate indicators show that the twentieth century is probably warmer than any other century since 1400, before which

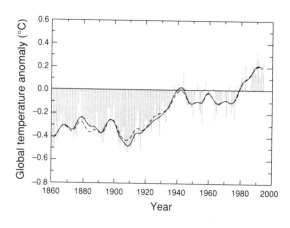

FIGURE 1.6. Combined land-surface, air, and sea-surface temperatures (°C), 1861 to 1994, relative to 1961 to 1990. The solid curve represents smoothing of the annual values shown by the bars to suppress subdecadal timescale variations. The dashed smoothed curve is the corresponding result from Houghton, Callander, and Varney (1992).

sufficiently reliable data were not available. Even within this century, some areas have recorded the warmest decade-long spells that have occurred in the past millennium. More specifically, the world's average surface temperature has increased between 0.3 and 0.6°C over the past hundred years, and by about 0.2 to 0.3°C since the 1950s, based on surface air temperature measurements over both land and ocean (Figure 1.6). The warming has not been uniform – for example, nighttime temperatures over land have increased more than daytime temperatures. The mean sea level has risen between 10 and 25 cm over the same period, due primarily to the warming and consequent expansion of the oceans (0.2 to 0.7 mm per year), retreat of glaciers (0.2 to 0.4 mm per year), and other temperature-related causes, including possible melting of the Antarctic and Greenland ice sheets.

Second, a number of sophisticated statistical tests indicate that the increasing trend in global mean temperature during the past hundred years cannot be attributed entirely to variations caused by natural phenomena. These tests have included more refined estimates of the expected range of natural temperature variations derived from instrumental measurements on air, land, and sea, paleoclimatic samples (e.g., tree rings), snow cover and glacier recession data, climate simulation models, and statistical models of data.

Third, the observed spatial variations in temperature (both horizontally across the globe and vertically through the atmosphere), as well as seasonal changes, are consistent with the pattern of increasing human intervention over time. There have been wide regional climatic variations – for example, greater warming over midlatitude continental areas during winter and spring, but more cooling over the North Atlantic. Precipitation has increased over land areas in the high northern latitudes, particularly during winter. The range of variation in the climate and extreme weather events at the regional level has changed also, although there is no clear link with human activities. One example is the massive ENSO in temperatures in the tropical East Pacific Ocean. The frequency of ENSO events has increased since the 1970s, and the sustained 1990–5 oscillation was never observed before, on the basis of records going back 100 years.

On the issue of detection, large numerical global climate models have been developed that seek to capture not only the heat-trapping effects of GHGs, but also the negative radiative forcing provided by sulfate aerosols, as well as the contribution from the coupling of the atmospheric–oceanic systems over timescales ranging from decades to centuries. These models have been used to simulate the global mean climate and even compare disaggregate model results with actual measurements at the regional level (for more details see Gates et al. 1996). The predictions made by such models correspond closely enough to the observed patterns of temperature change to confirm that the latter cannot be caused solely by natural phenomena.

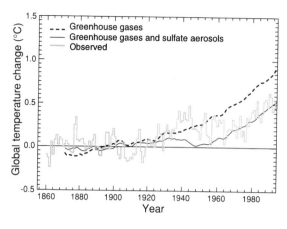

FIGURE 1.7. Simulated global annual mean warming from 1860 to 1990, allowing for increases in GHGs only (dashed line) and GHGs and sulfate aerosols (solid line), compared with observed changes over the same period. From Houghton et al. (1996).

Figure 1.7 provides one example of the agreement between past observations and simulated results of changes in global annual mean warming derived from a recent model. Moreover, model simulations of the more important large-scale features of the climate system (including seasonal, geographic, and vertical variations in temperature as well as precipitation) agree well with observations, although some discrepancies exist at regional levels. Such comparisons with empirical measurements help to confirm the validity of a given model as a sufficiently realistic representation of the global climate system and build greater confidence in predictions about the future (see Section 1.3). The inability of present models to incorporate the variability of solar radiation and volcanic dust released into the atmosphere is a major weakness to be remedied in future studies.

The IPCC Second Assessment Report of Working Group I (Houghton et al. 1996) concludes that "the balance of evidence suggests a discernible human influence on the global climate," despite the problems of measurement and analysis caused by natural variability, lack of knowledge of the complex systems at work (especially the radiative forcing effects of GHGs and aerosols and the role of clouds, oceans, and vegetation cover), and the inherent time lags in decade- to century-long processes. Figure 1.8 provides a concise summary of the main basis for this judgment.

1.3   LIKELY FUTURE CHANGES IN THE GLOBAL CLIMATE

This section, like the subsequent one, summarizes the main physical, ecological, and social changes that might occur in the context of global climate change. It deals primarily with the results produced by Working Groups I and

II of the IPCC (Houghton et al. 1996; Watson et al. 1996). Readers who are interested primarily in the economic and policy aspects dealt with in later chapters would still be well advised to persevere by at least glancing through the next two sections in order to develop a better understanding of the scientific facts underlying the global climate change phenomenon.

(a)   **Temperature indicators**

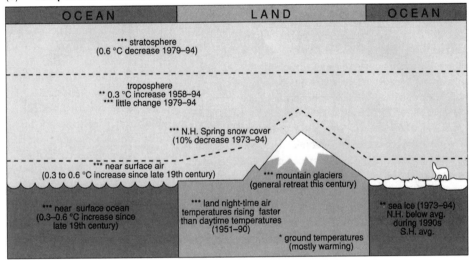

(b)   **Hydrological indicators**

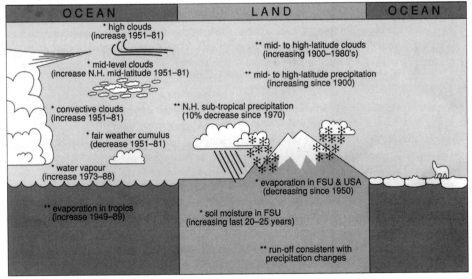

FIGURE 1.8. Summary of trends in observed climatic behavior during the past century. Asterisks indicate confidence level (i.e., assessment): ***high; **medium; *low. From Houghton et al. (1996).

1.3.1   The IS92 Scenarios (without Intervention)

Emissions and Concentrations   In 1992, the IPCC made a series of projections up to 2100 of emissions of the principal GHGs (CO_2, CH_4, N_2O, and halocarbons) and aerosols (from sulfate precursors and biomass combustion) that would most likely occur if no measures were taken to limit GHG emissions. We focus initially on the CO_2 component of these scenarios as shown in Figure 1.9a, because CO_2 has far more potent climate change effects and exhibits more complex behavior than the other GHGs. The emission projections for other GHGs are discussed later in this section.

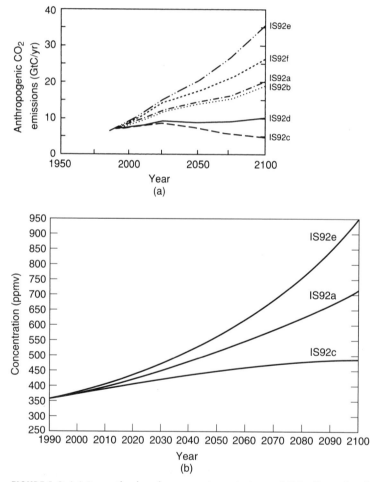

FIGURE 1.9. (a) Prescribed anthropogenic emissions of CO_2 (from fossil fuel use, deforestation, and cement production) for the IS92 scenarios. From Houghton et al. (1996). (b) Future atmospheric CO_2 concentrations based on the IS92 scenarios. Modified from Schimel et al. (1996).

While alternative projections exist, the IS92 curves provide an important baseline for thinking about climate change. IS92a is the middle case, which is closest to a "business-as-usual" scenario for growth in the future emissions of CO_2, using moderate assumptions about global economic growth (2.3 to 2.9%), population increase (11.3 billion by 2100), and mix of energy sources. IS92e is the highest emission scenario, based on rapid economic growth (3.0 to 3.5% per year), moderate population growth (11.3 billion by 2100), high availability of fossil fuels, and the phasing out of nuclear power. The lowest curve (IS92c), in which emissions grow initially but fall in later years, assumes reduced economic growth rates (1.2 to 2% per annum), a limited increase in global population (only 6.4 billion in 2100), and severe constraints on fossil fuel use.

Figure 1.9b indicates how corresponding atmospheric concentrations of CO_2 will increase over the same period based on the "Bern" model, which is fairly representative of the variety of carbon-cycle models used to link emissions and concentrations. By 2100, carbon levels in the atmosphere have soared above 600 ppmv (or more than twice the preindustrial level) for the midrange IS92a case. Even in the most optimistic IS92c scenario, the atmospheric CO_2 concentration is barely beginning to stabilize below 500 ppmv. Except for case IS92c, CO_2 levels will continue to grow for centuries, beyond 2100. Even if global emissions were to level off by the year 2000, CO_2 concentrations would continue to rise indefinitely beyond 2100 (Schimel et al. 1995, 1996).

The focus so far has been on CO_2. The IPCC augmented its 1992 projections in 1995 to incorporate the effects of aerosols and tropospheric ozone, as well as revised estimates of CH_4, N_2O, and halocarbon emissions. The following discussion refers to these augmented 1992 scenarios. As already shown in Figure 1.9a and summarized in Table 1.2, starting from the 1990 net emissions of just over 7 GtC per year from anthropogenic sources, CO_2 emissions in the year 2100 range from about 5 to 36 GtC per year (i.e., a sevenfold variation), depending on the IS92 scenario that is selected. Although the emission levels of the other GHGs projected under the different augmented IS92 scenarios also differ widely, the range of variation is far less than for CO_2. For example, anthropogenic CH_4 emissions increase from about 170 megatonnes of CH_4 ($MtCH_4$) per year in 1990 to 510 to 830 $MtCH_4$ per year in 2100. At the same time, anthropogenic N_2O emissions grow from around 8 MtN per year in 1990 to between 9 and 14 MtN per year in 2100. Meanwhile sulfur aerosol emissions (which have a net cooling effect) change from 25 MtS per year to 0 to 180 MtS per year over the same period. For comparison, the net emissions of these gases from purely natural sources in 1990 were estimated to be about 0 MtC, 5 MtN, 340 $MtCH_4$, and 75 MtS; these natural emission rates are assumed to remain unchanged in the future.

TABLE 1.2 Main GHG Emissions from Anthropogenic Sources in the Augmented IS92 Scenarios

| Year | Gas | | | |
	Carbon Dioxide (GtC)[a]	Methane (MtCH$_4$)[b]	Nitrous Oxide (MtN)[c]	Sulfur Aerosols (MtS)[d]
Annual emission in 1992 (actual)	7	170	8	25
Annual emission in 2100 (projected range)	5–36	210–830	9–14	0–180
Natural emissions (1990)	—	340	5	75

[a]10^{15} g carbon.
[b]10^{12} g methane.
[c]10^{12} g nitrogen.
[d]10^{12} g sulfur.
Source: Alcamo et al. (1995, Table 6.2).

Examining the problem from a different perspective, if annual anthropogenic CH_4 emissions were reduced immediately by 30 GtCH$_4$, the concentration of CH_4 would remain at the current level of about 1,700 ppbv. On the other hand, if future CH_4 emissions were to remain constant at present levels, the concentration would stabilize at over 1,800 ppbv within the next 40 years. In the case of N_2O, future concentrations could be stabilized at the current level of about 310 ppbv, provided that present anthropogenic emissions were approximately halved. Holding current emissions constant in the future would increase the concentration of N_2O to around 400 ppbv over several hundred years (because of its relatively long lifetime of 120 years).

In Figure 1.10, the estimated radiative forcing arising from the IS92a case illustrates the relative future global warming contributions of various gases and aerosols. Clearly, CO_2 will remain the dominant component as in the past (see also Figure 1.5), and therefore managing CO_2 emissions will be the key to any successful climate change mitigation strategy. For example, in the IS92a scenario, the contribution of CO_2 to global warming will increase from the current 60% to about 75% in 2100. The radiative forcing effects of CO_2 will almost quadruple over this same period, whereas the corresponding forcing from CH_4 and N_2O will increase by a smaller factor ranging between 2 and 3.

Modeling also indicates that concentrations of GHGs with lifetimes shorter than CO_2 respond more quickly to changes in emissions (relative to CO_2). For example, CH_4 emissions affect atmospheric CH_4 concentrations over a period

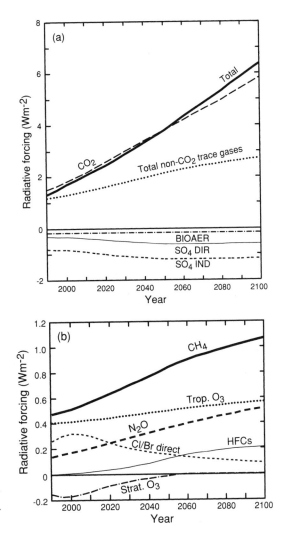

FIGURE 1.10. Radiative forcing components for augmented IS92a scenarios. From Kattenberg et al. (1996).

of 9 to 15 years. Tropospheric aerosol levels show an almost immediate response to changes in emissions. Since aerosols have highly variable regional effects, the globally averaged radiative forcing shown in Figure 1.10 may not adequately reflect their complete impact on climate.

Global Mean Temperature Increase   As indicated earlier, forecasting the global climate through the year 2100 requires the use of large and complex models.[5] Given the wide range of variation in the model parameters, the mod-

---

[5] General circulation models are the most complex now in use. However, simpler one-dimensional upwelling diffusion models have proved extremely useful in many of the IPCC calculations.

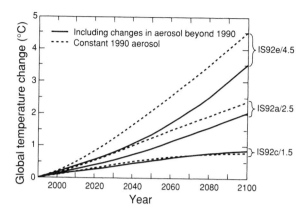

FIGURE 1.11. Temperature changes over 1990–2100 for climate sensitivities of 1.5, 2.5, and 4.5°C for the with-aerosol (solid lines) and no-aerosol (dashed lines) cases. From Hougton et al. (1996).

eling assumes three different levels of sensitivity of the global climate system to radiative forcing (high, medium, and low).[6] The moderate sensitivity assumption is considered the most likely by the IPCC (Kattenberg et al. 1996).

Figure 1.11 depicts future increases in the global mean temperature, for the IS92a or middle-emission scenario, for each of the three climate sensitivity levels (denoted by the labels 4.5, 2.5, and 1.5, respectively). For a given sensitivity level, the (higher) dashed line assumes that aerosol concentrations remain constant at 1990 levels, whereas the corresponding (lower) solid line reflects the additional cooling effect of rising aerosol concentrations consistent with the emissions of aerosol precursors in the IS92a projection. These values of increase in the global mean temperature do not capture significant regional and local temperature variations about the mean. Furthermore, a small increase in the average temperature is likely to cause relatively large increases in extreme events – for example, the number of "hot" days in which the maximum temperature exceeds 40°C.

The full range of potential temperature changes may be gauged from Figure 1.12, which illustrates the main uncertainties associated with both human actions (reflected in the emission forecast) and climate modeling (reflected in the sensitivity parameter). Thus, the steepest curve, which yields a temperature rise of 4.5°C by 2100, corresponds to the highest emission scenario (IS92e) and high climate sensitivity.[7] By contrast, the best outcome is an increase in temperature of less than 1°C by 2100 for the most optimistic emis-

---

[6] Presentation of the model results has been facilitated by using as a realistic benchmark the CO_2 concentration of 560 ppm in 2100 (or double the preindustrial level), denoted by $\Delta T(2x)$. For the high, medium, and low climate sensitivity levels associated with the $\Delta T(2x)$ benchmark case, the mean *long-term equilibrium* global temperature is estimated to rise about 4.5, 2.5, and 1.5°C, respectively.

[7] Note that in this case the temperature will continue to rise beyond 2100 toward a long-term equilibrium that will exceed 4.5°C, which is quite different from the $\Delta T(2x)$ high climate sensitivity case in which the temperature has stabilized in the distant future, *after* having risen by 4.5°C.

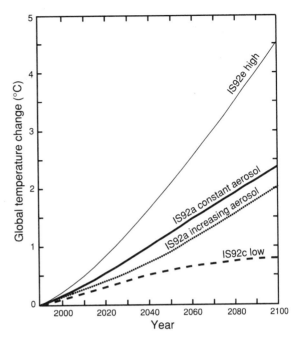

FIGURE 1.12. The extreme range of possible warming for the IS92 emission scenarios, together with results for the middle case (IS92a) with and without changing aerosol concentrations after 1990. From Kattenberg et al. (1996).

sion projection (IS92c) coupled with low climate sensitivity. All other combinations of assumptions give results that fall within these two extremes. One noteworthy feature is that, in almost all cases, global temperatures will continue to rise after 2100 (toward some long-term equilibrium several hundred years in the future), due to the inertia and time lags in the climate system. In one extension of the IS92a scenario, even when emissions were made to decline to zero linearly between 2100 and 2200, the temperature did not begin to stabilize until after 2200.

Global Mean Sea Level Rise and Other Effects   The rise in global mean sea level is the other major aggregate indicator of global warming, after temperature change. As mentioned earlier, an increase in the average global temperature will result in a corresponding rise in sea level, due mainly to the thermal expansion of the oceans and the melting of glaciers and ice sheets. The global mean sea level has already increased by 10 to 25 cm during the past century. Although there has been no recent acceleration in this rate of increase, it is significantly higher than the average rate of change recorded over several thousand years (Warrick et al. 1996).

In Figure 1.13, future increases in the mean sea level between 1990 and 2100 are shown, for the IS92a or middle emission scenario, for each of the three composite sensitivity levels (denoted by the labels high, mid, and low, respectively). These sensitivity levels combine the effects of climate sensitivity (or increase in equilibrium temperature because of radiative forcing, as described earlier) with the sensitivity of ice melting due to the temperature rise.

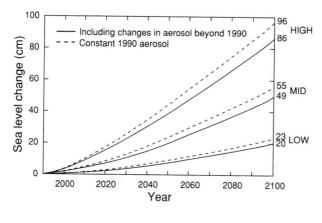

FIGURE 1.13. The 1990–2100 sea level rise for scenario IS92a. From Warrick et al. (1996).

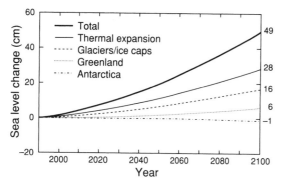

FIGURE 1.14. Contributions to the 1990–2100 sea level rise for scenario IS92a (including the effect of changing aerosol concentrations beyond 1990). From Warrick et al. (1996).

For a given composite sensitivity level, the (higher) dashed line assumes constant 1990 levels of aerosol concentration, whereas the corresponding (lower) solid line reflects the additional cooling effect of rising aerosol concentrations. In summary, these midrange results of sea level rise up to 2100 have a mean of around 50 cm and vary from 23 to 96 cm (or almost a factor of 4). They correspond to the temperature change estimates given in Figure 1.11.

The relative contributions of the various causes of sea level rise are shown in Figure 1.14 for the typical middle sensitivity and rising aerosol case associated with the IS92a emission scenario. The main underlying factor is seen to be thermal expansion of the oceans (more than 50% of the total). It is interesting that the net contribution from Antarctica is slightly negative, because global warming increases precipitation in the south polar regions by an amount sufficient to counterbalance the magnitude of the ice-melting effect.

Figure 1.15 (which is analogous to Figure 1.12) indicates the full range of potential global sea level variations up to 2100. The different cases in this figure attempt to capture the main uncertainties associated with both human actions (reflected in the emission forecast) and climate modeling (reflected in the composite sensitivity parameter). Thus, the highest projection that

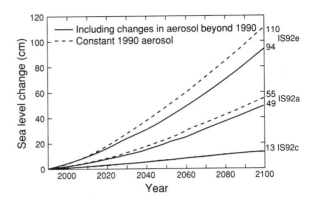

FIGURE 1.15. Extreme projections of sea level rise, 1990–2100. From Warrick et al. (1996).

yields a sea level rise of 110 cm by 2100 corresponds to the worst emission scenario (IS92e) and the high climate and ice-melt sensitivity levels. By contrast, the most optimistic result is an increase in mean sea level of only 13 cm by 2100, for the lowest emission projection (IS92c) combined with the low climate and ice-melt sensitivities. All other combinations of assumptions give results that fall within these two extremes. The sea level would continue to rise for several centuries after 2100, due to continued warming and expansion of the ocean depths and delayed melting of ice. Stabilization of the mean sea level would occur a long time after stabilization of the global temperature, which in turn would significantly delay the stabilization of GHG concentrations.

Although different models yield differing results about the geographic and seasonal variations in expected climate change, there are certain common features. Many of the weather changes will become apparent during the winter season, including more warming of the land surface than the oceans, maximum warming in the high northern latitudes, and increased precipitation and soil moisture in the high latitudes. In addition, there will be little surface warming in the Arctic regions during the summer (Ingram, Wilson, and Mitchell 1989).

The global average hydrological cycle will be enhanced. This is linked to the likelihood of more severe floods and droughts in some areas but less intense floods and droughts in others. There is also a possibility of higher levels of precipitation and more extreme rainfall events. At the same time, warmer areas could face the risk of more severe droughts (Nicholls et al. 1996). It is expected that small changes in the mean climate or in the variability of climate can produce much larger changes in the frequency of extreme weather events like typhoons and cyclones. However, present models are unable to confirm whether the occurrence of extreme weather events will increase or decrease.

1.3.2 Stabilization Cases (with Intervention)

Emissions and Concentrations   In contrast to the unconstrained IS92 emission scenarios discussed earlier (which helped to explore nonintervention outcomes up to about 2100), the CO_2 concentration stabilization profiles outlined in this subsection are useful in predicting possible consequences over a longer period of several centuries during which human intervention could lead to stabilization (for stabilization scenarios see Houghton et al. 1995). The same upwelling diffusion–energy balance model used to analyze the IS92 scenarios was applied to the stabilization cases to predict long-term climate change. Figure 1.16a illustrates some typical time paths for concentrations of atmospheric CO_2 that will terminate in specific, *stable* concentrations some time beyond 2100. These target levels range from 450 to 1,000 ppmv (designated by the respective curves S450 to S1000).[8] A doubling of preindustrial CO_2 concentrations, which is the benchmark used in many studies of the impact of climate change (see Section 1.4), amounts to 560 ppmv, or roughly the S550 case. More steeply rising concentration profiles will result in faster rates of climate change and more severe effects.

Figure 1.16b shows the CO_2 emission profiles that will permit stabilization at the various target levels indicated in Figure 1.16a.[9] The two extreme cases, S450 and S1,000, are indicated by solid lines. The pair of (solid and dashed) lines show two of the many alternative emission pathways to achieve the intermediate case, S650. In fact, each level of stabilized CO_2 concentration may be achieved by many different emission profiles, because concentration depends on total accumulations, which is measured by the area under any given emission profile rather than the shape of that profile. However, the steeper the increase in emissions, the faster the rate of climate change. For reference, emissions corresponding to the IS92a scenario are also shown in the figure. If we begin with the IS92a profile, even the highest stabilization target level of S1000 appears to be achievable only if fairly drastic emission abatement measures begin before 2050. The lowest IS92c emission projection is consistent with a CO_2 concentration that appears to be stabilizing around 500 ppmv, after 2100.

Global Mean Temperature Increase   Figure 1.17 shows the global mean temperature increase corresponding to two of the intervention cases – S450

---

[8] The stable CO_2 concentrations of 450, 650, and 1,000 ppmv would result in long-term equilibrium global temperature increases of 1, 2, and 3.5°C, respectively. These benchmark values exclude the effects of other GHGs and aerosols.

[9] These curves are derived from a typical midrange carbon-cycle model. Results from other models could differ from these values by ±15%.

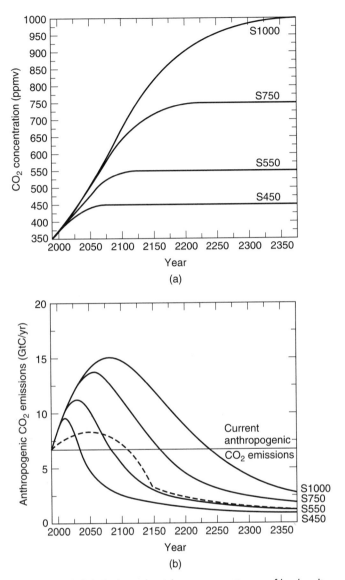

FIGURE 1.16. (a) Carbon dioxide concentration profiles leading to stabilization at 450, 550, 750, and 1,000 ppmv for pathways that allow emissions to follow IS92a until at least the year 2000. A doubling of the preindustrial CO_2 concentration of 280 ppmv would lead to a concentration of 560 ppmv, and a doubling of the current concentration of 360 ppmv would lead to a concentration of about 720 ppmv. (b) Carbon dioxide emissions (solid lines) leading to stabilization at concentrations of 450, 550, 750, and 1,000 ppmv corresponding to the profiles shown in Figure 1.16a, based on a midrange carbon-cycle model. The dashed curve shows one of many alternative emission paths to achieve the S550 stabilization objective. Modified from Schimel et al. (1996).

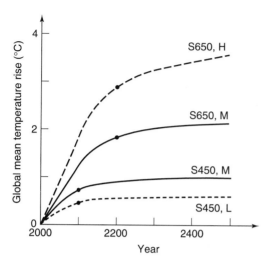

FIGURE 1.17. Global mean temperature increase corresponding to the pathways leading to stable CO_2 concentrations of 450 (S450) and 650 (S650) ppmv, for low (L), moderate (M), and high (H) climate sensitivities. The dots indicate date of concentration stabilization. The changes are due to increasing CO_2 concentrations alone (with aerosol concentrations held constant at 1990 levels). Modified from Houghton et al. (1996).

and S650.[10] The lowest (dashed) curve indicates the best case, which combines the S450 profile with the low (L) value of climate sensitivity. In this outcome, the temperature rise is only about 0.5°C in 2100 (which is a better result than the most optimistic IS92 scenario – IS92c, low, in Figure 1.12) and has stabilized below 0.7°C by 2500. The two solid curves in the middle of the graph show the results if the moderate (M) climate sensitivity is used with the S450 and S650 profiles. These cases yield global mean temperature increases in the range 0.8 to 1.3°C in 2100 (which is better than the midrange temperature rise exceeding 2°C for the IS92a scenario in Figure 1.12). By 2500, the corresponding range of temperature increase is 1.1 to 2.2°C. The uppermost (dashed) curve shows that for the S650 case with high (H) climate sensitivity, the global mean temperature rise is about 2°C in 2100 and 3.5°C by 2500 (which is far better than the worst IS92 outcome – IS92e, high, in Figure 1.12). Global temperatures stabilize centuries after atmospheric CO_2 concentrations have stabilized, due to the very long time lags inherent in the climate system (and especially the oceans). As noted earlier, very significant efforts will be required to reduce emissions in order to achieve even the S650 target profile.

Global Mean Sea Level Rise   Figure 1.18 indicates the global mean sea level changes corresponding to the S450 and S650 CO_2 concentration stabilization cases. In the best case (S450 combined with low climate and ice-melt sensitivities), the lowest (dashed) curve shows that sea level rise could be curtailed to about 10 cm up to and beyond 2500. The pair of solid lines show that

[10] These two cases bracket the value of 560 ppmv (or doubling of preindustrial CO_2 concentrations), which is the benchmark used in the assessment of climate change effects. In the S650 case, the CO_2 concentration rises above 560 ppmv in 2100, before stabilizing at 650 ppmv by about 2200.

FIGURE 1.18. Global mean sea level increase corresponding to the pathways leading to stable CO_2 concentrations of 450 (S450) and 650 (S650) ppmv, for low (L), moderate (M), and high (H) combined sensitivities for both climate and ice melt. The dots indicate dates of concentration stabilization. The changes are due to increasing CO_2 concentrations alone (with aerosol concentrations held constant at 1990 levels). Modified from Houghton et al. (1996).

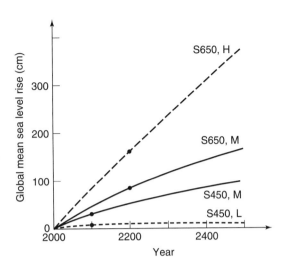

even with moderate climate and ice-melt sensitivities, the range of sea level rise for the S450 and S650 cases is about 25 to 30 cm in 2100 (which is better than the midrange expected sea level increase of 50 cm in the IS92a moderate scenario; see Figure 1.13) and around 85 to 145 cm by 2500. The figure also indicates that the S650 case combined with high climate and ice-melt sensitivities leads to a 70-cm mean sea level rise in 2100 (which is below the worst-case increase of 95 cm for the IS92e high scenario in Figure 1.13) and an alarming 325 cm increase by 2500. Once again, due to the very long time lags in the ocean system, the sea level will continue to rise for centuries after both CO_2 concentrations and global mean temperatures have stabilized.

Prospects for Stabilization   Table 1.3 represents another way of examining the emission–concentration link, and thereby assessing the prospects for achieving a given emission or concentration profile. It shows the cumulative carbon emissions between 1990 and 2100, corresponding to both the various stabilization cases and the IS92 curves. Two slightly different emission profiles are analyzed for each stabilization case, except S1000. For comparison, the emissions of carbon accumulated under the different IS92 scenarios are given in the lower portion of the table. Several interesting conclusions can be drawn from Table 1.3. First, clearly only emission profiles comparable to the rather stringent IS92c or d scenario provide any hope of stabilizing future equivalent CO_2 concentrations at or below the 550 ppmv level (or about double the preindustrial value). Second, while a given accumulation of atmospheric carbon may be achieved by following many different emission profiles, the stabilization of CO_2 concentrations anywhere below 1000 ppmv will require longer-term emissions to be reduced to well below current levels. Third, increasing the concentrations of other GHGs will significantly reduce the cumulative amount of CO_2 that could be emitted under any given stabilization case. In

TABLE 1.3 Emissions of Carbon Accumulated from 1991 to the End of the Twenty-first Century Leading to Stabilization of CO_2 Concentrations at 450, 550, 650, 750, and 1,000 ppmv

	Accumulated Carbon Emissions[a] (GtC)[b]
Stabilization case[c]	
S450	630–650
S550	870–990
S650	1,030–1,190
S750	1,200–1,300
S1000	1,410
IS92 emission scenario	
IS92c	770
IS92d	980
IS92b	1,430
IS92a	1,500
IS92f	1,830
IS92e	2,190

Note: For comparison, the accumulated emissions are also shown for the IS92 emission scenarios.

[a] The accumulated emissions during the period 1860–1994 were 360 GtC, roughly two-thirds of which were due to the burning of fossil fuels and the rest due to deforestation and changes in land use.

[b] 1 gigatonne (Gt) = 1 teragram (Tg) = 10^{18} grams; 1 Gt of carbon represents 3.67 Gt of CO_2.

[c] The range of values for accumulated carbon arises from the slightly different assumptions made for two alternative emission profiles corresponding to each stabilization case (except 1,000 ppmv).

this interpretation, each stabilization case is taken to represent the "equivalent" CO_2 concentration, which is defined as that concentration of CO_2 that will provide the same radiative forcing and climate change effects as any given mix of GHGs.

The likely constraints on future emissions required to achieve a reasonably stable CO_2 concentration can also be analyzed in terms of emissions either per capita or per unit of economic activity. For example, even to achieve S750 or S1000, global mean emissions per capita and per unit of output in the future would have to fall to under 50% of current values. Typically, the present average global per capita emissions are about 1.1 tonnes from fossil-based carbon and about 0.2 tonne from deforestation and land-use changes. Developing countries emit about 0.5 tonne of carbon per capita from fossil fuel burning (values range from 0.1 to 2), while the corresponding figure for industrialized countries (developed and transitional economies) is over five times more, or 2.8 tonnes per capita (values range from 1.5 to 5.5).

Similarly, stabilization of CO_2 concentrations at the higher levels (S750 or S1000) would require halving the present rates of energy use per unit of economic activity. For comparison, a thousand 1990 U.S. dollars' worth of output currently results in the emission of 0.3 tonne of carbon from energy use and a further 0.05 tonne from land-use changes, as a global average. Using market exchange rates, industrialized and developing countries emit 0.27 and 0.41 tonne, respectively, of energy-related carbon per thousand U.S. dollars of output.

In the next section, we summarize the likely ecological and social effects of climate change corresponding to a much more stringent benchmark value of CO_2 concentrations – that is, 560 ppmv, or a doubling of preindustrial levels. From the foregoing discussion, it is clear that the task of stabilizing future CO_2 concentrations at such a value will prove to be a very formidable challenge to decision makers. Thus, the effects described below are likely to occur anyway, and the hope is that policy measures will help us avoid even more severe consequences.

1.4   IMPACT ON ECOLOGICAL AND SOCIAL SYSTEMS

Both the magnitude and the rate of global climate change are likely to affect human well-being in numerous ways. In this section, we review the physical effects on ecological and social systems in a world with an atmospheric CO_2 concentration of 560 ppmv, while economic valuation or costing of such effects is discussed in Chapter 5. As indicated earlier, stabilizing the CO_2 concentration at 560 ppmv will require significant interventions to change current trends in human activity. Broadly stated, in such circumstances climate change will significantly alter weather patterns and terrestrial and aquatic ecosystems, which in turn will affect socioeconomic systems upon which human welfare depends. Major effects on human health will most likely occur as well. Most areas will experience adverse effects, but some will benefit from climate changes. Therefore, response strategies and the extent of adaptation will tend to vary widely across the globe.

Vulnerability to climate change will depend greatly on the economic circumstances and infrastructural capacity of nations. Climate-induced effects will impose significant additional stress on ecological and socioeconomic systems that are already burdened by pollution, natural resource scarcities, and other unsustainable practices. The more affluent and technologically advanced countries will be better prepared to respond to climate change, particularly by developing and establishing suitable institutional and social structures capable of dealing with the consequences. The poorer developing countries are likely to be the most severely affected, because of the socioeconomic, scientific, and technological limitations they already face.

Present assessments of the impact of climate change are qualitative rather than quantitative. Estimating the effects on a particular ecological or socioeconomic system is complicated by several factors. First, predictions on a regional scale are difficult and marked by high uncertainty. Second, the knowledge base and understanding of many critical functions are limited. Finally, ecological and socioeconomic systems are made up of numerous complex environmental as well as nonenvironmental components that are subject to multiple climatic and nonclimatic stresses, and their interactions are often nonlinear (or not additive).

Partly due to these complexities, most current models simply assess the consequences of a doubling of atmospheric CO_2 concentrations above preindustrial levels (i.e., up to 560 ppmv). Neither the dynamic effects from a steady increase in the concentration of GHGs over the same period of time nor the effects of more than a doubling of CO_2 concentrations have been systematically studied. The identification of a particular change as climate-induced will be very difficult, especially in the early stages of global warming. However, as future climate changes begin to exceed the typical parameters associated with past periods of climate change, human analysis will become increasingly limited by lack of knowledge, and the risk of unexpected and even catastrophic outcomes will grow.

1.4.1   Ecosystems and Natural Habitats

Terrestrial and aquatic ecosystems provide many of the goods and services upon which human societies depend. Major benefits that such natural habitats yield include (a) energy, food, fiber, and medicines; (b) recycling and storage of nutrients and carbon; (c) assimilation of wastes, purification of water, and regulation of water runoff; (d) control of floods, prevention of soil degradation, and limitation of coastal erosion; and (e) ecotourism and recreation. Ecosystems provide the principal reservoir of biological diversity. It is likely that global climate change will lead to a reduction in the goods and services that ecosystems provide, as well as a decline in genetic and species diversity.

As shown in Figure 1.19, mean annual temperature and annual precipitation can be correlated with the distribution of major biomes (or biological habitats) throughout the world. Fluctuations and changes in temperature and precipitation caused by climate change will affect the geographic distribution of these biomes. Adaptation to such change will vary among species of plants and animals, depending on their growth and reproductive cycles. As different species respond to climate change, the geographic location and composition

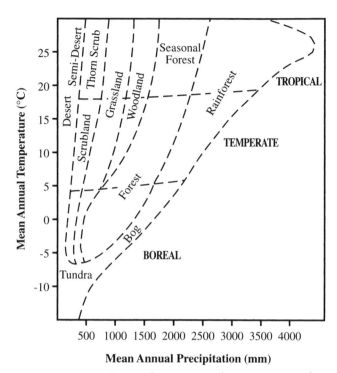

FIGURE 1.19. Correlation of mean annual temperature and mean annual precipitation with the distribution of the world's major biomes. While the role of these annual means appears to be important, the distribution of biomes may also strongly depend on seasonal factors such as the length of the dry season or the lowest absolute minimum temperature, on soil properties such as water-holding capacity, on history of land use such as for agricultural or grazing purposes, and on disturbance regime such as the frequency of fire. From Watson et al. (1996).

of entire ecosystems will shift. Some species will be able to migrate or adapt rapidly, while others will face extinction. However, it should be noted that non-climate-related conditions and factors such as soil properties and "disturbance regimes" also affect the distribution of biomes. Changes in the frequency of fires or the increasing presence of pests, which could also be associated with changes in the climate system, can affect the structure, functions, and productivity of ecosystems.

Both terrestrial and aquatic ecosystems are already under intense pressure due to increased human activity and a growing population base (Munasinghe and McNeely 1994). This will continue to be the major cause of damage to ecosystems. However, climate change could exacerbate the process by disrupting established ecosystem structures and functions, increasing their vulnerability, and inducing the formation of whole new systems. It may take several centuries for major ecosystems to return to equilibrium once they are disturbed (Karl, Meim, and Quayle 1991). Some typical effects on key ecosystems are reviewed in the following subsections.

Forests Forests affect the climate and are in turn influenced by the climate. The vegetation of forest ecosystems accounts for 80% of carbon above ground, and the soil beneath forests captures 40% of soil carbon (Melillo et al. 1990; Dixon et al. 1994; Watson et al. 1996). Therefore, forests are a crucial element of climate change in their capacity as both a carbon source (through deforestation and degradation of forests) and a carbon sink (via reforestation, afforestation, and enhanced growth caused by CO_2 fertilization). Forests also affect climate conditions extensively at the continental level due to the combined impact on ground temperatures, evapotranspiration, surface roughness, albedo, cloud formation, and precipitation (Henderson-Sellers, Dickinson, and Wilson 1988).

In general, the productivity of forests as well as the number of species inhabiting them depend on temperature, precipitation, and nutrient availabil-

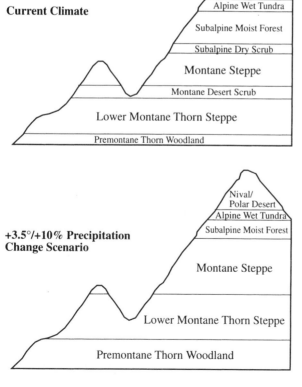

FIGURE 1.20. Comparison of current vegetation zones at a hypothetical dry temperate mountain site with simulated vegetation zones under a climate-warming scenario. From Beniston (1994).

ity – all of which will be affected by climate change. Significant variations in water availability (both drought or water logging) could result in the destruction of forests (Holridge 1967). Figure 1.20 illustrates how climate change can affect a change in vegetation zones on a dry temperate mountainside (Davis and Botkin 1985). Meanwhile, even a 1°C change in air temperature sustained over time is likely to change the growth and regeneration patterns of many forest species (Whitehead, Leathwick, and Hobbs 1993).

According to ecosystem models, on average about one-third of global forests (ranging from one-seventh to two-thirds, by region) will experience major changes in broad vegetation types if CO_2 levels double from preindustrial levels. Forests in high latitudes are likely to be most seriously affected, and those in the tropics, least changed (Greco et al. 1994). These models assume global temperature changes of between 1 and 4°C over the next century (as discussed in the preceding section), which could shift isotherms poleward by about 160 to 640 km. When compared with past migration patterns of forest species, which range from 4 to 200 km per century, it is clear that many forest species that are unable to migrate fast enough will probably disappear completely and be replaced by new ecosystems and groups of species (Kirschbaum and Fischlin 1996).

Forests have a large capacity to sequester carbon, thus serving as a vast carbon store. While the net primary productivity of forests will generally increase with increasing levels of carbon, some forests are also likely to disappear due to higher temperatures and an increase in the number of pests and pathogens. The net effect of these two phenomena on the level of carbon in the atmosphere is not clear from existing research. In the transition period, it is likely that carbon uptake through regrowth to maturity of some forests will be overtaken by the carbon released from other forests facing extinction, resulting in a large quantity of carbon being discharged into the atmosphere in the form of CO_2 – or worse, as CH_4.

Climate change will have the greatest impact on boreal forests. Warming effects will be most pronounced at high latitudes, where boreal forests are prevalent. Temperate forests will be affected to a lesser extent, and tropical forests will probably be least affected by climate change conditions. While increases in temperature and atmospheric carbon concentrations are likely to increase the net primary productivity of temperate forests, increased incidences of fire and pests will counter these effects. Tropical forests, which are sensitive to changes in the quantity and seasonality of rainfall, could be adversely affected by climate change.

Rangelands and Mountain Ecosystems Rangelands (i.e., unimproved grasslands, shrublands, savannas, deserts, and tundra) account for 61% of the earth's land surface, and support hundreds of millions of people (Lean, Hinrichsen, and Markham 1990; Prentice et al. 1992). About 50% of all livestock live off them (Solomon et al. 1993). The seasonal distribution and quantity of precipitation are crucial elements in the productivity of rangelands. Slight variations in extreme weather conditions and temperature can have a disproportionately large impact on their survival (Briske and Heitschmidt 1991).

As the global temperature rises, mountainous areas will become generally warmer and the distribution of vegetation will gradually shift toward higher elevations. Those species that are already limited to mountain peaks could face extinction as their habitats disappear and they are unable to migrate to a climate more conducive to survival.

Desertification and Land Degradation   Climate change is likely to make conditions in desert areas more extreme, as the weather will become hotter but not significantly wetter. Desert organisms that are already close to their heat tolerance thresholds will be threatened. Furthermore, lands that border on deserts – mainly arid, semiarid, and dry subhumid areas – will be most likely to undergo irreversible desertification, as the soils become drier and undergo further degradation due to compaction and erosion. Adaptation to desertification will depend on the development of crops and animals suited to such conditions (Noble and Gitay 1996).

Land degradation is likely to occur in some regions due to a greater fluctuation in weather patterns. Extended droughts followed by periods of heavy rainfall will degrade soil structure and reduce the soil's capacity to circulate organic matter and nutrients (Bullock and Le Houérou 1996). Even areas that receive more rainfall could be negatively affected by increased leaching, which causes acidification and nutrient loss.

Aquatic Ecosystems   Aquatic ecosystems include lakes and streams, nontidal wetlands, coastal environs, and oceans. Every sustained increase in air temperature of 1°C can shift the geographic range of species by about 150 km poleward. Temperature increases caused by climate change may alter the diversity and geographic distribution of species, the productivity of organisms in ecosystems, and the mixing properties of lakes – which would affect their primary productivity.

Although the impact of climate change on lake and stream ecosystems will vary by region, some broad generalizations can be made. The most favorable effects of warming will probably be felt at high latitudes, where biological productivity and species diversity are likely to increase. On the other hand, the most significant negative effects will be experienced by cold- or cool-water species in low latitudes, where extinction is likely to increase and biodiversity will decline. Runoff and groundwater flow into lakes and streams could also be altered by an increase in air temperature. This could change the clarity and productivity of water bodies. Such hydrological variations would be expected to have a greater impact on aquatic ecosystems than changes in mean flow rates or temperatures. Changes in the frequency and duration of flash floods or droughts could severely affect stream ecosystems.

Hydrological variations caused by climate change will affect the biological, biogeochemical, and hydrological functions of wetlands. About 4 to 6% of the earth's land surface is covered by wetlands, which serve several useful functions, such as the recharging of aquifers, sediment retention, waste processing, and carbon storage (Matthews and Fung 1987). The boreal and subarctic peatlands alone store about 20% of the global organic carbon captured by soils. An increase in air temperature could severely affect such wetlands by thawing the permafrost, which is crucial for maintaining high water tables in these ecosystems (Gorham 1994; Oechel and Vourlities 1994). For example, the drying of some arctic areas due to climate change has already caused them to shift from weak carbon sinks to weak carbon sources.

Coastal ecosystems include wetlands, coral reefs, estuaries, saltwater marshes, mangroves, and beaches. They are both economically and ecologically important. Research data and models indicate significant damage to these coastal ecosystems from climate change effects, such as sea level rise, changes in ocean temperature, and variations in rainfall patterns. Wetlands and lowlands would be displaced, shorelines eroded, estuaries and freshwater aquifers rendered more saline due to saltwater intrusion, and nutrient and sediment transport affected. Many of the valuable economic and ecological functions of such systems, including tourism, freshwater supplies, fisheries, storm and flood protection, and biodiversity, would be threatened by climate change.

Perhaps the greatest impact of climate change on many aquatic ecosystems would be the exacerbation of already existing stresses resulting from human activity. Over the past few decades coastal wetlands, saltwater marshes, and mangrove systems have disappeared at a rate of 0.5 to 1.5% per year in some regions. Temperature changes and sea level rise will accelerate these trends. If sea levels rise faster than sediment accretion rates, low-lying areas will become inundated. Thus, sandy beaches, coral reefs and atolls, and river deltas will be threatened. Of all marine ecosystems, coral reefs represent the greatest biological diversity. They are also very sensitive to environmental change. A temperature increase of just 1 or 2°C could not only cause "bleaching" of coral reefs and their eventual destruction, but also increase the risk of subsequent flooding.

1.4.2 Hydrology, Cryosphere, and Water Resources

As indicated earlier, a warming trend will intensify the global hydrological cycle and significantly affect regional conditions, especially the cryosphere (i.e.,

snow, ice, and permafrost). Between one-third and one-half of present mountain snow cover and glaciers will melt during the next century, thereby affecting the magnitude and seasonality of river flows that supply water to many hydropower, potable water, and irrigation schemes. Other hydrological shifts
and reductions in the area and depth of permafrost may cause heavy damage
to human infrastructure, release further CO_2 into the atmosphere, and also
alter the processes that contribute to natural CH_4 production. At the same
time, reductions in sea ice cover in the Arctic might open up sea lanes and
benefit shipping. The impact of temperature increases on polar ice sheets and
calving of icebergs is still being debated.

Like other effects, changes in the availability of water will exacerbate prevailing problems, such as pressure on limited water supplies caused by increasing population. Hydrological models show that water availability could vary
widely among nations and within nations. Experts differ in their projections as
to whether human water supply systems will advance sufficiently to counteract
the anticipated negative impact of climate change and increased demand.

Such factors as vegetation, projected demand for consumptive uses, and
population growth complicate efforts to assess the impact of climate change
on water resources. However, on the basis of general circulation models, it is
generally accepted that climate change will have a significant impact on regional water supplies. These models have so far been capable of providing
only large-scale geographic projections. Nevertheless, they indicate that even
minor changes in temperature and precipitation that are magnified by nonlinear effects involving evapotranspiration and soil moisture conditions can
have a significant impact on runoff into surface and groundwater tables, especially in arid and semiarid areas. Runoff may increase in high latitudes and decrease in lower latitudes due to the effects of evapotranspiration and precipitation.

In general, countries in arid and semiarid zones are at greater risk of expe-
riencing water shortages. They are typically characterized by low-volume total
runoff and infiltration. A widely used benchmark for water scarcity is 1,000 m³
per person annually. In several countries, including Kuwait, Jordan, Israel,
Rwanda, Somalia, Algeria, and Kenya, the availability of water will fall below
this level. In several other countries, including Libya, Egypt, South Africa,
Iran, and Ethiopia, availability is likely to fall below this benchmark in the next
two or three decades. Such countries are particularly vulnerable to reductions
in water supply. The quantity and quality of freshwater supplies in low-lying
coastal areas, river deltas, and small islands are also at considerable risk.

The impact of climate change on water resources will affect human well-
being to various degrees, depending largely on how country-specific water
management methods can accommodate such change. Wealthier countries

with sophisticated water management systems will be better prepared to deal with the consequences of climate change, whereas poorer nations that are more dependent on seasonal rainfall will be more vulnerable. In general, irrigation may be the first activity to be significantly affected in many countries facing water shortages.

1.4.3 Food and Agriculture

The overall impact of climate change on worldwide food production is considered to be low to moderate. On balance, with successful adaptation and adequate irrigation, global agricultural production could be maintained or increased – due to the CO_2 fertilization effect arising from the doubling of CO_2 concentrations. Agricultural pests and greater climatic variability may have negative consequences. Far greater and more localized risks are likely to arise from the regional effects. Temperate areas may actually benefit from climate change in terms of their food production. Tropical and subtropical areas are likely to suffer more negative consequences. Isolated agricultural systems located in arid and semiarid zones will be most seriously threatened by climate change. Global food supplies will become increasingly inadequate to meet human needs as climatic changes exacerbate existing stresses. The populations most at risk are those in sub-Saharan Africa, South, East, and Southeast Asia, and tropical areas of Latin America, as well as some Pacific island nations.

Climate change can affect food production systems in two ways: directly, through changes in temperature, water balance, and atmospheric composition, as well as extreme weather events; and indirectly, through changes in the distribution, frequency, and severity of pest and disease outbreaks, incidence of fire, weed infestations, or variations in soil properties. The impact of such direct and indirect effects on agricultural systems will manifest not only through the physiological responses of plants and animals responding to climate change, but also via fluctuating yields, which will have a negative impact on production and distribution systems.

Farming and other food production systems have historically had to adapt to changing economic, population, technological, and resource availability trends. It is unclear whether climate change will constitute a significant additional disruption to existing socioeconomic and environmental trends. In any event, adaptability to change will depend on access to advanced technology and skills, scientific research facilities, availability of water, soil characteristics, topography, and the genetic diversity introduced into crops. Many developing countries not only will be unable to afford the incremental cost of such adaptation, but also will have to cope with prevailing agricultural practices and policies that run counter to adaptation strategies.

Models suggest that intensively managed livestock systems will be more adaptable to climate change than crop systems. However, pastoral systems whose production is very sensitive to such change will be under greater threat.

The impact of climate change on fisheries will add to existing stresses caused by human-induced activities, such as overfishing, diminishing nursery areas, and inland and coastal pollution. IPCC projections indicate that if fishery management improves, climate change will emerge as the dominant negative impact on fishery production by the latter half of the next century. Freshwater fisheries in small rivers and lakes that lie in regions expected to experience larger temperature and precipitation changes are likely to suffer the greatest impact. Fishing communities restricted to exclusive zones and unable to adjust to changes in stock migration and regeneration would also be severely affected. In addition, fisheries in large rivers and lakes, in estuaries (especially those without migration paths or those affected by sea level rise or reduced river flows), and in high seas would be adversely affected in that order. The loss of wetlands might also have a negative impact on fishery production.

1.4.4 Human Health

Figure 1.21 illustrates the adverse effects on human health produced both directly and indirectly by climate change. Various adaptation techniques are available to counter them. However, very poorly developing countries in the tropics and subtropics, which would be most susceptible to such effects, might well be the least well equipped with the necessary technological, scientific and economic resources and the organizational capacity to implement such response strategies (Chatterjee, Munasinghe, and Ganguly 1997). It would be important to ensure that such capability were strengthened before the impact of climate change became significant.

One major direct consequence of climate change would be an increase in heat-related deaths and illnesses (primarily from cardiorespiratory failure). Studies have shown that heat-related deaths could increase severalfold if climate change projections hold. Correspondingly, deaths due to cold weather conditions would most likely decrease as a result of global warming. An increase in the frequency and magnitude of extreme weather conditions would also increase the number of fatalities, particularly in regions with high population densities and poor infrastructure (i.e., developing countries).

The extension of geographic ranges and seasons favoring vector-borne organisms would be likely to result in an increase in such diseases as malaria, dengue, yellow fever, and encephalitis. In particular, models show that changes in basic climate variables would result in an expansion of the areas under the influence of the malaria mosquito. This indirect effect of climate

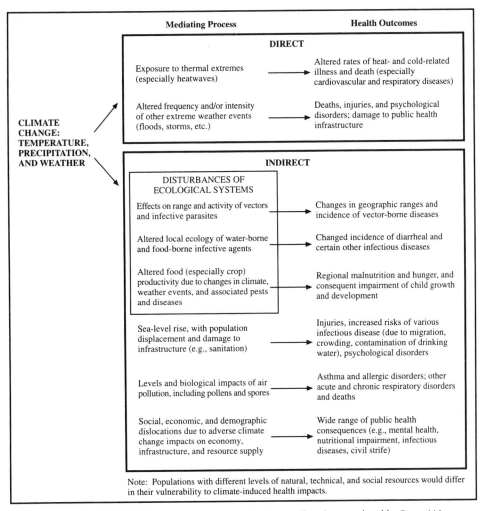

FIGURE 1.21. Ways in which climate change is likely to affect human health. From Watson et al. (1996).

change could increase the global population exposed to malaria from the current 45% to 60% by the latter half of the next century. Temperate regions would experience the greatest expansion in transmission area. However, actual increases in the number of people with malaria – estimated to be between 50 and 80 million – would occur mainly in the tropical, subtropical, and poorly protected temperate zones.

Increases in cases of asthma, allergic disorders, and cardiorespiratory diseases would probably also occur due to climate-induced changes in pollens and spores and to temperature increases that enhance the formation, persistence, and respiratory impact of certain air pollutants.

Finally, other effects of climate change, such as food and water shortages, would exacerbate hunger, malnutrition, and various health impairments, thereby making humans more vulnerable to non-climate-related causes of disease and death.

1.4.5 Human Infrastructure and Habitats

Climate change can have both direct and indirect effects on human infrastructure and habitats. Changes in temperature, precipitation, and sea level or extreme weather events can directly damage physical infrastructure and affect outputs. Indirect effects are likely to be felt via markets sensitive to climate change, such as energy for heating and cooling, and through changes in resources affected by climate change, such as agroindustries and biomass production. Effects on human infrastructure could be exacerbated by human migration caused by large-scale flooding, destruction of crops, droughts, or spread of disease.

Generally, the impact of climate change on hard infrastructure sectors such as energy, transport, and industry will be much milder than that on agricultural and ecological systems (discussed in the preceding section). Transportation infrastructure located in vulnerable coastal zones and permafrost regions will be the most seriously affected by climate change. The significant impact of such change on water supply systems has already been reviewed. In particular, stormwater management systems in industrialized (and urbanized) countries are likely to be at risk. Renewable energy sources such as hydroelectricity and biomass will also be affected. The impact on hydroelectric power generation will depend on the frequency and quantity of rainfall and evaporation. The overall impact on thermal power generation will be minimal. Deforestation and other effects of climate change on forests will reduce the availability of fuelwood. Many nations will face the risk of losing capital valued at over 10% of GDP.

Coastal communities will be particularly vulnerable to inundation and erosive land loss due to sea level rise, with consequent flooding of homes and places of work, as well as threats to health and safety. Currently, about 46 million people are at risk every year due to flooding. Assuming no increases in population and no adaptive behavior, a 50-cm rise in mean sea level (the midrange projection for 2100) would potentially affect more than 90 million individuals. Under the same conditions, a 1-m sea level increase (which is the upper end of the IPCC range of values predicted for 2100) would put almost 120 million people at risk. It is worth noting that this is a static estimate at a given point in time, under conservative assumptions. Since the global population will increase and the sea level is likely to continue rising beyond 2100 (for any reasonable scenario), the population affected by flooding will also increase unless adaptive or defensive measures are undertaken.

Developing nations with high population densities and relatively weak coastal defense systems are especially likely to face the problems of internal or even international forced migration of a large number of people – for example, about 70 million each in Bangladesh and China will be affected by a 1-m rise in mean sea level. Small islands and river deltas will also be seriously affected by such an increase in sea level. Estimated land losses as a fraction of total area would range from 80% for Majuro Atoll in the Marshall Islands to 17.5% for Bangladesh, 6% for the Netherlands, 1% for Egypt, and 0.05% for Uruguay.

1.5 SUMMARY AND RATIONALE FOR ACTION

If the enhanced greenhouse effect is indeed the outcome of human activities, then those activities might well be modified in order to curb the potential impact of global climate change. Deciding whether or not to act, and what types of response strategies might be adopted, depends crucially on the likelihood of significant changes in the global climate and the gravity of the consequences.

The evidence presented in this chapter indicates that human actions such as the burning of fossil fuels, deforestation, agricultural practices, and land-use changes have resulted in an unprecedented increase in the concentrations of GHGs in the atmosphere. The world's leading experts working under the aegis of the IPCC have recently concluded that increases in the global mean surface temperature during the past century are unlikely to have been caused entirely by natural effects, and that changes in both the average temperature and the geographic, seasonal, and vertical patterns of temperature indicate the influence of human actions on global climate. Climate models also predict that GHG emissions arising from any reasonable range of future economic activities will lead to increasing concentrations of atmospheric GHGs and result in significant changes in average global and regional temperatures, mean sea level, and related indicators such as precipitation and soil moisture content.

The effects of global climate change that could be potentially serious over the next century include regional increases in floods and droughts, inundation of coastal areas, high-temperature events, fires, outbreaks of pests and diseases, significant damage to ecosystems, and threats to agricultural production. Climate changes will also pose a major risk to human health and safety, especially among poorer communities with high population densities in areas like river basins and low-lying coastal plains, which are vulnerable to climate-related natural hazards such as storms, floods, and droughts.

The risk of catastrophic consequences, the complexity of the issues and analyses, and uncertainties in our knowledge suggest that, at the very least, considerable effort should be devoted to investigating the problem. Furthermore, uncertainty does not constitute valid grounds for inaction. The precautionary approach (enunciated in Article 3.3 of the UNFCCC) suggests that steps be taken to anticipate, mitigate, or minimize the causes of global climate change without undue delay. Common sense and simple economics also indicate that the least expensive measures should be adopted first. The remaining chapters of this book examine how humanity might collectively decide on a course of action to mitigate global climate change, what options seem the most promising, and how decisions could best be implemented.

REFERENCES

Alcamo, J., A. Bouwman, J. Edmonds, A. Grübler, T. Morita, and A. Sugandhy. 1995. An evaluation of the IPCC IS92 emission scenarios. In J. T. Houghton et al. (eds.), *Climate change 1994: Radiative forcing of climate change and an evaluation of the IPCC IS92 emission scenarios.* Cambridge University Press, pp. 247–304.

Beniston, M. (ed). 1994. *Mountain environments in changing climates.* London: Routledge.

Briske, D. D., and R. K. Heitschmidt. 1991. An ecological perspective. In R. K. Heitschmidt and B. Stoth (eds.), *Grazing management.* Portland, OR: Timber Press, pp. 11–27.

Bullock, P., and H. Le Houérou. 1996. Land degradation and desertification. In R. T. Watson et al. (eds.), *Climate change 1995: Impacts, adaptations and mitigation of climate change – Scientific-technical analyses.* Contribution of Working Group II to the Second Assessment Report of the Intergovernmental Panel on Climate Change. Cambridge University Press, pp. 191–214.

Chatterjee, M., M. Munasinghe, and M. Ganguly (eds.). 1997. *Health and environment in developing countries.* Delhi: Macmillan (India).

Davis, M. B., and D. B. Botkin. 1985. Sensitivity of cool-temperature forests and their fossil pollen record to rapid temperature change. *Quarternary Research,* **23,** 327–40.

Dickinson, R. E., V. Meleshko, D. Randall, E. Sarachik, P. Silva-Dias, and A. Slingo. 1996. Climate processes. In J. T. Houghton et al. (eds.), *Climate change 1995: The science of climate change.* Contribution of Working Group I to the Second Assessment Report on the Intergovernmental Panel on Climate Change. Cambridge University Press, pp. 193–228.

Dixon, R. K., S. Brown, R. A. Houghton, A. M. Solomon, M. C. Trexler, and J. Wisneiwski. 1994. Carbon pools and flux of global forest ecosystems. *Science,* **263,** 185–90.

Dutton, E. G., and R. Christy. 1992. Solar radiative forcing at selected locations and evidence for global lower tropospheric cooling following eruptions of El Chicon and Pinatubo. *Geophysical Research Letters,* **19,** 2313–16.

Eswaran, H., E. Van den Berg, and P. Reich. 1993. Organic carbon in soils of the world. *Soil Science Society of America Journal,* **57,** 192–4.

Gates, W. L., et al. 1996. Climate models – Evaluation. In J. T. Houghton et al. (eds.), *Climate change 1995: The science of climate change.* Contribution of Working Group I to the Second Assessment Report on the Intergovernmental Panel on Climate Change. Cambridge University Press, pp. 229–84.

Gorham, E. 1994. The future of research in Canadian peatlands: A brief survey with particular reference to global change. *Wetlands,* **14,** 206–15.

Graf, H. F., I. Kirchbner, A. Robock, and I. Schult. 1993. Pinatubo eruption winter climate effects. *Climate Dynamics,* **9,** 81–93.

Greco, S., R. H. Moss, D. Viner, and R. Jenne. 1994. *Climate scenarios and socioeconomic projections for IPCC WG II assessment.* Washington, DC: IPCC–WMO and UNEP.

Henderson-Sellers, A., R. E. Dickinson, and M. F. Wilson. 1988. Tropical deforestation: Important process for climate change models. *Climate Change,* **13,** 43–67.

Holdridge, L. R. 1967. *Life zone ecology.* San Jose, Costa Rica: Tropical Science Center.

Houghton, J. T., B. A. Callander, and S. K. Varney (eds.). 1992. *Climate change 1992: The supplementary report to the IPCC scientific assessment.* Cambridge University Press.

Houghton, J. T., et al. (eds.). 1995. *Climate change 1994: Radiative forcing and an evaluation of the IPCC IS92 emission scenarios.* Cambridge University Press.

Houghton, J. T., et al. (eds.).1996. *Climate change 1995: The science of climate change.* Contribution of Working Group I to the Second Assessment Report of the Intergovernmental Panel on Climate Change. Cambridge University Press.

Ingram, W. J., C. A. Wilson, and J. F. B. Mitchell. 1989. Modelling climate change: An assessment of sea-ice and surface albedo feedbacks. *Journal of Geophysical Research,* **94,** 8609–22.

Karl, T. R., R. R. Heim, Jr., and R. Quayle. 1991. The greenhouse effect in Central North America: If not now, when? *Science,* **251,** 1058–61.

Kattenberg, A., et al. 1996. Climate models – Projections of future climate. In J. T. Houghton et al. (eds.), *Climate change 1995: The science of climate change.* Contribution of Working Group I to the Second Assessment Report on the Intergovernmental Panel on Climate Change. Cambridge University Press, pp. 285–357.

Kirschbaum, M., and A. Fischlin. 1996. Climate change impacts on forests. In R. T. Watson et al. (eds.), *Climate change 1995: Impacts, adaptations and mitigation of climate change – Scientific-technical analyses.* Contribution of Working Group II to the Second Assessment Report of the Intergovernmental Panel on Climate Change. Cambridge University Press, pp. 95–130.

Lean, G., D. Hinrichsen, and A. Markham. 1990. *Atlas of the environment.* New York: Prentice-Hall.

Matthews, E., and I. Fung. 1987. Methane emissions from natural wetlands: Global distribution, area, and environmental characteristics of sources. *Global Biogeochemical Cycles,* **1,** 61–86.

Melillo, J. M., et al. 1990. Effects on ecosystems. In J. T. Houghton, G. J. Jenkins, and J. J. Ephraums (eds.), *Climate change: The IPCC scientific assessment.* Report prepared for IPCC by Working Group I. Cambridge University Press, pp. 283–310.

Munasinghe, M. 1996. Analyzing economic policy issues in climate change. Paper presented at the Yale–NBER Conference on Climate Change, Snowmass, CO, July.

Munasinghe, M., and J. McNeely (eds.). 1994. *Protected area economics and policy.* Geneva and Washington, DC: World Conservation Union (IUCN) and the World Bank.

Nicholls, N., G. V. Gruza, J. Jouzel, T. R. Karl, L. A. Ogallo, and D. E. Parker. 1996. Observed climate variability and change. In J. T. Houghton et al. (eds.), *Climate change 1995: The science of climate change.* Contribution of Working Group I to the Second Assessment Report on the Intergovernmental Panel on Climate Change. Cambridge University Press, pp. 133–82.

Noble, I. R., and H. Gitay. 1996. Deserts in a changing climate: Impacts. In R. T. Watson et al. (eds.), *Climate change 1995: Impacts, adaptations and mitigation of climate change – Sci-*

entific-technical analyses. Contribution of Working Group II to the Second Assessment Report on the Intergovernmental Panel on Climate Change. Cambridge University Press, pp. 159–70.

Oechel, W. C., and G. L. Vourlities. 1994. The effects of climate change on land–atmosphere feedbacks in arctic tundra regions. *Tree,* **9,** 324–9.

Potter, C. S., et al. 1996. Terrestrial ecosystem production: A process model based on global satellite and surface data. *Global Biogeochemical Cycles,* **7,** 811–41.

Prentice, K. C., W. Cramer, S. P. Harrison, R. Leemans, R. A. Monserud, and A. M. Solomon. 1992. A global biome model based on plant physiology and dominance, soil properties, and climate. *Journal of Biogeography,* **19,** 117–34.

Schimel, D., et al. 1995. CO_2 and the carbon cycle. In J. T. Houghton et al. (eds.), *Climate change 1994: Radiative forcing of climate change and an evaluation of the IPCC IS92 emission scenarios.* Cambridge University Press, pp. 35–72.

Schimel, D., et al. 1996. Radiative forcing of climate change. In J. T. Houghton et al. (eds.), *Climate change 1995: The science of climate change.* Contribution of Working Group I to the Second Assessment Report on the Intergovernmental Panel on Climate Change. Cambridge University Press, pp. 65–132.

Seigenthaler, U., and J. L. Sarmiento. 1993. Atmospheric carbon dioxide and the ocean. *Nature,* **365,** 119–25.

Solomon, A. M., I. C. Prentice, R. Leemans, and W. P. Cramer. 1993. The interaction of climate and land use in future terrestrial carbon storage and release. *Water, Air, and Soil Pollution,* **70,** 595–614.

Trenberth, K. E., J. T. Houghton, and L. G. Meira Filho. 1996. The climate system: An overview. In J. T. Houghton et al. (eds.), *Climate change 1995: The science of climate change.* Contribution of Working Group I to the Second Assessment Report of the Intergovernmental Panel on Climate Change. Cambridge University Press, pp. 51–64.

UNFCCC (United Nations Framework Convention on Climate Change). 1992. Articles. New York: United Nations.

Warrick R. A., C. Le Provost, M. F. Meier, J. Oerlemans, and P. L. Woodworth. 1996. Changes in sea level. In J. T. Houghton et al. (eds.), *Climate change 1995: The science of climate change.* Contribution of Working Group I to the Second Assessment Report on the Intergovernmental Panel on Climate Change. Cambridge University Press, pp. 359–406.

Watson, R. T., M. C. Zinyowera, R. H. Moss, and D. J. Dokken (eds.). 1996. *Climate change 1995: Impacts, adaptations and mitigation of climate change – Scientific-technical analyses.* Contribution of Working Group II to the Second Assessment Report on the Intergovernmental Panel on Climate Change. Cambridge University Press.

Whitehead, D., J. R. Leathwick, and J. F. F. Hobbs. 1993. How will New Zealand's forests respond to climate change? Potential changes in response to increasing temperature. *New Zealand Journal of Forestry Science,* **22,** 39–53.

2

METHODOLOGICAL AND CONCEPTUAL ISSUES

This chapter sets out the main methodological and conceptual difficulties that arise in the analysis of global climate change. The relationship between sustainable development and global warming, which is described in the next section, provides a useful contextual background. The chapter continues with a discussion of the chief complications that hinder the analysis of climate change, the problems created by uncertainty, and equity and social issues. Methods of dealing with these difficulties are described in subsequent chapters.

2.1 SUSTAINABLE DEVELOPMENT AND CLIMATE CHANGE

Sustainable development is an overarching objective for human society that has emerged at the end of the twentieth century (WCED 1987). The interaction between sustainable development and global climate change is especially important in view of the wide-ranging impact that the latter is likely to have (see Chapter 1). Furthermore, the UN Framework Convention on Climate Change (UNFCCC) has recognized this relationship explicitly in Article 2, which states (inter alia) that the stabilization of greenhouse gas (GHG) concentrations "should be achieved within a time-frame sufficient to . . . enable economic development to proceed in a sustainable manner" (UNFCCC 1992).

The state of the environment is a major worldwide concern today. Pollution in particular is perceived as a serious threat in industrialized countries, where the quality of life has hitherto been measured mainly in terms of growth in material output. Meanwhile, environmental degradation has become a serious impediment to economic development and the alleviation of poverty in the developing world.

2.1.1 The Economic, Social, and Environmental Dimensions of Sustainable Development

Humanity's relationship with the environment has gone through several stages. In primitive times, human beings lived in a state of symbiosis with nature. This was followed by a period of increasing mastery over nature up to the industrial age, culminating in the rapid material-intensive growth of the twentieth century, which adversely affected the environment in many ways. The initial reaction to such environmental damage was a reactive approach characterized by an increase in cleanup activities. Most recently, attitudes toward the environment have evolved to encompass the more proactive design of projects and policies that seek to anticipate and limit environmental degradation – though not always successfully. The ongoing effort to predict and address the future consequences of global climate change is undertaken very much in this spirit.

In this proactive context, human beings worldwide are currently exploring the concept of sustainable development – an approach that will permit continuing improvements in the present quality of life at a lower intensity of resource use, thereby leaving behind for future generations an undiminished or even enhanced stock of natural resources and other assets. While no universally accepted practical definition of sustainable development yet exists, there is increasing agreement that it should incorporate three critical elements – economic, social, and environmental – in a balanced manner (see Box 2.1).

BOX 2.1   Approaches to Sustainable Development

Current approaches to the concept of sustainable development draw on the experience of several decades of development efforts. Historically, the development of the industrialized world focused on material production. Not surprisingly, therefore, the model followed by developing nations in the 1950s and 1960s was dominated by output and growth, based mainly on the concept of economic efficiency. By the early 1970s the large and growing number of poor in the developing world, and the dearth of "trickle-down" benefits accruing to them, led to greater efforts to improve income distribution directly. The development paradigm shifted toward equitable growth, where social (distributional) objectives, especially poverty alleviation, were recognized as distinct from, and as important as, economic efficiency.

Protection of the environment has now become the third major objective of development. By the early 1980s, a large body of evidence had accumulated that environmental degradation was a major barrier to devel-

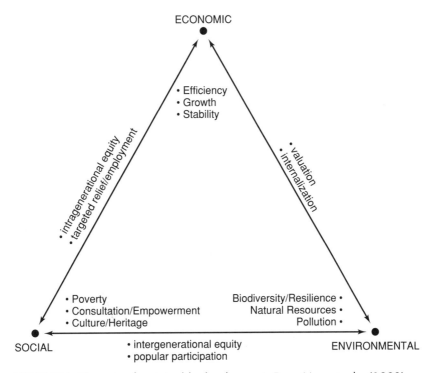

FIGURE B2.1. Elements of sustainable development. From Munasinghe (1993).

opment. The concept of sustainable development has therefore evolved to encompass three elements: economic, social, and environmental, as shown in Figure B2.1 (Munasinghe 1993).

The economic approach to sustainability is based on the concept of the maximum flow of income that could be generated while at least maintaining the stock of assets (or capital) that yield these benefits. (This concept is attributed to Hicks and Lindahl.) An underlying concept of optimality and economic efficiency is applied to the use of scarce resources. Problems of interpretation arise in identifying the kinds of capital (e.g., manufactured, natural, and human) to be maintained and their substitutability, as well as in valuing these assets, particularly ecological resources. The issues of uncertainty, irreversibility, and catastrophic collapse pose additional difficulties.

The social concept of sustainability is people-oriented and seeks to maintain the resilience of social and cultural systems and their capacity to withstand shocks. Greater equity and the reduction of destructive conflicts are important aspects of this approach. Preservation of cultural diversity and cultural capital across the globe, and better use of knowledge concerning sustainable practices embedded in less dominant cultures, are desirable. Modern society must encourage and incorporate

(continued)

BOX 2.1 Approaches to Sustainable Development *(continued)*

pluralism and grassroots participation into a more effective decision-making framework for socially sustainable development.

The environmental view of sustainable development focuses on the resilience of biological and physical systems. Of particular importance is the viability of subsystems that are critical to the global stability of the overall ecosystem. Furthermore, "natural" systems and habitats may be interpreted broadly to include man-made environments like cities. The emphasis is on preserving the resilience and dynamic capacity of such systems to adapt to change, rather than the conservation of some "ideal" static state. Natural resource degradation, pollution, and loss of biodiversity reduce system resilience.

Reconciling these various concepts and operationalizing them as a means to achieve sustainable development is a formidable task, since all three elements of sustainable development must be given balanced consideration. The interfaces among the three approaches are also important. Thus, the economic and social elements interact to give rise to such issues as intragenerational equity (income distribution) and targeted relief for the poor. The economic–environmental interface has yielded new ideas on valuation and internalization of environmental effects. Finally, the social–environmental linkage has led to renewed interest in areas like intergenerational equity (rights of future generations) and grassroots participation.

In seeking to integrate the three approaches in a practical way, it is useful to recognize that most development decisions continue to be based on the economic efficiency criteria. Thus, it is useful to turn to the relatively new area of environmental economics as a starting point for developing a broader conceptual framework that integrates the economic, sociocultural, and ecological approaches (Munasinghe 1993). For example, economists attempt to incorporate environmental concerns into decision making by valuing environmental resources in monetary terms and ensuring that resource prices reflect their scarcity values. Similarly, economists have addressed social equity concerns by placing special emphasis on costs and benefits accruing to the poor, by ensuring that those who impose costs on others pay commensurate charges, and, more recently, by seeking to protect productive assets for future generations.

The foregoing suggests a broad integrated conceptual approach in which the net benefits of economic activities are maximized, subject to maintaining the stock of productive assets over time, and providing a social safety net to meet the basic needs of the poor. Some analysts support a "strong sustainability" rule, which requires the separate preservation of each category of critical assets (e.g., manufactured, natural, sociocultural, and human capital), assuming that they are complements rather than substitutes. Other researchers favor a "weak sustainability," which seeks to maintain the aggregate monetary value of the total stock of assets, assuming a high degree of substitutability among the various types of asset. At the same time, the underlying basis of economic valuation, optimization, and efficient use of resources may not be easily applied to ecological objectives like protecting biodiversity or to social goals such as promoting public participation and empowerment – thereby forcing reliance on noneconomic indicators of social and environmental status, as well as on techniques like multicriteria analysis (see Chapter 4) to facilitate trade-offs among a variety of such noncommensurable objectives. Furthermore, uncertainty about the future will require the use of methods based on decision analysis.

Source: Munasinghe (1993).

2.1.2 How Climate Change Might Affect Prospects for Sustainable Development

The climate change problem fits quite handily within the conceptual framework for sustainable development just described. In fact, the discussion in Chapter 1 clearly indicates that global warming will have serious implications for all three elements. Global warming poses a significant threat to future economic activities and the well-being of a large number of human beings (see Chapters 5 to 8 for details of potential economic damage and costs of abatement). The economic approach is to maximize the net benefits from use of the global resource represented by the atmosphere. The stock of atmospheric assets that provide a sink function for GHGs has to be maintained at a level that ensures future benefits (in terms of avoiding damage due to climate change) that equal or just exceed the costs of measures required to restore the sink function to that level. The underlying principles are based on optimality and the economically efficient use of a scarce resource, that is, the global atmosphere.

The ecological approach is to ensure the stability and viability of the global ecosystem. Increasing anthropogenic emissions and accumulations of GHGs will significantly perturb one of the subsystems (the atmosphere) to the point where the global mean temperature and climate will change at a rate that might threaten the stability of the overall global ecosystem (see Chapter 1). An important goal will be to determine the limits of climate change within which the resilience of the global system can be maintained at an adequate level (Munasinghe and Shearer 1995). In turn, the accumulation of GHGs in the atmosphere will have to be sufficiently constrained to prevent climate change from exceeding safe margins.

Climate change could also undermine the social elements of sustainable development in an unprecedented manner. Both intra- and intergenerational equity (i.e., spatial and temporal fairness) could worsen due to the uneven distribution of the costs of damage due to climate change, as well as of required adaptation and mitigation efforts. Some social aspects worth considering include (a) the establishment of an equitable and participative global framework for making and implementing collective decisions about climate change; (b) the reduction of the potential for social disruption and conflicts arising from climate change effects; and (c) the protection of threatened cultures and the preservation of cultural diversity (particularly in small islands, where the impact of a sea level rise will be greatest).

The different approaches to sustainable development often employ different criteria, which could result in potentially conflicting decisions. The advantages and disadvantages of a variety of decision-making criteria are discussed further in Chapter 4. In the context of sustainable development, both public and private sector decision makers in most countries still rely on a basically economic-financial framework to manage their activities and make development decisions. Furthermore, the techniques of environmental economics, which focus on the impact of human activities on natural resources, ecosystems, and social systems, provide an important bridge among the three approaches (for details see Munasinghe 1993).

These considerations suggest that an integrated and comprehensive framework for analyzing the climate change problem might begin with the concept of maximizing net benefits of development, subject to maintaining the services provided by the stock of economic, social, and environmental resources, over time. In this context, net benefits are defined as the benefits derived from development activities minus the costs incurred in carrying out those actions (see also Chapter 3, especially Box 3.1). This approach implies that wastes should be generated at rates less than or equal to the assimilative capacity of the environment – in particular, the global atmosphere. Renewable resources, especially if they are scarce, should be utilized at rates less than or equal to the natural rate of regeneration. The efficiency with which nonrenewable resources are used should be optimized subject to substitutability between these resources and technological progress. Finally, both intra- and intergenerational equity (especially the alleviation of poverty), as well as a pluralistic and consultative social framework, are important additional considerations. Such an integrative framework would also help to incorporate climate change response measures within a national sustainable development strategy (see Chapter 4).

2.2 **PRINCIPAL COMPLICATIONS**

All approaches to understanding and dealing with climate change are hampered by formidable complications, which are explored in this section.

2.2.1 Causal Chain

The chain of causality that is associated with global climate change is a convenient starting point for understanding the magnitude of the climate change phenomenon and attendant complications. The main relationships – each of which is characterized by significant uncertainty – were presented in Figure 1.1 (see also Munasinghe et al. 1996). The causal chain leads from GHG emissions and concentrations to damage caused by climate change, followed by a succeeding set of linkages from proposed abatement measures to net reductions in GHG emissions.

In Figure 1.1, linkage 1 (which connects modules A and B) shows how net emissions of GHGs lead to increasing atmospheric concentrations. There are major scientific uncertainties associated with determining the quantitative relationship between emissions and concentrations, including the multitude of sources and sinks and their interactions, and the number of GHGs with their varying lifetimes and global warming potentials. The second link (connecting modules B and C) relates atmospheric concentrations of GHGs to geophysical effects such as changes in temperature and precipitation, soil moisture content, sea level rise, and the frequency and severity of extreme weather events. Once again, there are formidable problems in accurately predicting such effects – for large and complex systems, over long periods of time, and in specific regions and locations of the world.

Linkage 3 (which connects modules C and D) explores the extent of economic, social, and ecological damage caused by geophysical changes. While Article 2 of the UNFCCC has highlighted potential effects on ecosystems and food production, the full range of such effects is much larger and difficult to enumerate taxonomically. As discussed in Chapter 1, the broadest categories of impact concern ecosystems and natural habitats, hydrology and water resources, food and agriculture, human infrastructure and habitats, and human health. The complexity of relatively unknown relationships within and among large and often closely linked physical, ecological, and socioeconomic systems, as well as the multidisciplinary nature of the required analysis, poses great problems. Economic valuation of the many potential consequences or the assignment of comparative weights to such effects is important in establishing both the overall magnitude of damage and relative priorities. This step also has many theoretical and practical pitfalls.

Linkage 4 (connecting modules D and E) is the critical one. It determines the response strategy that will address the potential damage from global climate change. Before this step is taken, however, the range of feasible technological options, policy instruments, and adaptive responses must be identified and well understood. The costs of implementing such measures must also be determined. Linkage 5 (which connects modules E and F) helps to establish how responsibilities for abatement measures will be assigned, how emission rights will be allocated, and what the international and national implementing mechanisms might be. A related question involves the extent to which those responsible for GHG emissions (both past and present) might compensate others who suffer the costs of climate change and how such transfers could be ensured. The implementing mechanisms will have a feedback effect on the response strategy, as shown by the reverse linkage 5. The main difficulties associated with the decision-making process related to linkages 4 and 5 have to do with sociopolitical and equity issues. While there are scientific and technical questions at this stage as well, the uncertainty and lack of knowledge associated with them are significantly less pronounced than in the earlier stages.

The sixth and final linkage closes the chain in two ways, through the implementation of the response strategy. First, the response strategy will result in the abatement of GHG emissions, which will eventually reduce global climate change. Second, it will lead to the adoption of adaptive measures that will directly reduce vulnerability to climate change. There will be many uncertainties (both scientific and socioeconomic) at this stage, as relatively new and untried policies and technologies are implemented. Typical questions include how well the technological options might work, how quickly GHG emissions will respond to abatement measures, and how individuals will react to the policies.

For clarity it is useful to group into several categories the multitude of difficulties identified in the foregoing discussion. We begin in the following subsection with an explanation of several generic issues that complicate climate change analysis. These issues compound the difficulties described in subsequent sections, under two main categories: (a) uncertainty and (b) equity and social issues.

2.2.2   Generic Issues

The generic problems that pervade climate change analysis include the very large spatial and temporal scales of events, complexity, irreversibility, nonlinear behavior, and catastrophic collapse. Both the worldwide spatial extent and centuries-long temporal dimension are considerably greater than the normal range of magnitudes for which conventional techniques of analysis have been designed. The physical, ecological, economic, and social systems and subsystems involved are extremely large and complex. Their interactions are not well known and require multidisciplinary analysis. Another unprecedented complication is the fact that every human being, present and future, has a stake in the outcome. Determining who will bear the costs and who will benefit, allocating rights and responsibilities for damage and abatement, and establishing collective decision-making procedures and implementation mechanisms for agreed-upon response strategies are typical tasks to be undertaken on a global scale. At the same time, every link in the causal chain from emissions to impact involves decades- or centuries-long physical and ecological processes and corresponding time lags. Predictions about the structure of socioeconomic systems and human behavior for at least the next hundred years are equally problematic, but are essential for assessing the consequences of climate change and the appropriate responses.

The potential irreversibility of some effects is another major complication, especially in the context of the long gestation periods and inertia of the systems involved. For example, since the effective agent is the cumulative stock of GHGs, past emissions may already have committed us to an unacceptably high risk of global warming. Furthermore, we could well discover too late that the consequences of ongoing or future emissions are irreversible. Thus, a failure to reduce emissions in the short to medium term might have an irreversible impact in the sense that once the effects of climate change become apparent, it will be too late to do anything about them. If some of these consequences turn out to be catastrophic as well, human beings will literally have no other planet to migrate to. Catastrophic outcomes are more likely to occur if potential nonlinearities are involved in the entire causal chain – that is, the responses of systems are not proportional to the stimuli to which they are subjected. If global-scale systems (climate, ecological, or social) are perturbed outside their normal or stable states of equilibrium, positive feedbacks could lead to instability and unpredictable "flipping" of these large systems to new states, which may have unpleasant consequences for humanity. In other words, we are dealing with planetary-scale mechanisms that are little understood and may become uncontrollable. The foregoing indicates the merits of a cautious approach, especially if humanity is averse to taking risks on a global scale (see Chapter 3 for a related discussion of the precautionary principle).

2.3 UNCERTAINTY

In this section we discuss the different forms of uncertainty in two broad categories: (a) scientific uncertainty, which arises due to limited knowledge, mainly in the case of physical and natural systems; and (b) socioeconomic and technological uncertainty, which is inherent in predictions about human systems and infrastructure. Much of this uncertainty is linked to the generic issues mentioned earlier, that is, the large spatial and temporal scales, as well as the complexity of the systems involved. Key related concepts such as aversion to risk, the precautionary approach, and the importance of robust policies that are sound under a variety of circumstances are taken up in Chapter 3, and Chapter 4 outlines techniques such as decision analysis for dealing with uncertainty.

2.3.1 Scientific Uncertainty

To begin with, the processes involved in the biological and geochemical cycling of GHGs, aerosols, and aerosol precursors, as well as their rates of accumulation, are all uncertain. Since many GHGs and multiple sources and sinks are spread across the world, GHG emissions are subject to much uncertainty (except for a few cases, like CO_2 from fossil fuel use). Moreover, distinguishing between anthropogenic and natural causes of GHG emissions is more difficult than in the case of other important local or global pollution issues (such as those involving nuclear wastes and chlorofluorocarbons [CFCs]).

Next, the link between emissions and ambient concentrations of most GHGs is far from clear because of complex chemical transformations in the atmosphere. Even the relatively well understood case of CO_2 is complicated by the operation of several processes that remove this gas from the atmosphere over a range of timescales. Many poorly understood interactions are involved. For example, the presence of CFCs could affect climate change not only directly through their greenhouse warming potential, but also indirectly by their impact on biota like nanoplankton – which in turn influences oceanic CO_2 uptake. Similarly, reliance on fossil fuels affects CO_2 and SO_2 emissions directly and CO_2 absorption indirectly – via the effects of acid rain on forests and biomass.

The impact of atmospheric GHG concentrations on the climate is subject to even greater scientific uncertainty. Current climate models do not adequately capture the effects of clouds, sea ice, vegetation, and oceans. Some of these key feedbacks involve changes in albedo, or reflection of solar radiation (e.g., from clouds or ice), while others concern long term equilibrium exchanges of GHGs and heat (e.g., between the atmosphere and oceans). Other types of uncertainty that affect greenhouse warming include variations in solar output, hydrological cycles, and ecosystem changes. Present models are even less well equipped to predict how regional precipitation patterns, including the spatial and temporal distribution of rainfall, might change. Uncertainty about changes in the patterns of extreme weather events will be of even greater concern, especially to vulnerable developing countries.

Quantifying the impact of climate change on flora, fauna, and human beings is another area of great scientific uncertainty. Problems arise in such analyses because of the sheer number of effects to be considered, the complexity of biological and ecological systems and subsystems, and the existence of many nonlinear feedbacks. Disentangling the effects of climate change on these systems from those of other nonclimatic factors will be especially difficult.

2.3.2 Socioeconomic and Technological Uncertainty

The future emissions of GHGs, aerosols, and aerosol precursors from anthropogenic sources are subject to significant uncertainty, because they depend on human actions and policies that are themselves unpredictable. This form of uncertainty compounds the problems raised by scientific uncertainty (see preceding subsection). Another major source of economic uncertainty stems from the inability to estimate the magnitude of certain types of impact in terms of monetary value, especially those involving assets that are not valued directly in markets (see Chapter 5). For example, the loss of rain forests will entail a reduction in biological diversity and the degradation of ecological functions like watershed protection, which are very difficult to value. Another problem concerns the valuation of effects on human health, especially the loss of life. Both ethical and equity issues add to the complications of determining the value of a human life.

The manner in which human communities will adapt to climate change is not well known. Simple extrapolation of existing information based on the impact of disasters on social systems will be inadequate, because the scale of global climate change is so much greater. For example, the destabilizing effects of impoverishment and the displacement of a large number of environmental refugees, including the potential for destructive conflicts over diminishing resources, are not well known. Another source of uncertainty is the reaction of individual human beings and economic agents to some of the strategic response options.

Finally, there is considerable uncertainty concerning the availability, reliability, and costs of various technological options now under review. Timing is particularly important, and the long lead times required for the development of new technologies will influence decisions regarding response strategies. The interactions among new inventions, markets, venture capital, and government incentives for research and development are also critical (see Chapters 6 and 7).

2.4   EQUITY AND SOCIAL ISSUES

Equity in the context of a social decision means that the outcome is fair and just. It is an important element of the collective decision-making framework needed to respond to the challenge of global climate change (see Box 2.2).

BOX 2.2   Why Is Equity Important?

Equity considerations are important in addressing global climate change for a number of reasons, including (a) moral and ethical concerns, (b) facilitating effectiveness, (c) sustainable development, and (d) the UNFCCC itself.

First, the principles of justice and fair play are important in all types of human interaction. In particular, most modern international agreements, including the UN Charter, enshrine moral and ethical concerns relating to the basic equality of all human beings and the existence of inalienable and fundamental human rights. Equity is also embodied, explicitly or implicitly, in many of the decision-making criteria used by policy makers (see Chapter 4).

Second, equitable decisions generally carry greater legitimacy and encourage parties with differing interests to cooperate more fully in carrying out mutually agreed-upon actions. The successful implementation of a collective human response to the problem of global climate change will require the sustained collaboration of all sovereign nation-states and many billions of human beings over long periods of time. While penalties and safeguards will play a role, decisions that are widely accepted as equitable are likely to be implemented with greater willingness and goodwill than ones enforced under conditions of mistrust or coercion. In other words, cooperation and effective outcomes are more likely to occur when all parties to a decision feel that it is fair.

Third, as explained in Box 2.1, equity and fairness are extremely important elements of the social dimension of sustainable development. Thus, the impetus for sustainable development provides another crucial reason for finding equitable solutions to the problem of global warming.

Fourth, there are several specific references to equity in the UNFCCC's substantive provisions (UNFCCC 1992). To begin with, Article 3.1 states, "The Parties should protect the climate system for the benefit of present and future generations of humankind, on the basis of equity and in accordance with their common but differentiated responsibilities and respective capabilities. Accordingly, the developed country Parties should take the lead in combating climate change and the adverse effects thereof." Other equity-related principles emphasized in Article 3 include (a) the right to promote sustainable development, (b) the need to take into account the specific needs and special circumstances of developing-country and vulnerable parties, (c) the commitment to promote a supportive and open international economic system, and (d) the precautionary principle (to protect the rights of future generations).

According to Article 4.2(a), all developed-country parties, including those with economies in transition, are required to take the lead in mitigating climate change. Furthermore, they are required to transfer technology and financial resources to developing-country parties that are particularly vulnerable to the adverse effects of climate change in meeting the costs of adaptation (Article 4.4). Another reference to equity in Article 4.2(a) requires developed-country parties to commit themselves to "adopt[ing] national policies and tak[ing] corresponding measures on the mitigation of climate change. . . . These policies and measures will demonstrate that developed countries are taking the lead in modifying longer-term trends in anthropogenic emissions consistent with the objective of the Convention . . . taking into account the difference in the Parties' starting points and approaches, economic structures, available technologies and other individual circumstances, as well ás the need for equitable and appropriate contributions by each of the Parties to the global effort regarding that objective." Finally, Article 11.2 requires the Convention's financial mechanism to "have an equitable and balanced representation of all Parties within a transparent system of governance."

The foregoing provisions of the UNFCCC provide important guidance on how equity considerations should influence or modify the achievement of the Convention's objectives. While protecting the climate system is considered to be a "common concern of humankind," developed countries (and transition economies) are expected to take a lead in initiating actions and to assume a greater share of the burden. Furthermore, in burden sharing, emphasis is placed on applying equity considerations among developed countries as well. The responsibilities of the present generation with respect to those of future generations are also referred to. Finally, equity is mentioned in the context of governance, to emphasize the importance of including procedural elements that guarantee distributive outcomes perceived to be equitable.

2.4.1 Procedural and Consequential Equity

The requirements of the UNFCCC indicate that equity principles must apply to (a) procedural issues – how decisions are made; and (b) consequential issues – the outcomes of those decisions (Banuri et al. 1996). Both aspects are important because equitable procedures need not guarantee equitable decisions, and conversely, equitable outcomes could well arise from quite inequitable decision-making processes. Support for the Convention and acceptance of its recommended courses of action will depend largely on widespread participation by the global community and on how equitable it is perceived to be by all participants.

Procedural equity itself has two components. First, pertaining to participation, equity implies that those who are affected by decisions should have some say in the making of these decisions, through either direct participation or representation. Second, pertaining to the process, equity must ensure equal treatment before the law – similar cases must be dealt with in a similar manner, and exceptions must be made on a principled basis.

Consequential equity also has two elements, relating to the distribution of the costs and benefits of (a) the effects of and adaptation to climate change and (b) mitigating measures (including the allocation of future emission rights). Both elements have implications for burden sharing among and within countries (intragenerational and spatial distribution) and between present and future generations (intergenerational and temporal distribution). The equity of any specific outcome may be assessed in terms of a number of generic approaches explained in Chapter 4, including parity, proportionality, priority, classical utilitarianism, and Rawlsian distributive justice (Young 1994). Societies normally seek to achieve equity by balancing and combining several of these criteria. Self-interest also influences the selection of criteria and the determination of equitable decisions. Consequential equity as applied in the international arena is derived largely from these principles, which were developed originally in the context of human interactions within specific societies.

A human response to climate change requires the application of equity at an even more elevated (global) level, where there is far less practical experience. Cultural and societal norms and views about ethics, the environment, and development complicate efforts to achieve a worldwide consensus on matters of both procedural and consequential equity. Even the urgency of responding to climate change is subject to dispute. Given the different meanings, philosophical interpretations, and policy approaches associated with equity, judgment plays an important role in resolving potential conflicts. Ultimately, any global response strategy will be a compromise among different worldviews, each of which is also influenced by self-interest and attempts to shift the compromise in one's favor. As an example, the practical difficulties of allocating future emission rights among nations are explored in Box 2.3.

Nevertheless, from a pragmatic viewpoint significant progress toward reaching a global consensus would be made if the decision-making framework could harness enlightened self-interest to support equitable or ethical goals. For example, developed countries are likely to have a self-interest in taking the lead and shouldering the major burdens of addressing climate change issues, because their own citizens have shown greater willingness to cover the costs of solving environmental problems. Similarly, developed nations would enjoy greater opportunities for trade and export if developing-country markets grew without being disrupted by climate change, and the former could also avoid the significant negative spillover impact of worldwide instability arising from disasters associated with climate change. At the same time, the higher risks and vulnerability faced by developing countries would provide them an incentive to seek common solutions to the climate change problem.

BOX 2.3 How Might GHG Emission Rights Be Allocated Fairly?

Suppose that an analysis of climate change (e.g., see Box 3.1 on the global optimization process) yielded a target level of desirable worldwide GHG emissions in the future. To illustrate the issue, we will take a single constant level of emissions that will achieve some desired stabilization case (e.g., S550 discussed in Chapter 1). The principles of allocation discussed here would apply in exactly the same way to any other case involving an alternative emission profile, such as IS92c. One method of allocating constant emissions might be based on ethics and basic human rights – that is, equal per capita (EPC) emission rights for all human beings. The total national "right to emit" would then be the product of the population and the basic per capita emission quota.

Figure B2.2 illustrates the dynamics of this allocation issue in simplified form. Line EPC indicates the constant level of per capita emissions if the total global emission target is allocated equally to all human beings during the decision-making time horizon. If we assume a total permissible accumulation of 800 GtC during the 100-year period from 2000 to 2100 corresponding to the S550 case (see Table 1.3), shared equally among the global population of about 6 billion people (in 2000), then the constant average per capita emission right would amount to 1.33 tonnes of carbon (TC) per year, up to 2100 – as shown by the solid line EPC in the figure. A more precise calculation might aggregate both past and future emissions (using discounting techniques that further penalize near-term emissions, which would cause damage over a longer period), to yield the grand total over any given period of time.

Points IC and DC represent the average current per capita GHG emissions of the industrialized (i.e., OECD nations, Eastern Europe, and former Soviet Union) and developing countries, respectively. Although the figure is not exactly to scale, IC (about 3.5 TC per capita annually) is both above EPC and considerably larger than DC (about 0.5 TC per capita). Thus, industrialized countries would have to cut back GHG emis-

(continued)

BOX 2.3   How Might GHG Emission Rights Be Allocated Fairly? *(continued)*

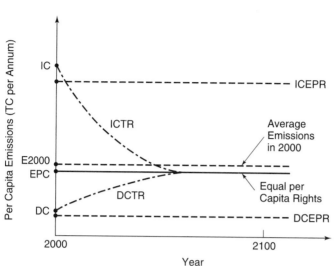

FIGURE B2.2. Allocating emission rights. From Munasinghe (1996).

sions significantly if they were to meet the EPC criterion, which would entail economic costs (depending on the severity of the curtailment in each country). On the other hand, developing countries would have a considerable amount of room to increase their per capita emissions as incomes and energy consumption grew.

An alternative allocation rule is based on equiproportional reductions (EPR) of emissions. In this case, all countries would reduce emissions by the same percentage relative to some pre-agreed-upon baseline year to achieve the desired global emission target. Assuming a global average emission rate of about 1.47 TC per capita annually in 2000 (indicated by the dashed line E2000 in the figure), this implies that all countries would have to curtail carbon emissions by about 10% to meet the EPR criterion (as shown by the dashed lines ICEPR and DCEPR). Clearly, given that energy is a vital prerequisite to economic development, such a solution would severely restrict growth prospects in the developing world, where per capita energy consumption is quite low to begin with (Munasinghe 1995).

Thus, the EPC and EPR approaches would result in some hardship and inequity for developed and developing countries, respectively. Another related equity issue is whether past emissions should also be considered or ignored in making decisions concerning the current and future quotas. Suppose we assume that the future global atmospheric concentration of CO_2 must be stabilized at 550 ppmv. More than 80% of carbon accumulated up to 1990 has resulted from fossil fuel use in the industrialized world. Clearly, the industrialized countries have used up a significant share of the "global carbon space" available to humanity while driving up atmospheric CO_2 concentrations from the preindustrial norm of 280 ppmv to the current level of about 360 ppmv. Therefore, the developing countries argue that responsibility for past emissions should be considered when future rights are allocated. Correspondingly, it would be in the industrialized countries' interest to use a fixed base year population (e.g., in the year 2000) as the multiplier of the per capita emissions right (e.g., EPC) in determining total national emission quotas. This would effectively penalize those countries with high population growth rates, since their allowed national quota (determined by the base year population) would have to be divided among more people in the future.

In practice, it is possible that some intermediate requirement falling between EPC and EPR might eventually emerge from the collective decision-making process. For example, EPC may be set as a long-term goal. In the shorter run, pragmatic considerations might suggest that both industrialized and transition countries be given a period of time to adjust to the lower GHG emission level, in order to avoid undue economic disruptions and hardship – especially among poorer groups in those countries (see transition emission paths ICTR and DCTR in Figure B2.2). Even if some industrialized nations argue that the goal of EPC emission rights for all individuals is too idealistic or impractical, the directions of adjustment are clear. Net CO_2 emissions per capita in industrialized countries should decline, while such emissions in developing countries will increase with time. This result emerges even if the objective is a more equitable distribution of per capita emissions rather than absolute equality of per capita emissions.

Another adjustment option might be the facilitation of an emission trading system. For example, once national emission quotas have been assigned, a particular developing country may find that it is unable to utilize its allocation fully in a given year. At the same time, an industrialized country might find it cheaper to buy such "excess" emission rights from the developing nation rather than undertake a much higher cost abatement program to cut back emissions and meet its own target. More generally, the emission trading system would permit emission quotas to be bought and sold freely on the international market, thereby establishing an efficient current price and even a futures market for GHG emissions (see the discussions of joint implementation and North–South transfers in Chapters 3 and 8).

Note: Numerical values in this box have been chosen for illustrative purposes only and are based on rounded-off data from several sources (Houghton et al. 1995; Raskin 1995).

2.4.2 Equity and Economics

The preceding section reviewed some arguments for reconciling equity and economic self-interest; however, conflicts between economic efficiency and equity may arise among nations due to assumptions about the definition, comparison, and aggregation of the welfare of different individuals or nations (Banuri et al. 1996). For example, efficiency often implies maximization of output subject to resource constraints (see Chapter 3). This approach can potentially result in an inequitable income distribution. Overall, human welfare could drop depending on how welfare is defined in relation to the distribution of income. Conversely, total welfare might increase if appropriate institutions can ensure appropriate resource transfers – usually from the rich to the poor.

In the same context, aggregating and comparing human welfare across different countries is a disputable issue. Gross national product is simply an indication of the total measurable economic output of a country and does not represent welfare directly. Aggregating GNP across nations is not necessarily a valid measure of global welfare levels. However, national economic policies frequently focus more on the growth of GNP than on its distribution, indirectly implying that additional wealth is equally valuable to rich and poor alike, or that there are mechanisms for redistributing wealth in a way that satisfies equity goals. Attempts have been made to incorporate equity considerations within a purely economic framework by the weighting of costs and benefits so as to give preference to the poor. Although systematic procedures exist for determining such weights, often the arbitrariness of assigning weights has caused many practical problems. At the same time, it should be recognized that all decision-making procedures do assign weights (arbitrarily or otherwise). For example, approaches based on economic efficiency that seek to maximize net benefits (see Box 3.1 and Chapter 4) assign the same weight of unity to all monetary costs and benefits, irrespective of income levels. More pragmatically, in most countries the tension between economic efficiency and equity is resolved by keeping the two approaches separate – for example, by maintaining a balance between maximizing GNP and establishing institutions and processes charged with redistribution, social protection, and the provision of various social goods to meet basic needs.

The lack of appropriate institutions to carry out such a redistributive role on an international scale raises concerns over how – if at all – national welfare levels can be compared internationally. The extreme viewpoints are that (a) welfare levels should be compared if all countries valued each other's welfare equally (i.e., as if equivalent welfare functions existed across countries and equal weights could be assigned to each), and (b) each country is concerned primarily with its own welfare and bears no responsibility for the welfare of any other (i.e., welfare cannot be aggregated and compared across countries). Since climate change involves situations in which the activities of one country affect other countries, a convention on climate change must arrive at some compromise between these two extremes.

2.4.3 Intragenerational (Spatial) Equity

While equity is not synonymous with equality, differences among countries clearly affect issues of international equity. International response strategies will eventually translate into actions adopted at the national level, and therefore should reflect equity concerns within countries as well. Several categories of differences among countries that are relevant to the question of equity are discussed next.

Wealth and Consumption   Wealth is perhaps the most obvious and prevalent difference among (and within) countries. Measured in terms of GNP, the World Bank's 1994 *World Development Report* states that more than half the world's population (58.7%) live in countries classified as "low-income" (p. 162). These countries have an average per capita GNP of $390. In contrast, 15.2% of the world's population live in "high-income economies," which have an average per capita GNP of $22,160. The remaining 26.1% of the population live in "middle-income economies," which have an average per capita GNP of $2,490. Such wide variations in per capita income among countries imply that simply comparing this measure of welfare may be inappropriate (as explained in the preceding section).[1]

These differences have direct implications for the way climate change is addressed. For instance, activities in developing countries that produce GHGs are generally related to fulfilling "basic needs." They may include generating energy for cooking or keeping warm, engaging in agricultural practices, and consuming energy to provide barely adequate lighting and occasionally travel by public transport. In contrast, GHG emissions in developed countries are likely to result from the operation of personal vehicles and central heating or cooling, as well as the energy required to produce and utilize a wide variety of manufactured goods. Therefore, the level of personal wealth is directly re-

---

[1] One method of comparing incomes across countries is to use purchasing power parities (PPPs) instead of market exchange rates. PPPs are used to adjust exchange rates, such that the monetary value of a standard basket of commodities (typically including food, clothing, and shelter) is equalized across all countries. Such a correction tends to provide a better assessment of the ultimate welfare provided by income levels in different nations. However, even when incomes are adjusted on the basis of PPPs, wide differences in real per capita income are evident among countries.

lated to the present and future concentrations of GHGs (WCED 1987). Furthermore, wealth has a direct bearing on vulnerability to the consequences of climate change. By virtue of being richer, some countries will adapt more effectively than other countries to climate change. A similar relationship between the poor and the rich prevails within countries.

Poor countries may be less prepared than wealthier ones to adopt mitigation and adaptation strategies, for several reasons. First, poverty influences the degree of urgency of other national priorities and of timescales used in policy planning. Wealth has a direct correlation to personal discount rates (i.e., discount rates decline with rising wealth). The affluent have a greater share of disposable wealth to invest in the future and therefore are able to conceptualize longer planning time horizons. The poor are forced to focus on short-term objectives such as basic survival necessities.

A similar phenomenon applies to national-level economic and political systems. Consequently, interest rates are higher in poor countries, capital is more scarce, and the emphasis in policy planning is on short-term needs, such as the alleviation of poverty and the generation of employment. The focus of government may be to keep up with infrastructure needs due to rapidly rising demands. These countries may not have the luxury of considering optimal development strategies as some richer countries may be able to. Thus, national wealth affects both actual investment decisions and broader public policy planning capability (Mathur 1991).

The Report of the IPCC Special Committee on Developing Countries (1989) addresses this concern by stating that "the priority for the alleviation of poverty continues to be an overriding concern of the developing countries; they would rather conserve their financial and technical resources for tackling their immediate economic problems than make investments to avert a global problem which may manifest itself after two generations." Similarly, Article 4.7 of the UNFCCC states that "economic and social development and poverty eradication are the first and overriding priorities of the developing country Parties" and thus their commitment to implementing climate change responses will be influenced by these considerations. Even though concerns about climate change are likely to grow in developing countries (especially those that consider themselves the most vulnerable), they are likely to lack the resources to address the issue.

Contributions to Climate Change   Countries vary in the nature and degree of their contribution to climate change. Many different gases and sources contribute to such change. The capacity of sinks to absorb carbon emissions also differs widely among countries. The range of sources and sinks may not be an issue of equity, but different ways of aggregating and presenting the data can have implications for equity considerations. In particular, developing countries emit much less per capita and have contributed less to past emissions than developed countries. In this context, some authors have argued that the industrialized countries owe the developing world a "carbon debt," due to disproportionately high GHG emissions in the past (see, e.g., Jenkins 1996).

The developing nations also need considerable "headroom" to allow for the growth of future economic output and energy consumption, since they are starting from a much lower base (see Section 1.3 and Box 2.3). At the same time, there are many variations within developed and developing countries that must be acknowledged as well. An analysis that involves simply differentiating along the lines of developed and developing countries will exclude many important issues. As discussed in Chapter 3, incorporating into the decision-making process equity issues associated with variations in contributions to climate change will be critical both in facilitating the adoption of a worldwide consensus on burden sharing and in subsequently implementing difficult mitigation and adaptation measures.

Incidence of and Vulnerability to Effects The incidence of adverse effects may bear no relationship to the pattern of GHG emissions. This violates equity principles and is inconsistent with the "polluter pays" and "victim is recompensed" approach that has already been applied to local environmental pollution problems. In particular, the negative impact of climate change is likely to be most pronounced in tropical regions typically occupied by developing countries. In addition to asymmetries in the incidence of effects, many developing countries are more vulnerable to the consequences of global warming because of fewer resources, a weaker institutional capacity, and a smaller pool of skilled human resources to draw on in times of crisis. The plight of poor and subsistence-level communities, or of low-lying small island nations subject to sea level rise, will be quite bleak. Therefore, both humanitarian and equity principles, along the lines of the principles and procedures established during the UN International Decade for Natural Disaster Relief, must be invoked to provide them some relief (see, e.g., Munasinghe and Clarke 1995).

Equity Within Countries Almost all of the arguments mentioned in the context of equity across countries also apply to equity within individual nations. Fortunately, many mechanisms exist within countries (such as subsidized food, health care, and schooling, social security, and progressive taxation) to ensure action consistent with what is considered acceptable and proper, and to achieve proper redistribution of resources. Equity issues, especially in the form of views about what constitutes justice, will influence the formation, decision making, and credibility of these institutions. Although their capacity and legitimacy may vary, they will provide a useful framework within which climate change issues can begin to be addressed at the national and subnational levels.

2.4.4 Intergenerational (Temporal) Equity and Discounting

Most of the points enumerated earlier with respect to spatial equity also affect equity across time, and in very similar ways. First, future generations may be richer or poorer than the present generation. Second, those living in the past and the present will undoubtedly contribute to future climate change effects. Third, while future generations will have to bear the consequences of past GHG emissions, they will also benefit from sacrifices and investments made by their forbears. At the same time, it is unclear whether our descendants will be more or less vulnerable to the consequences of climate change.

Two fundamental issues require us to pay special attention to intergenerational equity. First, all decisions relating to climate change are made by the generation living at that time. To the extent that future generations are not represented in the decision-making process, particular care must be exercised to ensure that their rights are protected (Broome 1992; Munasinghe 1993). Second, once a chain of events unfolds, it will be difficult to compensate future generations for past mistakes or miscalculations. Again, extra prudence is required to avoid imposing future burdens that are both irreversible and impossible to compensate. Nevertheless, generations do overlap in practice (e.g., parents and children), and this is likely to result in the automatic incorporation of some intergenerational concerns into the discount rate and decision-making in general.

Social Rate of Discount The various equity-related decision criteria discussed in Chapter 4 may be used to ensure a desirable measure of temporal equity. From an economic viewpoint, one of the principal instruments available for influencing the allocation of resources across time is the social rate of discount (see Box 2.4). Indeed, the conclusions derived from any long-term analysis of climate change policy will depend crucially on the numerical value of the discount rate that is selected (Arrow et al. 1996). It is important to bear in mind that we are discussing the real discount rate from which the effects of inflation are netted out. Furthermore, conceptually the interest rate (at which present-day capital will grow into the future) is the exact mirror image of the discount rate (at which future expenditures should be discounted to the present date).

Since discounting is a method for comparing economic costs and benefits that occur at different times, it will have a direct bearing on intergenerational equity. In the case of climate change analysis, the effects of discounting will be especially pronounced, for two reasons: (a) the relevant time horizons are extremely long, and (b) many of the costs of mitigation occur relatively early, while potential benefits lie in the distant future. In brief, as far as present-day decisions are concerned, a higher discount rate will reduce the importance of future benefits (of avoided climate change damage) relative to the near-term costs (of mitigation measures); see Figure B2.3 in Box 2.4.

BOX 2.4 Discount Rate

Basic Concepts

The social rate of discount is defined as the discount rate used by decision makers in determining public policy. The main text indicates that some fundamental issues of value and equity are involved in the choice of such a social discount rate. In addition to the technical aspect of comparing economic costs and benefits over time, the sustainable development dimension described earlier provides a more overarching guideline – that each generation has the right to inherit a set of economic, social, and environmental assets that are at least as good as the one enjoyed by the preceding generation (see Box 2.1). In subsequent discussions, mention of "discount rate" refers to the social rate of discount, unless otherwise specified.

Even in traditional cost–benefit analysis used for project evaluation, which is far less complicated than climate change decision making (see Chapters 3 and 4), the choice of a discount rate is not clear-cut. Discount rates vary across countries, depending on behavioral preferences and economic conditions. Furthermore, it is considered prudent to test the sensitivity of the results by using a range of discount rates (usually about 4 to 12% per annum), even for a project within a given country.

Starting from the theoretically ideal (or first-best) situation of perfectly functioning, competitive markets and an optimal distribution of income, it is possible to show that the discount rate should be equal to the marginal returns to investment (or marginal yield on capital), which will also equal the interest rate on borrowing by both consumers and producers (Arrow et al. 1996). More specifically, there are three conditions to be met in order to ensure an efficient (or optimal) growth path. First, the marginal returns to investment between one period and the next should equal the rate of interest (i) charged from borrowing producers. Second, the rate of change of the marginal utility of consumption (or satisfaction derived from one extra unit consumed) from one period to the next should be equal to the interest rate (r) paid out to lending consumers. Third, the producer and consumer rates of interest should be equal (i.e., $i = r$), throughout the economy and over all time periods.

As we deviate from ideal market conditions and optimal income distribution, determining the discount (or interest) rate becomes less clear. For example, taxes (subsidies) may increase (decrease) the borrowing

(continued)

BOX 2.4   Discount Rate *(continued)*

rate to producers above (below) the interest rate paid to consumers on their savings (i.e., $i \neq r$). More generally, if the three conditions do not hold because of economic distortions, then efficiency may require project- or sector-specific discount rates that would include so-called second-best corrections to compensate for the various economic imperfections. In extreme cases, there is no theoretical basis for linking observed market interest rates to the social rate of discount. Nevertheless, market behavior would still provide useful information for estimating the social rate of discount.

Numeraire

One important issue that precedes discounting is the question of measuring the aggregate quantity that one chooses to discount. The accepted approach is to convert all costs and benefits into a common numeraire, or measuring yardstick. For example, if one were trading bananas (b) against pineapples (p), one basis for comparison might be a set of parameters that include taste, color, texture, and nutritional content. However, such a comparison would most likely prove highly subjective and debatable. A more stable comparison might be made by trading the two types of fruit according to their weight; in this case, the numeraire would be kilos. Another alternative would be a comparison based on the relative prices of the two fruits. Thus, if a pineapple cost 10 rupees and a banana only 1 rupee, the agreed-upon rate of exchange would be $p = 10b$. Furthermore, if one were asked to aggregate the value of 2 pineapples and 10 bananas, the response would be 30 rupees (or, equivalently, 3 pineapples or 30 bananas). In the latter case, using currency units as the numeraire would be more convenient and natural, but the analysis would be just as rigorous if it were carried out consistently in units of bananas (or pineapples).

With economic value as the numeraire, the situation could become more complicated, because the same nominal unit of currency might have a different value depending on the economic circumstances in which it is used. For example, a peso's worth of a certain good purchased in a duty-free shop at the border is likely to be more than the physical quantity of the same good obtained for 1 peso from a retail store, after import duties and taxes have been levied. Therefore, it is possible to distinguish intuitively between the border-priced peso, which is used in international markets free of import tariffs, and a domestic-priced peso, which is used in the domestic market subject to various distortions.

Thus, in conventional project evaluation, domestic market prices may be adjusted by netting out such taxes and duties through a process called "shadow pricing." (Chapter 3 provides details of the valuation and pricing of environmental assets.) A more sophisticated example of the differences in the value of a currency unit in various uses concerns countries in which the current investment level is considered inadequate. In these instances, a peso's worth of savings that could be invested now to increase the level of future consumption may be considered more valuable than a peso devoted to current consumption.

Whatever the problems of choosing a numeraire for valuing goods and services, we may conclude that once it is agreed upon, this selection (like the choice of a currency unit) should not influence the economic criteria for decision making – provided that the same consistent framework and assumptions are used in the analysis. In the case of climate change analysis, the most appropriate numeraire would be "consumption equivalents"; that is, all costs and benefits need to be transformed into monetary units used to measure current household consumption. Even this choice poses problems in the case of climate change effects that are not normally traded in the market and cannot be easily valued for ethical or practical reasons (e.g., human life or biodiversity; see Chapter 5).

Illustrative Numerical Results

Figure B2.3 indicates the present value of U.S.$1,000 worth of climate change damage occurring at 1, 10, 20, 50, and 100 years in the future, and discounted at rates of 1, 5, and 10%. The general formula for this

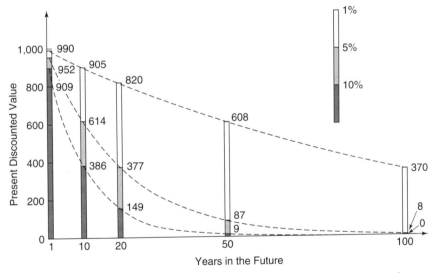

FIGURE B2.3. Value of U.S.$1,000 worth of damage in the future, discounted to the present day. Damage occurs 1, 10, 20, 50, and 100 years in the future and is discounted at 1, 5, and 10%.

(continued)

BOX 2.4   Discount Rate *(continued)*

present value would be $1,000/(1 + r) T$, where r is the discount rate (per year) and T is the time (in years) in the future when the damage occurs. For example, suppose that the damage occurs 10 years in the future. Assuming a discount rate of 1%, the present value of this cost would be $1,000/(1.01)10$, or about U.S.$905. Clearly, the present value falls off very sharply with time, especially for the higher discount rates. For example, if we consider a time 50 or 100 years in the future (which is quite modest on the timescale of climate change) and a discount rate of 5% (which is comparable to estimated values of the social discount rate in the main text), then the present value of U.S.$1,000 would be only about U.S.$87 and 8, respectively. For higher discount rates, the present value of future damage would become even less significant, while even at the rather low value of $r = 1\%$, the value of U.S.$1,000 would shrink to around U.S.$370 after 100 years.

Two Approaches to Estimating the Social Rate of Discount   There are two main approaches to practically determining a value for the social rate of discount for climate change analysis, one based on the social rate of time preference (SRTP) and the other on market returns to investment (MRI). While the concepts underlying the two approaches may appear to diverge, when practical adjustments are made they tend to produce estimates for the social discount rate that are comparable – typically with SRTP varying from 1 to 4% and MRI lying in the range 3 to 6%.

Social Rate of Time Preference.   This parameter is defined as follows: SRTP = a + ($b \times g$). Here, a is the pure rate of time preference, which basically reflects the impatience of consumers and is normally positive. In other words, individuals tend to value consumption today more highly than consumption next year, simply because current consumption is preferred (myopia), while future consumption may be considered less certain (risk aversion). In the second term, g represents the growth rate of per capita consumption, which is normally positive. The parameter b is the "elasticity of marginal utility," which is also a positive number that reflects how rapidly the welfare (or satisfaction) provided by each successive unit of consumption declines as the consumption level rises. To summarize, even if individuals were not impatient (i.e., $a = 0$), the term ($b \times g$) tells us that one unit of current consumption will still provide greater satisfaction than a unit of future consumption, to the extent that succeeding generations are likely to be richer and more satiated.

Since both a and b are normally positive, if the growth rate g is high, then SRTP will also be large. This is consistent with an optimistic scenario in which technological and other advances will ensure high levels of future per capita consumption. Therefore, future benefits will carry a smaller weight relative to current costs. On the other hand, if we assume a gloomy future in which global climate change has depressed production and consumption levels, g might become sufficiently negative to make SRTP zero or even negative. In this case future benefits would have exaggerated importance compared with present-day costs. The main conclusion is that the choice of a discount rate will also depend on the future scenario that is assumed (Munasinghe 1993). Given the uncertainties indicated earlier, it would be rather difficult to predict which outcome will predominate.

In any comparison of competing investment alternatives, all expenditures should be converted to the same numeraire (see Box 2.4), based on consumption equivalents (i.e., the value of consumption that could be provided by a given expenditure). Environmental impact would also be valued in units of this agreed-upon consumption-based numeraire and discounted at the SRTP. This approach obviates the need for a special (lower) discount rate for environmental effects. In other words, to the extent that environmental assets become scarcer in the future, their value will tend to rise faster than the social discount rate, which is equivalent to using a lower discount rate for environmental assets while keeping relative prices unchanged. Uncertainty about future effects must be handled by converting the value of such effects into certainty equivalents – that is, the certain result that would be equivalent in value to the uncertain actual outcome. Finally, in GHG mitigation projects, the opportunity cost of capital must be taken into account to adjust costs, which reflect the forgone benefits from alternative uses of the same investment resources.

Market Returns to Investment.   This approach focuses explicitly on the opportunity costs of capital adjusted for risk (i.e., the MRI capital measured in terms of the certainty equivalents just mentioned). The basic argument is that investments in GHG mitigation measures could well have been applied to other projects (both public and private) that would themselves have yielded returns in the future. Therefore, a discount rate that reflects the yields on such alternative projects (i.e., the opportunity cost of capital) provides a good basis for comparing whether future generations would be better off with or without climate change response measures. Furthermore, returns to current projects and yields in financial markets reveal the actual preferences of society today concerning trade-offs between current and future consumption.

One criticism of the MRI approach is that some alternative schemes may in fact ignore the impact of climate change. For example, investments in coastal zones that are normally expected to provide high yields (based on present-day conditions) may not be feasible if there is a major sea level rise in the future.

More broadly, if the effects of global warming are severe enough to disrupt human society significantly, the general instability of national economies and the business climate may undermine returns to all activities.

Additional problems arise in attempting to deal with taxes and uncertainty. Thus, the extent to which the MRI should reflect the producer rate of interest (or returns to private investment before taxes) or the consumer rate of interest (or returns to private investment after taxes) will depend on the extent of distortion introduced by the tax system. Uncertainty has to be dealt with by adjusting gross yields according to risk. While this is routinely done in financial markets for short- to medium-term investments (e.g., the risk rating system for bonds), the long-term perspective required in the context of climate change decisions poses additional problems. The yield on a longer-term government guaranteed bond is often used as an example of a relatively risk-free rate of return on investments.

Practical Estimates.   While actual estimates of SRTP are scarce, it is possible to begin with the assumption that the pure rate of time preference is rather low (say, $0 < a < 2$). Reasonable estimates for the elasticity of marginal utility b might be in the range 1 to 1.5. The long-term growth rate of per capita consumption is likely to be rather modest (say, $0.5\% < g < 1.5\%$, where the higher figure is close to the value of 1.6% used in the IPCC IS92 emission scenarios). After adjusting for forgone investments and depreciation, it is possible to arrive at a range of about 1 to 4% for the SRTP.

A logical first step in estimating the MRI is to review current returns to various types of market investments. A selected sample of yields is provided in

TABLE 2.1   Estimated Returns on Financial Assets and Direct Investments

Asset	Period	Real Return (%)
High-income industrial countries		
Equities	1960–84 (a)	5.4
Bonds	1960–84 (b)	1.6
Nonresidential capital	1975–90 (b)	15.1
Government short-term bonds	1960–90 (c)	0.3
Developing countries		
Education projects	1975–90 (d)	13–26
Energy projects	1980–90 (e)	15–30
Water supply projects	1980–90 (f)	10–25

Sources: (a) Ibbotson and Brinson (1987), updated by Nordaus (1994); (b) UNDP (1992, table 4, results of G7 countries); (c) Cline (1992); (d) Psacharopolous (1985); (e) Munasinghe (1990); (f) Munasinghe (1992).

Table 2.1. Adjustment for risk will tend to lower expected returns. Furthermore, historic rates of economic growth may have been higher than those expected for the long-run future, which could reduce projected yields on investment. When these considerations are taken into account, a range of 3 to 6% appears reasonable for the MRI, which is also consistent with the long-term returns on risk-free public investments.

2.5 SUMMARY

The issues presented in this chapter underline the formidable difficulties faced by climate change analysts and decision makers. First, climate change issues must be examined in relation to the three main elements of sustainable development (economic, social, and environmental). Second, our understanding of the chain of causality from emissions of GHGs to ultimate effects on natural and human systems is hampered by the large spatial and temporal scale of events, their complexity and irreversibility, and the risks of nonlinear responses and catastrophic collapse. Third, scientific uncertainty arising from lack of knowledge about natural systems, and socioeconomic and technological uncertainty associated with human systems, are difficult to deal with. Fourth, equity issues occur in many forms and are often intertwined with efficiency issues. In decision making, both the procedures and their consequences must be equitable to all stakeholders. Intragenerational equity involves trade-offs among affected groups, especially between richer and poorer nations. Intergenerational equity also requires a careful balance to be maintained between the rights of current and future generations, with the discount rate playing a crucial role. Two different estimates of the social rate of discount – based on the social rate of time preference and the market rate of interest – yield comparable results of 1 to 4% and 3 to 6%, respectively.

REFERENCES

Arrow, K. J., W. R. Cline, K.-G. Maler, M. Munasinghe, R. Squitieri, and J. E. Stiglitz. 1996. Intertemporal equity, discounting, and economic efficiency. In J. P. Bruce, H. Lee, and E. F. Haites (eds.), *Climate change 1995: Economic and social dimensions of climate change.* Contribution of Working Group III to the Second Assessment Report of the Intergovernmental Panel on Climate Change. Cambridge University Press, pp. 125–44.

Banuri, T., K. Goran-Maler, M. Grubb, H. K. Jacobson, and F. Yamin. 1996. Equity and social considerations. In J. P. Bruce, H. Lee, and E. F. Haites (eds.), *Climate change 1995: Economic and social dimensions of climate change.* Contribution of Working Group III to the Second Assessment Report of the Intergovernmental Panel on Climate Change. Cambridge University Press, pp. 79–124.

Broome, J. 1992. *Counting the costs of global warming.* Cambridge, MA: White Horse Press.

Cline, W. R. 1992. *The economics of global warming*. Washington, DC: Institute for International Economics.

Houghton, J. T., et al. 1995. *Climate change 1994: Radiative forcing of climate change and an evaluation of the IPCC IS92 emission scenarios*. Cambridge University Press, 1995.

Ibbotson, R. G., and G. P. Brinson. 1987. *Investment markets*. New York: McGraw-Hill.

Mathur, A. 1991. India: Vast opportunities and constraints. In M. Grubb et al. (eds.), *Energy policies and the greenhouse effect, Volume 2: Country studies and technical options*. London: Royal Institute for International Affairs, p. 78.

Munasinghe, M. 1990. *Electric power economics*. London: Butterworths.

Munasinghe, M. 1992. *Water supply and environmental management*. Boulder, CO: Westview Press.

Munasinghe, M. 1993. *Environmental economics and sustainable development*. Washington, DC: World Bank.

Munasinghe, M. 1995. *Sustainable energy development: Issues and policy*. Environment Dept. Paper no. 16. Washington, DC: World Bank.

Munasinghe, M. 1996. Analyzing economic and policy issues in climate change. Paper presented at the Yale–NBER Conference on Climate Change, Snowmass, CO, July.

Munasinghe, M., and C. Clark. 1995. *Disaster prevention for sustainable development*. Geneva and Washington, DC: International Decade for Natural Disaster Reduction and World Bank.

Munasinghe, M., P. Meier, M. Hoel, S. W. Hong, and A. Aaheim. 1996. Applicability of techniques of cost–benefit analysis. In J. P. Bruce, H. Lee, and E. F. Haites (eds.), *Climate change 1995: Economic and social dimensions of climate change*. Contribution of Working Group III to the Second Assessment Report of the Intergovernmental Panel on Climate Change. Cambridge University Press, pp. 145–78.

Munasinghe, M., and W. Shearer (eds.). 1995. *Defining and measuring sustainability*. Tokyo and Washington, DC: United Nations University and World Bank.

Nordhaus, W. D. 1994. *Managing the global commons: The economics of climate change*. Mimeo, MIT, Cambridge, MA.

Psacharopoulos, G. 1985. Returns to education: A further international update and implications. *Journal of Human Resources*, **20**(Fall), 583–604.

Raskin, P. D. 1995. Methods for estimating the population contribution to climate change. *Ecological Economics*, **15**(3), 225–34.

Report of the IPCC Special Committee on the Participation of Developing Countries. 1989. Geneva, September.

UNDP (United Nations Development Program). 1992. *Human development report 1992*. Oxford: Oxford University Press.

UNFCCC (United Nations Framework Convention on Climate Change). 1992. Articles. New York: United Nations.

World Bank. 1994. *World development report*. Washington, DC: World Bank.

WCED (World Commission on Environment and Development). 1987. *Our common future*. London: Oxford University Press.

Young, P. 1994. *Equity in theory and practice*. Princeton, NJ: Princeton University Press.

3

DECISION-MAKING FRAMEWORK

One logical starting point for making decisions about the global climate change problem is the UN Framework Convention on Climate Change (UNFCCC 1992). Following the initial signing of the Convention by the heads of state who attended the UN Conference on Environment and Development in Rio de Janeiro in June 1992, its ratification by 165 nations (as of January 1997) made it legally binding internationally. Especially relevant is the wording of Article 2, which advocates "stabilization of greenhouse gas concentrations in the atmosphere at a level that would prevent dangerous anthropogenic interference with the climate system." To begin with, an effective decision-making framework should provide policy makers with the means to assess the available scientific, technical, economic, and social information, and thereby determine what range of future human activities might constitute such "dangerous anthropogenic interference" and how remedial measures could be identified. The basis for making immediate decisions concerning the climate change issue has already been provided in the context of present international discussions pertaining to the UNFCCC. However, the most important decisions lie in the future. Thus, the decision-making framework presented in this chapter outlines the process by which future measures may be developed in order to limit global climate change. Unfortunately, the complications set out in Chapter 2, especially uncertainty, impose limits on the precision of long-term optimizing solutions. However, uncertainty is not a valid reason for inaction, given the seriousness of the potential effects of global climate change. Therefore, the decision-making framework should facilitate consensus building on prudent short-term responses (e.g., the "no-regrets" measures, like energy conservation, which are robust over a wide range of future scenarios). At the same time, the longer-term strategy should ensure the consistency and smooth implementation of the near-term decisions already under way. It must certainly be flexible in the face of uncertainty and contain "heuristic" elements that permit the systematic future revision of both decision-making processes and objectives, in keeping with knowledge that may emerge periodically.

83

In the next section, we define the key questions that require responses from decision makers. Then we discuss the main elements of a decision-making framework used to respond to the problem of climate change.

3.1 DECISION-MAKING OBJECTIVES AND ISSUES

As we saw in the first two chapters, global climate change poses especially difficult analytical problems characterized by unprecedented scale (both spatial and temporal), complexity, and uncertainty. Thus, the objectives of decision making (at all levels of society) must be clearly specified in order to analyze the problem and arrive at effective decisions.

3.1.1 Causal Linkages

We begin by reviewing the chain of causality associated with the global climate change. The main relationships already set out in Chapters 1 and 2 are discussed here with reference to Figure 1.1 – but this time, from a decision-making viewpoint.

Linkages 1 and 2 concern the emissions and concentrations of GHGs, and the effects of such anthropogenic activities on climate. These links depend primarily on relationships within the physical world. Therefore, they are not directly responsive to human decisions or interventions. The most important step at this stage is to develop a more informed understanding of these two linkages by gathering knowledge and reducing the uncertainties that lead from cause to effect. The decisions involved are scientific ones, typically concerning the need for better research and information gathering.

Linkage 3 concerns the extent of economic, social, and ecological damage caused by global climate change. One key issue that arises in damage assessment is the choice of an anthropocentric (i.e., from a human perspective) versus ecocentric (i.e., from an ecological viewpoint) approach (see also Chapter 5). A related question involves the relative weights to be applied to diverse effects, including the role of economic valuation in establishing a common numeraire both to rank the severity of various effects and to compare them with the costs of corresponding measures to abate climate change. All these questions require making choices – mainly concerning impact assessment criteria and techniques, as discussed in Chapter 4. The same observations made in relation to linkages 1 and 2 apply here as well.

Linkage 4 involves the critical decisions that will determine the bulk of the human response to the threat of global climate change. In view of the potential impact and damage, decision makers must identify the feasible menu of mitigatory, adaptive, and other options, as well as policy instruments, from which a response strategy can be determined.

Linkage 5 helps to determine abatement responsibilities, emission rights, and implementation mechanisms. Clearly, the abatement measures and the implementation process are interrelated and depend on each other. The decision-making framework and the decision tools that play an especially important role in linkages 4 and 5 are discussed in greater detail in this chapter and the next one. The information required and decisions relating to these linkages are primarily of a technological, socioeconomic, or political nature.

Linkage 6 represents the effects of the human response strategy on climate change effects, via two paths. First, mitigation actions will affect net GHG emissions and ultimately change the consequences. Second, adaptive measures could influence damage more directly. In view of the many uncertainties surrounding the implementation and effectiveness of many new and untried measures, some key decisions will involve the extent of resources to be devoted to further research, data gathering, and modeling in the scientific, technological, and socioeconomic areas.

3.1.2 Key Issues

Within the broader objective of the UNFCCC – which seeks to "stabilize the concentrations of GHGs" – more specific decision-making objectives can be defined in the form of a set of issues that must be addressed by all of humanity:

1. Determining the acceptable concentrations of GHGs in the atmosphere at different times in the future
2. Determining the target levels for reduction of GHG emissions that are necessary to achieve the desired atmospheric concentrations, including the future time path of such emissions
3. Identifying the combination of measures that will bring about the desired reductions of emissions and, consequently, the ambient concentrations of GHGs
4. Allocating the appropriate distribution of emission reduction responsibilities among different nations, taking into consideration their past emissions and future development needs
5. Establishing mechanisms and institutions (both international and national) for making collective decisions and implementing them effectively
6. Formulating measures that also meet the criteria for sustainable development
7. Developing methods of dealing with special difficulties, especially the issue of uncertainty, which manifests in different forms with respect to each of the first five issues

The first issue, determining desirable GHG concentrations, depends on the entire chain of cause and effect reviewed earlier, including GHG emissions, atmospheric concentrations, temperature increases, biogeophysical effects, and socioeconomic factors. One important subsidiary issue is the relationship between ultimate damage and possible time profiles of GHG concentrations in the future. Since a major objective of this analysis is to confine socioeconomic effects to acceptable levels, methods of valuing such effects and trading them off against the costs of abatement are particularly important (see Chapter 4).

Regarding the second issue, Chapter 1 showed clearly that the primary focus should be the GHG concentrations that cause global climate change, whereas net GHG emissions are important only to the extent that they lead to increases in concentration. Furthermore, the same level of GHG concentration may be achieved at a given time in the future by following an infinite variety of time paths of GHG emissions – each having a different risk and abatement cost. For stabilization of GHG concentrations, annual emissions must eventually equal annual removals, very likely at a level significantly below current emission levels. At the same time, the actual emission trajectory is likely to be important also, in two ways. First, the extent of damage due to climate change will depend on the rate of climate change (which will be greater if emissions rise more sharply in early years), in addition to the magnitude of climate change (which will depend on the ultimate concentration of GHGs). Second, the counterbalancing argument is that earlier emission cutbacks will require greater mitigation costs, especially since energy-using capital stocks may have to be replaced more prematurely. As discussed further in Box 3.1 and Chapter 4, the ultimate selection of the best time profile of emission reductions will depend on the estimated trade-off between the magnitude and pattern of climate change damage, and the costs of abatement and adaptive measures associated with each such profile.

An adequate response to the third issue can be made only after carefully examining the practical feasibility and cost effectiveness (i.e., minimization of the cost of achieving the desired emission abatement targets) of the different response measures (see Chapter 7). The second and third issues are related, depending on the extent to which one believes that more stringent GHG mitigation measures and emission reductions will accelerate research and application of new mitigation technologies – thereby reducing the costs of further emission reductions. Associated issues to be considered include the timing of measures in relation to available information and the "political will" required to implement unpopular policies. The portfolio of response options may be grouped into three categories: mitigation, adaptive, and indirect. *Mitigation options* are engineering solutions focused primarily on limiting GHG sources and enhancing sinks. *Adaptive strategies* involve technological measures as well as changes in economic structure and behavior in order to limit the vulnerability of human habitats, social systems, infrastructure, and managed ecosystems (like farms) to the potential damage caused by global climate change. *Indirect options* are the vast range of actions (e.g., trade practices) that have little to do with climate change but nevertheless have implications for global warming. These options may be implemented through market-based, regulatory, and other policy instruments (see Chapters 6 and 7).

The fourth issue follows immediately from the third, and will require a collective decision-making process to ensure that a workable consensus is reached on the assignment of differentiated responsibilities to implement abatement measures and thereby achieve agreed-upon emission targets. Such a process will have to embody a variety of concerns, including poverty and equity (both geographic and intergenerational), economic and institutional capability, ecological and environmental vulnerability, and ethics (see Chapter 2). In practice, voluntary or ad hoc mitigation measures may be undertaken by individual countries (especially in the OECD group), before either global GHG reduction targets or allocation of responsibilities are agreed upon.

The fifth issue concerns the need for mechanisms and institutions by which the collective decisions described earlier can be reached and implemented over a long period of time. In particular, the development and extension of existing frameworks such as the UNFCCC and Montreal Protocol will require early attention. These issues are explored in greater detail later in this chapter (Section 3.2.4).

The sixth issue arises from the emerging consensus on the economic, social, and environmental dimensions of sustainable development – especially as it applies at the local and subnational levels (see Chapter 2). The UNFCCC specifies that the stabilization of GHG concentrations "should be achieved within a time-frame sufficient . . . to enable economic development to proceed in a sustainable manner." Thus, the complementarities and trade-offs between global climate change concerns and the narrower (but more immediate and politically weighty) sustainable development goals of individual nations must be resolved in order to develop a mutually consistent set of long-term measures.

Finally, the seventh issue requires us to deal with a number of complicating factors that are specific to the climate change problem. While these elements were set out as characteristics of the problem in Chapter 2, we discuss ways of dealing with them in the present chapter. Of particular relevance are the issues of scale, complexity, and uncertainty. An appropriate decision-making framework for climate change policies must successfully address the global dimension and long-time horizon of both the problem and the corresponding corrective measures. The complexity of the climate change problem also requires a framework that is sufficiently comprehensive to incorporate a multitude of relevant factors. Finally, decisions must reflect the high degree of uncertainty that affects the scientific, technical, social, and economic under-standing of climate change effects. The cumulative impact of these uncertain-ties can be greater than that of the individual components. Moreover, uncer-tainty and risk are amplified by the long time lags between the causes and effects of climate change; possible nonlinearity and irreversibility of climate change effects; the long planning horizon for policy decisions; the element of risk associated with policy decisions, including the decision to do nothing; the potential for "free riding" (or enjoying the benefits of abatement measures undertaken by others without doing anything oneself); the need to accommo-date new technology and information on climate change; and building collec-tive and preferably market-oriented strategies for sharing the risks (insurance and markets for risk) and the mutual benefits.

How much emphasis to place on research, data gathering, and modeling is a question that is common to all of the foregoing issues. The corresponding decision-making process must make use of a variety of decision tools (de-scribed in Chapter 4). For example, cost–benefit analysis might be of help in determining the magnitude of resources to be devoted to improving the state of scientific knowledge and analytical capability in relation to the payoff or benefits of better predictions and the consequent ability to mitigate or avoid damage caused by climate change. Future expectations concerning the emer-gence of better scientific information or improved technological options will certainly affect the timing of response strategies and measures.

3.2 FRAMEWORK FOR DECISION MAKING

3.2.1 Main Elements

The issues set out in the preceding section indicate that an effective decision-making framework must encompass three complementary elements. While economic efficiency and equity issues are often intertwined and difficult to separate, we seek a practical decision-making framework that might initially separate these two types of problems and allow us to proceed in a stepwise fashion (adapted from Arrow, Parikh, and Pillet 1996).

The first element in such a framework addresses questions of *global opti-mization*. More specifically, the desirable overall target GHG concentration and net emission levels must be determined on a global basis, since the total accumulation of GHGs is far more important than the location of specific sources and sinks. To make this type of decision, we first seek "efficient" solu-tions that maximize net benefits (i.e., benefits minus costs, broadly defined) for all of humanity, without necessarily being concerned about the incidence or allocation of such costs and benefits. However, the correspondence between global optimization and efficiency is not necessarily absolute. Equity considerations could also be important, to the extent that costs and benefits, as well as income levels, are unequally distributed. In the final analysis, it is more practical to bring in equity issues after both scientific and efficiency criteria have been used to arrive at a broad global consensus concerning desirable GHG concentrations and emissions.

The second element focuses on a *collective decision-making process* that will provide "equitable" solutions to problems involving the distribution and allocation of costs and benefits associated with global climate change that were raised in Chapter 2. Questions such as who should bear differentiated responsibilities for agreed-upon abatement response measures fall into this category. The key complication arises from the fact that the entire global population is potentially at risk. Thus, the stakeholders (or those who have an interest in the outcome) who should be involved in making decisions range from single individuals and small firms to nation-states and multinational groupings (e.g., the small island nations, which are particularly concerned with the inundation of land with a rise in sea levels). Furthermore, many selection criteria may be applied, including those concerning social equity, economic efficiency, sovereignty, ethics, and ecology. Clearly, a broadly participative process is essential for finding answers to this type of question (Arrow et al. 1996). Collective decisions could also be oriented toward efficiency rather than equity, although such a formulation might not be effective or practical enough to implement.

The third element concerns *procedures and mechanisms* for addressing climate change issues. Rules governing decision-making processes and behavior, as well as implementing mechanisms and structures, fall within this category. The UNFCCC has already provided many of the initial rules and mechanisms. The procedural element will be strongly influenced by the framework adopted for the global optimization and collective decision-making elements, and vice versa. Procedures should also facilitate abatement actions. For example, the application of decision tools discussed in Chapter 4 to the problems reviewed in Chapter 2 (especially uncertainty) indicates that such a set of response actions is likely to be (a) sequential (or step-by-step); (b) based on a preidentified portfolio of the most desirable mitigation, adaptation, and research measures (e.g., one would initially adopt "win–win" strategies or measures that would be taken anyway, even in the absence of global warming); (c) capable of providing a hedge against uncertainty and risk; and (d) adjusted and updated systematically to take new knowledge into account. The decision-making process must reflect these requirements. Finally, the process of formulating and implementing globally agreed-upon abatement decisions inside a specific country must take place within a national framework for sustainable development.

These three decision-making elements will now be discussed in further detail.

3.2.2 Global Optimization and Efficient Solutions

Global optimization is based on the rather basic concept of maximizing the aggregate net benefits (i.e., benefits minus costs) that are associated with different future climate change scenarios. This approach is fundamental to one of the main tools of decision making – cost–benefit analysis, which is described in Chapter 4 (Section 4.2). When such costs and benefits can be valued economically, the quantity to be optimized, or the "objective function," is defined in monetary units (ignoring distributional aspects). The resulting solution is termed *economically efficient*, as explained in Box 3.1.

BOX 3.1   Global Optimization Based on Economic Efficiency

An economically efficient policy for emission reduction is one that maximizes the welfare or net benefits (NB) of that policy – that is, it maximizes the global benefits (B) of reduced climate change net of the costs (C) associated with GHG abatement efforts.* Thus, simple arithmetic indicates that we seek to maximize the quantity $NB = B - C$. If we think of benefits in terms of the avoided costs (D) of greenhouse damage, then maximizing net benefits is equivalent to minimizing total costs (TC) or the sum of the costs of damage and abatement. In other words, since we can also write $NB = -D - C = -TC$, then maximizing NB is equivalent to minimizing TC.

Figure B3.1 illustrates the concept of global optimization, using total as well as marginal costs (and benefits) in simplified form. Curve C indicates that the costs of abating GHG emissions will rise at an ever-increasing rate as worldwide emission reduction efforts are intensified. Corresponding curve D shows that the costs of damage caused by climate change will decline (but at an ever-decreasing rate) when GHG emissions are reduced. The global total cost (TC) curve is simply the sum of curves C and D.

The shapes of the abatement and damage cost curves reflect the idea of diminishing returns. Thus, implicit in curve C is the fact that we start with the cheapest methods of emission reduction and progressively resort to more expensive techniques at later stages. Furthermore, each additional unit of emission reduction will require a higher expenditure: the first 10% of abatement can be achieved cheaply, but the next 10% will cost more, and so on. For example, the simplest energy conservation measures (that will also reduce GHG emissions) are relatively costless, whereas more sophisticated technologies, like combined-cycle electric

---

* The aggregation of individual benefits and costs to derive the global benefits and costs gives rise to significant theoretical and practical problems, as discussed in Chapter 4.

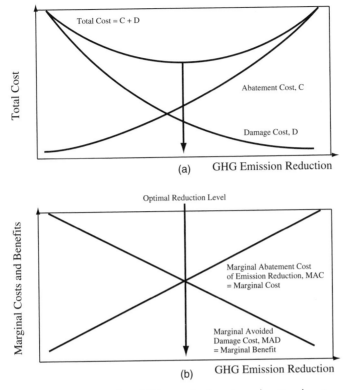

FIGURE B3.1. (a) Total and (b) marginal costs and optimal emissions reductions. From Munasinghe et al. (1996).

power plants (which emit less carbon per unit of energy produced), are more expensive. Thus, as the level of emission reduction increases, the abatement cost curve in the figure rises (with increasing slope), while the damage cost curve declines (with decreasing slope). One consequence of the foregoing is that the total cost has a minimum, as shown. This is the global optimum, because overall costs to humanity will be minimized.

An equivalent and perhaps conceptually clearer definition of the globally optimal level of emission reduction is based on marginal costs and benefits (or the costs and benefits associated with one additional unit of GHG emissions). Basic economic theory indicates that the abatement of emissions should be pursued up to the level where the marginal benefit of reducing emissions of GHGs by one additional unit is equal to the marginal cost of curbing such emissions. In Figure B3.1b, the MAC curve is derived by calculating the slope (or rate of change) of curve C in Figure B3.1a. In other words, the marginal cost at any level of emission reduction is equal to the slope of the corresponding cost curve at the same level. Similarly, the MAD curve represents the (negative) slope of curve D; the negative sign appears because D is defined as a cost, while MAD is a marginal benefit (or avoided cost).

(continued)

BOX 3.1 Global Optimization Based on Economic Efficiency
(continued)

It is quite easy to show mathematically that total cost is minimized at the point where the slope of the abatement cost curve equals the negative slope of the damage cost curve (or avoided costs) – that is, at the level of abatement where the MAC and MAD curves intersect. Although, for convenience, the marginal costs and benefits are shown as linearly proportional to emission reduction, this linearity is not strictly necessary to prove the point. For example, where abatement costs are subject to economies of scale, sections of the MAC curve might have a nonlinear form. In fact, empirical studies of marginal cost curves frequently do exhibit the stylized shapes shown in the figure.

Using the same notation, and introducing the symbol R to represent the level of emission reduction, we may write $TC(R) = -[C(R) + D(R)]$. In terms of simple calculus, TC is minimized when its first derivative equals zero, or $dTC/dR = [(dC/dR) - (dD/dR)] = 0$. This equation yields the condition that equalizes marginal costs and benefits: $dC/dR = -dD/dR$, or MAC = MAD.

Setting Targets An important practical handicap is the likelihood that the cost and benefit curves shown in Figure B3.1 will exhibit great uncertainty (for the reasons set out in Chapter 2) – especially the marginal avoided damage (MAD) costs. Figure 3.1 indicates how such uncertainty will affect the global optimization process. In Figure 3.1a, both the MAD and abatement costs are economically undefined. Nevertheless, it may be possible to use scientific judgment to determine the target level of emission reduction R_{AS}, beyond which the risk of damage is unacceptably high. The cross-hatching shows the extent of uncertainty in defining the two zones. Indeed, the dividing line at R_{AS} has been drawn closer to the error margin on the right to indicate a cautious viewpoint (see also the discussion below on the precautionary approach). The target level R_{AS} is based on an *absolute standard,* because the obligation to avoid harm is absolute. It implies that the underlying MAD curve (if available) would be quite low in the acceptable risk zone to the right of R_{AS} but would rise sharply in the zone of unacceptable risk. In other words, the potential for damage is so high in the unacceptable risk zone that the cost of abatement carries very little weight in this decision.

In Figure 3.1b, the marginal abatement costs (MAC) are available, while the MAD costs are still undefined. The cross-hatching on the MAC curve indicates the degree of uncertainty on either side of its expected (or mean) value. In this case, the target level of emission reduction R_{AM} reflects the judgmental balance between the affordable level of abatement costs and the acceptability of damage risk. This approach is termed the *affordable safe minimum standard.*

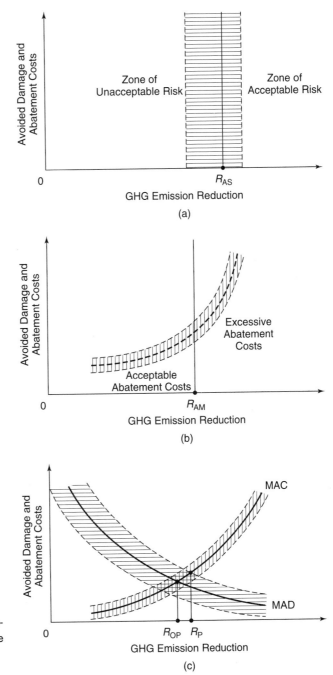

FIGURE 3.1. Determining abatement targets: (a) absolute standard; (b) affordable safe minimum standard; (c) cost–benefit optimum.

The relevant total affordable cost is the area under the MAC curve up to the vertical line. One of the difficulties is that the concept of affordability is not well defined and is subject to different interpretations. There is also an implication that the underlying MAD curve is still quite low in the zone to the right of R_{AM}.

Finally, in Figure 3.1c, the optimal level of emission reduction is defined in terms of the cost–benefit analysis framework presented in Box 3.1. In this case, either both the MAD cost and MAC functions are known with certainty, or the expected values of these curves are used (in the absence of risk aversion), with the cross-hatching again indicating the margin of error around the mean value. Then the globally desirable degree of emission reduction is indicated by R_{OP}, at the point where MAC = MAD.

The benefits of having greater information are apparent, to the extent that the abatement target may become less stringent as the decision making progresses through the three cases, in which (a) both MAC and MAD are undefined, (b) only MAD is undefined, and (c) both curves are defined. In other words, since $R_{AS} > R_{AM} > R_{OP}$, the level and costs of abatement would be progressively reduced as one approached the optimal point – reflecting greater confidence in the information available.

Attitude toward Risk and the Precautionary Approach Attitude toward risk plays a key role in decision making under uncertainty, as illustrated by the following simple example. Suppose that an individual places a bet in which he (or she) has an equal chance of gaining or losing 1,000 pesos. In this symmetrical case, the expected value of the outcome is zero, since (under normal circumstances) the bettor will expect neither to gain nor to lose.[1] Next, the individual is asked to name the certain (or guaranteed) cash payment that will be equivalent to the outcome of the bet. A risk-averse person will seek a positive cash payment (say, 100 pesos). One reason may be that the extra 1,000 pesos gained may not be considered as valuable as the 1,000 pesos that might be lost – a phenomenon known as the diminishing marginal value of income (or consumption). Equivalently, 1 peso may be worth more to a poorer person than to a richer one. A risk-neutral individual will be satisfied with a zero cash payment, while a risk lover (gambler) may be willing to pay more (say, 100 pesos) in the hope of gaining the 1,000 pesos. Lotteries are run on such an assumption – that most people have a gambling streak, especially if the sum at risk is small and the potential gain is large (although the expected value of the gain is much smaller). The *precautionary approach* to decision making is likely to be adopted (and was in fact endorsed by the UNFCCC) when risk aversion interacts with the uncertainty associated with potentially irreversible and catastrophic climate change effects. More specifically, Article 3.3 of the UNFCCC seeks to provide guidance to decision makers when uncertainty is present, through the "precautionary principle." The article states that the parties to

---

[1] More specifically, the expected value of a group of alternative outcomes is equal to the sum of each result weighted by its corresponding probability of occurrence. In this example, the outcomes are +1,000 pesos and −1,000 pesos, while the probabilities are .5 each. Therefore, the expected value is given by $(1,000 \times .5 - 1,000 \times .5) = 0$.

the UNFCCC should "take precautionary measures to anticipate, prevent or minimize the causes of climate change and mitigate its adverse effects. Where there are threats of serious or irreversible damage, lack of full scientific certainty should not be used as a reason for postponing such measures, taking into account that policies and measures to deal with climate change should be cost effective so as to ensure global benefits at the lowest possible cost. To achieve this, such policies and measures should take into account different socio-economic contexts, be comprehensive, cover all relevant sources, sinks and reservoirs of greenhouse gases and adaptation, and comprise all economic sectors. Efforts to address climate change may be carried out cooperatively by interested Parties" (UNFCCC 1992).

As an example of the precautionary approach, consider Figure 3.1c. Here, although the uncertainty in the avoided damage curve is likely to be large, the decision need not be delayed. A risk-averse decision maker would select the more stringent target emission reduction level R_P (lying to the right of R_{OP}, which is presently unknown). The relevant point B is determined roughly by the intersection of the MAC curve and some estimate of the upper envelope of the avoided damage (MAD) line. Furthermore, in the face of the greater level of uncertainty shown in Figure 3.1a, a precautionary approach might result in the even more stringent emission reduction target R_{AS}, leaving a smaller margin for error on the right side.

Concluding Remarks The foregoing analysis gives rise to several corollaries: (a) uncertainties in determining costs and benefits would require a great deal of judgment in determining target levels of emission reduction; (b) the decision criterion could progressively evolve as the quality of information improved – from the absolute standard, through the affordable safe minimum standard, to the precautionary and optimal approaches; and (c) there may be significant returns to investing in better research and information gathering on climate change.

As mentioned earlier, many difficulties would complicate the analysis. For example, the emission of a unit of GHG may give rise to a varying stream of environmental costs that must be discounted over time to yield a present value aggregate. The environmental damage function may be discontinuous and nonlinear. Abatement costs may change over time, depending on when the technologies are applied, because of technological progress. Similarly, abatement costs may exhibit economies of scale (e.g., mass production of solar photovoltaic cells), resulting in a marginal cost curve that actually declines beyond a certain point. Such costs may also differ across countries, for various reasons. Moreover, the abatement costs are net costs to the extent that certain technologies (e.g., renewables) may produce other (non-climate-related) benefits and costs; this is the so-called joint products complication. Finally, costs and benefits accrue to so many diverse individuals, groups, and nations that simple aggregation raises equity issues (see Chapter 2). It is possible to incorporate elements of "equity" into the globally optimal solution – for example, by weighting costs and benefits in inverse proportion to the income levels of the respective victims and beneficiaries. However, determining such weights has proved to be highly contentious, even in the much simpler context of development projects. To conclude, it is more practical to focus on an efficient solution (subject to some judgment) at the global optimization stage and introduce equity in the collective decision-making stage, discussed next.

3.2.3 Collective Decision Making and Equitable Solutions

One collective decision-making framework that already exists at the political level is the Conference of Parties (CoP) to the UNFCCC, which could evolve into an effective "apex forum" responsible for formulating global environmental policies (see Box 3.2 and the next section). Another international group set up before the CoP, primarily to make scientific assessments about climate change, is the Intergovernmental Panel on Climate Change (IPCC). The IPCC has been quite successful in reaching a worldwide consensus among scientists and experts working on the climate change problem. The present volume is based mainly on this consensus. Currently, the CoP is discussing the structure of a Subsidiary Body on Science and Technology Advice (SBSTA) under Article 9 of the UNFCCC. Meanwhile, the CoP will continue to rely on the IPCC for scientific input that could facilitate its negotiations. Ultimately, a comprehensive collective decision-making framework, as described in this subsection, may be the most appropriate setting for formulating practical climate change policies. While the focus here is on decision making at the global level, primarily among sovereign nation-states, many of the arguments are equally applicable to decision making and implementation at various levels within countries (as discussed in Section 3.2.4).

BOX 3.2 Evolution of a Climate Change Institutional Framework

The main bodies concerned with climate change and their areas of responsibility are set out here as a useful reference for much of the discussion on institutional issues throughout this volume. Among the first official bodies to be set up was the Intergovernmental Panel on Climate Change (IPCC), which was established in 1988 as a joint initiative of the World Meteorological Organization and the United Nations Environment Program. It was charged with the mandate to "(a) assess the scientific information that is related to the various components of the climate change issue, such as emission of major greenhouse gases and modification of the Earth's radiation balance resulting therefrom, and

that is needed to enable the environmental and socio-economic consequences of climate change to be evaluated; and (b) formulate realistic response strategies for the management of the climate change issue."

At the inaugural meeting of the IPCC in 1988, three working groups
(WGI, WGII, and WGIII) were created to (a) assess available scientific information on climate change, (b) assess the environmental and socioeconomic effects of climate change, and (c) formulate response strategies.
The first IPCC report, produced in 1990, was a compilation of the individual reports of these three working groups, which were composed of
several hundred scientists and experts from around the world. In 1990
the Second World Climate Conference was convened under the sponsor-
ship of several UN organizations. The conference was conducted as a
three-part series consisting of scientific, technical, and ministerial sessions. While the scientific and technical sessions presented the IPCC conclusions and recommendations on climate change, the ministerial ses-
sion acknowledged the importance of addressing greenhouse warming
as an urgent international concern.

A working group of government representatives that was formed in
1990 prepared the United Nations Framework Convention on Climate
Change (UNFCCC). The Convention is internationally binding since its
ratification by 165 states (as of January 1997).

The objective of the UNFCCC is as specified in Article 2 of the treaty:
"The ultimate objective of this Convention . . . is to achieve . . . stabilization of greenhouse gas concentrations in the atmosphere at a level that
would prevent dangerous anthropogenic interference with the climate
system. Such a level should be achieved within a timeframe sufficient to
allow ecosystems to adapt naturally to climate change, to ensure that
food production is not threatened and to enable economic development
to proceed in a sustainable manner" (UNFCCC 1992). The emphasis on
collective decision making reflected in the UNFCCC is operationalized
by means of the International Negotiations Committee (INC), originally
established to address environmental treaties, and the Conference of
Parties (CoP), which has replaced the INC. The activity following the
original UNFCCC agreement has been devoted to resolving issues that
were difficult to resolve in the initial phase. In particular, at the Berlin
meeting of the CoP in March–April 1995, a number of important decisions were made (see Box 3.4). Meanwhile, the IPCC was expanded in
1992. A new round of studies was launched, culminating in the Second
IPCC Report, which was presented to and approved by UN member governments in Rome, in December 1995. The results of this report (particularly the sections prepared by Working Group III) form the basis for
much of the material presented in the present volume.

The CoP has designated the Global Environment Facility as the interim operator of the financing mechanism for supporting the work of
the UNFCCC (see Box 3.5).

First, the requirements of procedural equity (introduced in Chapter 2) suggest that most of the key questions set out at the beginning of this chapter should preferably be resolved through a collective process with fair representation and treatment of all parties (mainly sovereign nations) and transparent procedures. The global optimization approach helps to establish desirable future GHG concentrations and emission profiles. However, other questions that are more equity-dependent, such as the allocation of emission rights and abatement responsibilities among countries, require special attention in the collective decision-making process. Thus, to the extent that the developing countries feel that they have not had an adequate influence on past international negotiating processes, special attention must be paid to procedural equity in the CoP. In particular, developing countries argue that they ought to be entitled to special considerations because they (a) historically have contributed less to the problem than developed countries, (b) have fewer resources, and (c) are more vulnerable to the effects of global warming.

Collective decisions should especially address international distributive issues. Thus, when responsibilities for implementing abatement measures are assigned, an individual country will be concerned about how emissions from other countries will affect it and how the benefits from its own emission reduction measures might accrue to all countries. More generally, climate change has different implications for different nations, raising issues of consequential equity (as outlined in Chapter 2). Countries listed in Annex 1 of the UNFCCC (i.e., the industrialized nations), which will bear the major financial burden, will be more concerned about the cost of abatement. At the same time, non–Annex 1 countries, which are poorer, will be more likely to worry about the impact of climate change on economic growth and development, the costs of adaptation, vulnerability to changes, increases in the frequency of extreme weather events, and irreversible damage. Meanwhile, more specific groups like the island states have already expressed fears about the inundation of land due to sea level rise, while the oil-exporting countries are particularly concerned about the potential effects of carbon taxes on their oil incomes and the drop in demand for petroleum fuels.

Second, the decision-making process must incorporate mechanisms for meeting the criteria of both equity and efficiency. Geographic or intragenerational equity is a prime concern. For example, the poorer nations require sufficient freedom to expand their economic activities. They are unwilling to curtail their development due to constraints imposed by emission reduction strategies. Generally, the welfare of poorer communities must be especially protected – even in industrialized countries. Another important example involves intergenerational equity. To paraphrase the Bruntland Commission: the interest of future generations must be considered, but without significantly compromising the well-being of present generations (WCED 1987).

The application of cost–benefit- or utility-based decision-making techniques to climate change decisions would adequately cover the efficiency aspects, but leave the bulk of the equity issues unaddressed. As described earlier, such approaches could help determine global optima for collective action, such as target GHG concentrations. However, using cost–benefit-based methods to deal with equity questions (e.g., to allocate responsibilities and to distribute costs and benefits associated with climate change) would require a comprehensive global welfare function with equity weighting that fairly incorporates the well-being of all stakeholders (i.e., affected parties). Such an objective function is virtually impossible to formulate.

Delays in the collective decision-making process also have equity implications. Typically, international negotiations and agreements take a long time to be implemented and may not be adopted by all parties. These "slippages" occur at the expense of increasing cumulative concentrations, thus reducing the potential GHGs that could be emitted in the future. For instance, after having ratified the UNFCCC, Annex 1 countries will emit 40 to 50 billion tonnes of CO_2 during the 1990s. This would be sufficient for the South to continue development activities for a quarter of a century. In the absence of consensus on an abatement strategy, Annex 1 countries could well be held accountable for emissions since 1990.[2]

Allocation of Emission Rights As mentioned in Chapter 2, an explicit or implicit allocation of emission rights does raise significant equity concerns. Once the desired global concentrations are established (using the global optimization principles discussed in Section 3.2.2), the permissible volume of emissions can be distributed among sovereign states in an equitable manner as decided by collective choice. Annex 1 countries, which have historically emitted more GHGs than developing countries, may feel entitled to a larger share of the rights – for example, by relying on a "grandfathering" allocation rule with existing country emissions as the starting point. Developing countries, however, would prefer an allocation rule based on an equal worldwide per capita emission rate weighted by population (Arrow et al. 1996). The implications of the foregoing allocation approaches have been explored in Chapter 2 (see Box 2.3). Meanwhile, the UNFCCC provides only general guidance on selecting an equitable allocation, by suggesting that nations act to protect the climate system in accordance with their "common but differentiated responsibilities."

---

[2] That is, in any global agreement, the share of emission rights assigned to Annex 1 countries might be adjusted accordingly (see also the section entitled "Contributions to Climate Change" in Chapter 2). Holding countries accountable for emissions after an agreed-upon baseline year (say, 1990) could help to reduce wasteful consumption in the scramble to beat regulations or manipulate rules.

The allocation of emission rights will affect the level of emission reduction that a country must adopt. Monitoring and enforcement are critical requirements for implementing any agreement on allocation of emissions (see Section 3.2.4). If rights to emit are clearly defined and enforceable, nations that wish to exceed their allocated emission rights could do so by agreeing to purchase or lease rights allocated to another nation – for example, by using a mechanism for trading emission quotas. This concept is discussed further in the subsection on North–South transfers and more extensively in Chapter 7. Here it is sufficient to note that the adoption of such measures imposes additional responsibilities on the collective decision-making process (i.e., in resolving such issues as how contracts might be arranged and implemented, how much should be paid, and how the revenues should be allocated).

Response Strategies Identifying the relevant portfolio of options and instruments that form the basis for a response strategy is an important requirement of collective decision making. Once globally efficient concentrations are decided, the decision process and analytical tools (discussed in Section 3.2.4) may be used to determine a range of response strategies. Specific response measures that include mitigation, adaptation, and indirect options, as well as market-based, regulatory, and other policy instruments, are described in detail in Chapters 6 and 7. For convenience, we briefly summarize the most important measures below under four broad categories: mitigation, adaptive, indirect, and cooperative responses. The joint implementation approach is presented as a hybrid response involving both technical and policy elements. These response options may be implemented through an appropriate mix of economic and noneconomic policy instruments.

Mitigation Responses. These include technological measures undertaken to reduce sources of GHG emissions into the atmosphere, as well as measures that increase the capacity of sinks to absorb CO_2 and other GHGs. Given that the great bulk of total CO_2 emissions originates from the burning of fossil fuels, energy efficiency and fuel switching are important options; removal of CO_2 from the atmosphere through carbon sinks such as new forests is also a priority. As discussed in Chapter 6, a range of technical responses can be adopted both collectively at a global level and also individually at the level of sovereign nations (Pearce et al. 1996).

Another key element of mitigation measures is the transfer of technology and training to developing countries, an objective spelled out in the UNFCCC. It embodies both efficiency and equity criteria. Efficiency is achieved by first selecting emission reduction strategies with higher marginal returns to the investment – that is, greater emission reduction per dollar of investment. Such opportunities are more likely to be found in developing countries and economies in transition, where technological improvements would be more cost-effective. Technology cooperation can be equitable when the social welfare of such countries is improved by assistance from developed countries. Ideally, the type of technology cooperation selected could achieve both efficiency and equity goals (a "win–win" outcome).

Adaptive Responses. In the absence of mitigation measures, countries must adapt to the effects of climate change. Thus, the cost of delay in dealing with the problem are externalized in the implementation of various adaptation measures. In general, the negative effects of adaptation are most strongly felt by the South and particularly by the poor within these countries. Many developing countries are ill-equipped to adapt to climate change. Adaptive responses generally result in increased human misery – for example, in terms of the homeless and poor who suffer and even die from starvation, heat exhaustion, or extreme weather events. It is fair to suggest that the cost of adaptation measures should be internalized by nations that have contributed to the problem. To some extent, the cost should also be internalized by parties that obstruct the adoption of corrective measures. However, it must be noted that adaptive responses are not a solution to the problem. They serve as corrective policy measures based on the polluter-pays principle, to the extent that affected parties are compensated for the costs of forced adaptation (Pearce et al. 1996).

Indirect (or Other) Responses. These measures, though designed to address priority development issues other than climate change, indirectly help to reduce the effects of climate change. Among the most important indirect responses are economic policies aimed at realigning resource allocation that has been distorted by market and institutional failures and making it consistent with the broad objective of improving social well-being. In general, economic measures serve several functions: (a) correcting market distortions, (b) encouraging cost minimization, and (c) providing flexibility in selecting strategies. A larger menu of instruments also provides more flexibility in order to encourage economic adjustments in the face of changing environmental conditions.

Cooperative Responses and Activities Implemented Jointly. Some of these approaches may be implemented at the global level, whereas others are more appropriate at the level of sovereign states. Cooperative programs are encouraged in the UNFCCC (e.g., in Article 4.2[a]). One such global policy instrument currently undergoing pilot testing is the activities implemented jointly (AIJ). It is a generic approach that provides opportunities for interactions between countries – especially between countries of the North and South. In particular, the North is encouraged to invest in emission reduction strategies in the South (through the process of AIJ) if this is less costly than adopting equivalent abatement measures in the North (see Box 3.3). Therefore, AIJ is a hybrid mechanism that essentially includes both technical and policy elements discussed so far.

BOX 3.3   Economic Rationale for North–South Transfers

The coordinated implementation of financing and technical cooperation mechanisms is essential to reduce GHG emissions in developing countries (see Munasinghe and Munasinghe 1993). Figure B3.2 elucidates the basic economic rationale for greater North–South resource transfers and technical cooperation, and also highlights the complex interplay of efficiency and equity considerations in addressing the climate change problem. Curve *ABCDE* indicates the combined marginal abatement costs (MAC) for a pair of countries (one developing, or Southern, and the other industrialized, or Northern). In other words, the graph shows the additional costs (over and above the costs of conventional technologies) of adopting various GHG-reducing schemes, plotted against the amount of avoided emissions. Portion *AB* indicates a region of negative costs, which includes so-called win–win or no-regrets options such as energy efficiency schemes for which cost–benefit analysis will show a net economic gain even before GHG abatement benefits have been considered (i.e., where the value of conventional energy savings exceeds project costs; see Chapter 6). Other measures like fuel switching,

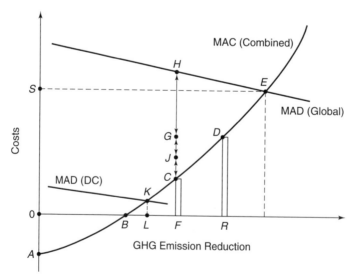

FIGURE B3.2. Economic rationale for North–South transfers and interplay of efficiency and equity. From Munasinghe and Munasinghe (1993).

new and renewable technologies, carbon sinks, and advanced energy technologies are likely to appear on the rising portion (*BCDE*) of the curve. Most of the lower cost options for GHG emission reduction, such as *CF*, would be in the developing country, whereas the more costly alternatives would lie in the industrialized nation. Ideally, all options should be pursued in both countries, up to the point *E*, where the additional costs (MAC combined) of the marginal unit of emissions curtailed are equal to the corresponding benefits (MAD) of avoided global warming effects.

First, we explore the broad rationale for Northern assistance to the South. In this context, consider a representative project, wind power generation, in the developing country, with additional costs *CF*. It would be *economically efficient* for the global community to finance these costs (on a grant basis) in the developing country, because it would thereby realize the global net benefits *HC* (i.e., *HC* = *HF* − *CF*). Without such a transfer, the developing country would be willing to pursue abatement measures only up to point *K*, where MAC is equal to the benefit of avoided climate change costs, or MAD (*DC*) accruing purely to that country.

Second, we make the case for a bilateral transfer of resources from the industrialized to the developing country. Consider the cost of the project *DR*, which seeks to reduce GHG emissions in the industrialized country. This country could realize a cost saving *GC* by transferring an amount *CF* to the developing country, while still achieving the same emission reduction. This would be the basis for so-called activities implemented jointly and similar bilateral cooperative schemes. To the extent that net benefits *HC* and cost savings *GC* are significant, it would be both *equitable* and *efficient* for the developing country to be given more resources than the barely break-even reimbursement *CF*. Thus, a share of the net profit (e.g., *JC*) would act as an incentive for the developing country to accelerate its GHG emission reduction activities. Basically the same argument has been made to accelerate implementation of the Montreal Protocol in order to reduce emissions of ozone-depleting substances (Munasinghe and King 1992).

Third, we examine how North–South transfers might help to overcome barriers to pursuing win–win options in developing countries. Consider the portion of the cost curve *AB* in Figure B3.2, where one might expect the developing country to undertake measures (such as energy conservation) without external inducements, because they are economically justified in themselves. Figure B3.3 shows the cash flow pattern over time for a typical energy-saving project. Although the net present value of benefits will be positive (i.e., negative costs, as shown in Figure B3.2), the initial investment costs *I* (horizontal stripes) are quite high, while en-

(continued)

BOX 3.3   Economic Rationale for North–South Transfers *(continued)*

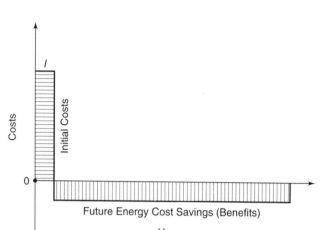

FIGURE B3.3. The barrier of initial costs. From Munasinghe and Munasinghe (1993).

ergy cost savings or benefits occur only in the future (vertical stripes). As indicated earlier, investment funds are scarce in developing countries, while perceived opportunity costs are high because of other uses for such funds (e.g., to overcome poverty, malnutrition, and poor health). Therefore, unless external funds are provided to surmount this initial investment barrier, many such cost-effective (or win–win) GHG abatement reduction schemes will not be undertaken.

The question of whether the North "earns" additional emission rights in exchange for emission reduction strategies that it promotes in the South has yet to be determined. Even if this principle were agreed upon, negotiations would be necessary to determine the amount of such credits and methods for monitoring joint implementation (JI) schemes. Clearly these issues should be discussed in a collective decision-making forum. A series of related issues must also be resolved. For instance, if developing countries are granted very large GHG emission rights (in recognition of their lesser contribution to the problem in the past), industrialized nations may find it difficult to justify such investments to taxpayers or shareholders who perceive the South to be receiving a free deal. Nevertheless, it would be possible for the North to devise methods for granting incentives or credits internally to recognize JI contributions – for example, in the form of carbon tax rebates. In the event that carbon taxes were not recognized by a country, rebates could be issued for other categories (such as research and development or charity). Even if international credits were not established, the North might benefit from participating in JI. Doing so would enhance the mitigation capacity of the South and reduce the pressure for mitigation in the North in the long term.

Policy Instruments. A range of both noneconomic and economic policy instruments (see Chapter 7) are available for implementing the response strategies discussed so far. Noneconomic policies are primarily of the command and control type, involving standards, regulations, and legislation. Economic policy instruments suitable for addressing climate change can be categorized according to the type of response generated by their implementation. These include (a) defining and allocating property rights (ownership rights, rights to use and develop resources, (b) creating new markets (to trade or exchange scarce environmental resources), (c) providing economic incentives (such as taxes on polluting emissions), (d) designing financial instruments (to make additional funds available for environmental protection), and (e) making better use of liability instruments (environmental damage liability, liability standards, liability insurance, etc.).

North–South (and West–East) Transfers and Technology Cooperation

The JI approach just discussed provides opportunities for North–South transfers of resources.[3] The tradable emission permit mechanism also shows promise, and we shall discuss some key issues concerning this approach. If national "rights to emit" can be defined, a mechanism may be developed to enable nations to transfer or trade such rights. For example, Southern countries that have more rights than they can presently utilize may decide to sell or lease them to Northern countries that may wish to exceed their allocated rights. One argument advanced by Southern nations is that many Northern countries currently utilize a larger share of the scarce global atmospheric commons, or "environmental space," than can be justified by their size (either geographic or population size) (Munasinghe 1992; Parikh 1994). Furthermore, the use of environmental space for the emission of GHGs has implications for intergenerational equity as well (Smith 1991; see also Chapter 2). Under the polluter-pays principle, countries (or communities) whose use of environmental space exceeds their fair share would be required to pay.

Poor nations may be forced to spend an increasing share of their resources to undertake adaptation measures against climate change effects. This may constrain development in these countries (Pearce et al. 1996). Transfer of financial and technical resources to affected countries could be justified on these grounds. Countries that have emitted more than their fair share of GHGs may become liable to other countries for the damage that may result from climate change. Finally, delay in adopting abatement strategies has im-

---

[3] The same arguments concerning North–South transfers apply to West–East transfers.

plications for the South in terms of lost opportunities for developing alternative energy sources, land use, and crops. The South may wish to be compensated for these forgone development opportunities. Technology cooperation between North and South is a key complement to financial transfers. While additional funds can enable developing countries to utilize the latest technologies, the long-term effectiveness of such investments will depend crucially on the improved knowledge and skill development that are essential elements of technology cooperation. Over the past decades, many studies of industrial productivity (including the efficiency of energy use) have sought to isolate the critical factors underlying long-term efficiency improvements. In particular, researchers agree that the greater bulk of efficiency increases may be attributed to "disembodied technical change" (i.e., basically the aspects linked to increased human skills and knowledge), rather than "embodied technical change" (i.e., mainly improvements in the physical capital stock or machines themselves). The key conclusion for those seeking to improve energy management and reduce GHG emissions in the developing world is that, to achieve maximum effectiveness, the *financing* of new investments must be synchronized with *human resource development* and *institution building*. Important barriers to improving the application of technology to environmental protection include (a) economic problems, like capital shortages and policy distortions; (b) inadequate environmental regulations and weak enforcement; (c) insufficient technological skills and knowledge; and (d) poor decision making and lack of political will. Implementing a climate change response strategy will require addressing these difficulties, as discussed in later chapters.

3.2.4 Procedures and Mechanisms

The UNFCCC broadly outlines some procedural guidelines for addressing climate change on a global scale. These pertain primarily to the collective decision-making framework – especially the negotiation and arbitration processes. At the outset, the UNFCCC states that collective decision making (in negotiating treaties and agreements) must be based on "procedural rationality," which places emphasis on the manner in which decisions are made. This stands in contrast to "substantive or consequential rationality," where the emphasis is on the outcome of the choice, irrespective of the decision-making process. A collective decision-making process is essential for negotiating international treaties and agreements, establishing decision procedures, and determining appropriate weights, decision criteria, or rules (Arrow et al. 1996).

The significant scientific, economic, and policy uncertainties surrounding the climate change problem compel international negotiators first to initiate environmental treaties on a more general basis and then to develop the details incrementally. Thus, once the broad outline of a treaty (like the UNFCCC) is agreed upon during the main negotiation phase, additional and specific details are addressed in the postnegotiation stage. Such discussions can be time-consuming because of the difficulty of arriving at a consensus among such a large number of stakeholders. In addition, environmental negotiations are complex because they must incorporate other policy concerns, such as international competitiveness and trade, as well as different opinions about what constitute fair and equitable solutions (particularly between developed and developing countries). Therefore, new and innovative approaches are required to improve the effectiveness and enforceability of international environmental negotiations.

Further work on procedural issues is being undertaken in view of the following factors. Postagreement negotiation processes that will critically influence the success of GHG abatement implementation need to be strengthened. In particular, some basic guidelines may be useful: involving domestic or local stakeholders and eventual implementers (including firms, consumer representatives, and energy producers) from the earliest stages of negotiations; restructuring agreements to reflect simplicity and transparency; and improving processes to educate the public about international environmental problems, which generally tend to be more abstract than local environmental issues.

In the absence of a global social welfare function, which could be maximized using methods of economic analysis (see Section 3.2.2 and Box 3.1), the resolution of global environmental problems rests squarely on the ability to reach consensus through a negotiation process and on the political will of countries to implement agreed-upon measures. The UNFCCC recommends a flexible procedure for collective decision making (recent developments are set out in Box 3.4). Governments are requested to submit national communications that will ensure their compliance with their commitments. Flexibility also facilitates modifying and changing the agreement on the basis of new information.

Countries are required by the Convention to coordinate their economic and administrative instruments. Coordination among countries is also necessary to ensure that measures adopted under international treaties are not misused in order to reinforce protectionist measures and other trade distortions or do not negatively affect third parties (IIASA 1993).

Financing Mechanisms for Climate Change Responses As mentioned throughout this book, the consequences of climate change are marked by great uncertainty, which will affect the nature and scale of technical and economic response strategies and will ultimately determine the extent of financial resources required. One primary objective of the international effort to mitigate the effects of climate change is to provide financial and other incentives to both developing and industrialized countries to implement agreed-upon measures for reducing emissions and increasing the absorption of GHGs. Activities implemented jointly, an approach described earlier, is just one example of a cooperative mechanism to provide such financial incentives (see Box 3.3).

BOX 3.4 The Berlin Mandate: The First Meeting of the CoP

The first meeting of the CoP (CoP-1), in Berlin in March–April 1995, helped maintain the momentum of the UNFCCC process, despite the economic downturn and scientific uncertainty concerning climate change. CoP-1 set a deadline of 1997 for negotiating a protocol or other binding legal instrument that would cover developed-country commitments after 2000. The other main decisions made at the meeting dealt with process issues involving (a) the arrangements for reviewing how parties to the UNFCCC meet their commitments and (b) the system for financial transfers. Developed and transition economies were to submit their second "national communications" by April 15, 1997, describing their attempts to implement the Convention and expected future emissions. Guidelines are being drawn up to help developing countries prepare their own national communications, which were due to start in early 1997.

A Subsidiary Body for Scientific and Technical Advice was set up to provide scientific and technical information that would assist the CoP in determining policy. The IPCC would continue to be a major source of such information. A Subsidiary Body for Implementation was established to assist the CoP in addressing implementation issues. The Convention Secretariat was made permanent on January 1, 1996, and was located in Bonn.

A set of pilot activities was initiated to explore possibilities for AIJ, as described in the main text and Box 3.3. However, it was agreed that the investing country could not claim future credit for reduced emissions during this phase of AIJ. The Berlin CoP also agreed that the Secretariat should produce an inventory of economically viable and environmentally sound current technologies. Further progress on the Berlin mandate was made at the second CoP meeting in Geneva in July 1996. The third meeting was due to take place in Kyoto in November 1997.

Given the massive potential costs of future GHG abatement measures worldwide (ranging from tens to hundreds of billions of U.S. dollars), a variety of avenues are being explored to secure funding for such programs. Specific mechanisms include general taxes and other sources of revenue not usually associated with climate change, as well as specific taxes targeting activities that discharge GHGs (e.g., taxes or fees levied on those using fossil fuels, like the carbon tax mechanism discussed in Chapter 7).

The trading of emission rights among nations (through a system for trading marketable emission permits) is another source of financing already discussed. Other creative methods for securing funding are also being explored. Using undisbursed funds left over from dormant projects or overestimated budget items, cost savings from increased energy efficiency, additional taxes on airline tickets, or introducing a worldwide lottery are some suggestions. Another possibility is to impose a noncompliance fee on countries that are unable to meet their GHG abatement obligations – this may not be appropriate for non–Annex 1 countries (i.e., developing nations), especially since they have no binding commitments at present. In the shorter term, multilateral development banks, bilateral assistance programs, UN organizations (especially those concerned with development, science, and technology), and other research and academic foundations are being encouraged to expand their capacity for addressing climate change issues.

More specifically, such institutions are expected to provide incentives in the form of external assistance to those developing countries that will accommodate their national development priorities to climate change considerations. Win–win strategies that simultaneously achieve internal sustainable development goals as well as GHG emission abatement objectives have the highest priority. Additional and complementary actions by groups of nations are being similarly promoted at the regional level. Regional development banks are already ensuring that their projects are environmentally sound. To the extent possible, such projects could be made more compatible with the climate change decisions emerging from the UNFCCC negotiations.

On such matters as assessing the financial assistance for climate change strategies, cooperation between nations, and development of mechanisms to be utilized as climate change strategies, a progressive approach is recommended. This mirrors the process described earlier in the development of international treaties. One example is the Global Environment Facility, which went through a five-year pilot phase before being accepted as an "interim" financing mechanism to support GHG mitigation efforts (see Box 3.5). Another case is the step-by-step approach being adopted in implementing the Montreal Protocol for reducing emissions of ozone-depleting substances like chlorofluorocarbons and halons.

Initially, a "twin response track" could help to ensure well-coordinated and cost-effective disbursement of resources to address climate change. The first track focuses on existing institutions, such as the World Bank, UN organizations, multilateral organizations, and some regional banks. Global climate change issues are being integrated into the agendas of these institutions, and they will each develop action plans while also identifying opportunities for future action. These organizations will reexamine the nature of ongoing development assistance in the context of this integration process. Funding priorities may be altered to reflect a greater emphasis on environmental programs.

The second response track refers to a parallel development of new institutional mechanisms, expressly created to facilitate the implementation of climate change conventions and related protocols. The Global Environment Facility is such an initiative (see Box 3.5).

BOX 3.5   Global Environment Facility

The Global Environment Facility (GEF) is a source of funding for global environmental actions. As of February 1997, there were 160 member countries. It is responsible to the global conventions (on climate change and biodiversity); it has its own council, and projects are implemented by the UN Development Program, the UN Environment Program, and the World Bank. The focal areas covered by the GEF include climate change, biodiversity, and international waters. The Montreal Protocol, a related multilateral fund for financing the phasing out of ozone-depleting substances, is managed by an executive committee and its secretariat.

The GEF has been identified by the UNFCCC as the operator of an interim financing mechanism (up to four years after CoP-1 in 1995; see Box 3.4) to support climate change initiatives. After restructuring in 1994, it was given a three-year budget of about U.S.$2 billion. The fundamental basis for financing projects through the GEF is the concept of "agreed-upon full incremental costs." This refers to a criterion whereby the incremental cost of a project, including components that are expressly designed to provide global benefits (e.g., through the reduction of GHG emissions or an increase in GHG absorption capacity), will be financed through the GEF. The GEF has established a policy on incremental costs and already financed more than 200 projects. Policy research initiatives have also been launched. For example, the Program for Measuring Incremental Costs for the Environment (PRINCE), which was financed by the GEF (during the pilot phase), is continuing to collaborate with "regional centers of excellence" in developing countries. PRINCE coordinates inputs to assist the policy-making process through research in developing countries, as well as to disseminate the results of such research.

On matters relating to climate change, the GEF responds to policies, eligibility criteria, and program priorities provided by the CoP. Furthermore, these guidelines must be transparent, consistent with the latest analysis and country-specific studies, and consistent across countries, sectors, project types, and focal areas. In general, projects selected for funding should be (a) country-driven and consistent with the national development priorities of the country; (b) consistent with the objectives of sustainable development as reflected in the Rio Declaration, Agenda 21, and other UN Conference on Environment and Development–related agreements; (c) sustainable and provide opportunities for "wider applications"; and (d) cost-effective. While striving to reach decisions through consensus, the GEF has adopted a "double majority" voting system to resolve issues that elude universal agreement. More specifically, GEF decisions require a 60% majority of all countries, as well as approval by donor nations representing more than 60% of contributions. Furthermore, the council comprises 32 representatives from 18 constituencies (or groupings) from recipient countries and 14 constituencies from donor nations.

The IPCC has discussed the merits of individual developing countries undertaking country-specific studies on current and projected emission levels along with estimates of financial and technical resources required to meet objectives of GHG mitigation or sink-enhancing strategies. Energy, forestry, and agriculture are the most important sectors to be examined in this regard. Such studies are to be conducted expeditiously, in order to determine the magnitude of the financial needs of developing countries. A similar approach was adopted in addressing issues and needs faced by countries in adhering with the Montreal Protocol.

Insurance and Related Market Mechanisms Insurance is closely linked with financial markets and provides an important means of dealing with risk and uncertainty using information derived from such markets. It provides a mechanism for agents (i.e., both among and within countries) to share the risk of adverse climate change outcomes. In general, the literature indicates that risk sharing through insurance tends to improve social welfare in an economy (see, e.g., Eeckhoudt and Gollier 1995). Concern about global climate change is concentrated not merely in rich nations, but in upper-income groups in those nations, which are the same groups that already purchase insurance to protect themselves against other losses through poor health, flooding, and earthquakes. Persuading poor people to make cash outlays for flood or earthquake insurance can be difficult, even in industrialized countries. However, less formal insurance schemes have been developed even in simple economies. For example, peasant farmers reduce their individual exposure to misfortune (e.g., bad weather or ill health) by collectively sharing farm work and outputs. In summary, to the extent that the lack of information about the impact of climate change constitutes a risk to society, this risk could be spread more widely through markets that trade securities that will pay off contingent on the occurrence of climate change effects.

Insurance schemes at the global level need to take some special concerns into account. These issues are linked to the "polluter-pays" and "victim-is-compensated" principles. First, to the extent that the bulk of past and present contributions to carbon emissions have been made by industrial countries, developing countries feel that the industrialized nations should bear a special responsibility for the costs of uncertainty imposed on the rest of the world. Second, developing countries are likely to be less able to protect themselves from the potential effects of climate change and more vulnerable to the risks.

Financial markets and private economic agents could serve as useful mechanisms for assessing the true importance that individual countries place on the climate change issue, as well as their publicly stated position (Chichilnisky and Heal 1993). When countries express their estimates of climate change risk through market commitments or payments, the latter are more likely to reflect actual views than are the mere public pronouncements of these governments.

A community's willingness to pay for reducing the risks from climate change depend on two factors: the degree of risk aversion and the discount rate. Differences in policy positions among countries can be attributed to these two causes and also to varying interpretations of the available evidence. The efficient global solutions mentioned in Chapter 3 do not necessarily have to be compromised by such differences in perceptions among countries. The differences in the degree of risk aversion, discount rate, and seriousness of the perceived outcomes can be addressed by creating markets in which the different risk positions can be traded, leading to a more efficient outcome.

Creating financial markets can be justified from an ecological-economic standpoint as well. Thus, the uncertainty over (a) the regional distribution and (b) overall intensity of effects merits the examination of two broad categories of financial instruments as a means of sharing the risks of both biological and cultural diversity (Ayres and Weaver 1994). For example, the alliance of small island states (AOSIS) has proposed a *mutual insurance* scheme to deal with the regional variations in the effects of climate change. The AOSIS proposal first pools nations (or communities) that face somewhat similar risks and then provides payoffs to the more adversely affected regions from those that suffer lesser effects. More specifically, AOSIS envisages an "international insurance pool" consisting of two groups: (a) low-lying coastal nations and small island states, which would receive insurance coverage from the pool, and (b) industrialized, donor countries, which would contribute funds to the pool. The size of the transfers between the two groups would depend on the relative magnitudes of adverse effects. In contrast to the approach based on regional variations, the overall or global incidence of effects could be insured against by so-called *Arrow securities*. Such an instrument would pay off only if the worldwide intensity of negative effects reached a predetermined trigger level (Weilenmann 1994).

Clearly, the insurance industry and financial community need to adopt new analytical methods and explore new instruments for dealing with climate change risks, especially since past experience is no longer a helpful guide for the future. One encouraging sign is the innovative reaction of some members of the insurance industry to recent extreme weather events (see, e.g., Munasinghe and Clarke 1995). Only in the 1990s have insurance losses associated with natural disasters exceeded the U.S.$1 billion mark – the most costly up to now being Hurricane Andrew in August 1992, which resulted in U.S.$15.5 billion in insurance payouts.

Making and Implementing Decisions Within Countries Any GHG reduction measures have to be designed and implemented within a national framework. Thus, individual projects must be generally consistent with the national economic and energy strategy, and also fit more specifically within the investment program of the energy system. It is necessary to briefly review the complexity of the overall energy decision-making process in order to identify barriers and opportunities for GHG emission reductions.

More effective policies may be designed by using a holistic framework that fully accounts for key macroeconomic and intersectoral linkages as well as energy–environment interactions – an approach that is more comprehensive than the narrower, intrasector analysis used in conventional analyses. Whatever the prevailing political system, market failures and policy distortions give rise to unsustainable practices. The decision-making framework described in this subsection helps formulate policies and provide decentralized market signals and information to economic agents that will encourage more sustainable energy production and use.

Climate change decision making is an iterative process. Figure 3.2 shows how global decisions are linked with national policies through the collective decision-making process to achieve sustainable energy development (SED) within a country. An effective SED process must deal with a multiplicity of actors, criteria, levels, policy tools, and impediments. Turning to the first column in the figure, there is an increasing need to ensure multiactor participation in energy decision-making (especially by the environmentally concerned public), but this involvement must be effectively structured to avoid a paralysis of decision making in the sector, which could result in costly energy shortages. The existence of many traditional and often conflicting policy criteria or goals (shown in the second column) is now further complicated by pressing new environmental considerations – of both a local and global nature. The core of the decision-making process is the integrated multilevel analysis shown in the

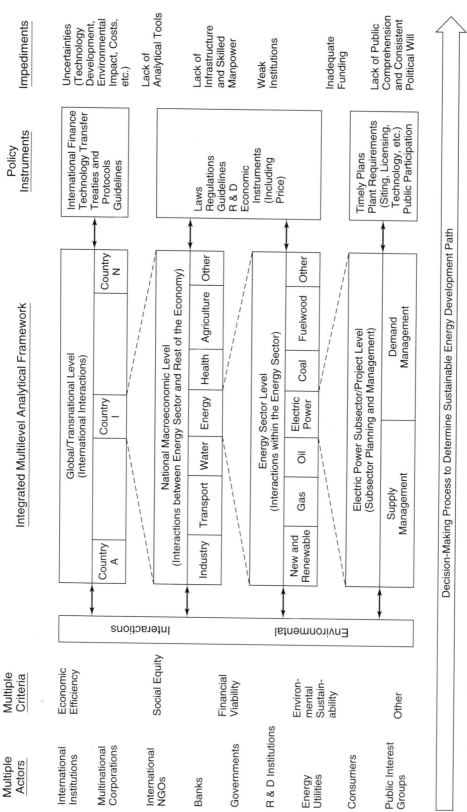

FIGURE 3.2. Conceptual framework for sustainable energy development. From Munasinghe (1991).

middle column. Within a given country, such an analysis can be carried out using a hierarchical framework for integrated national energy planning (INEP), policy analysis, and supply–demand management (Munasinghe 1990).

Although the INEP framework is primarily country-focused, we begin at the global level by recognizing that there are many transnational energy–environmental issues. Thus, individual countries are embedded in an international matrix, and economic and environmental conditions (e.g., global warming) at this level will impose a set of exogenous inputs or constraints on decision makers within countries. The next hierarchical level in Figure 3.2 focuses on the multisectoral national economy, of which the energy sector is a part. Therefore, energy planning requires analysis of the links between the energy sector and the rest of the economy. The intermediate level of the integrated approach treats the energy sector as a separate entity composed of subsectors such as electricity, petroleum, and coal. This permits detailed analysis, with special emphasis on interactions among the different energy subsectors, substitution possibilities, and the resolution of any resulting policy conflicts. The final, or micro, level pertains to analysis within a given subsector. It is at this most disaggregate level that most of the detailed energy resource evaluation, planning, and implementation of projects are carried out.

In practice, the many levels of INEP merge and overlap considerably. Thus, the interactions of electric power problems and linkages at every level must be carefully examined. Energy–environment interactions (represented by the vertical bar) tend to cut across all levels and have to be incorporated into the analysis. Finally, regional and spatial disaggregation may be required also, especially in larger countries. To reach the desired goals of sound energy management, a variety of policy instruments are available to decision makers, as summarized in the fourth column of the figure. Since these tools are interrelated, their use should be closely coordinated for maximum effect. Finally, the fifth column indicates the most important impediments to effective policy formulation and implementation. The practical application of the SED framework is described for the case of Sri Lanka in Box 3.6.

BOX 3.6 Application of the Sustainable Energy Development Methodology in Sri Lanka

As indicated in the main text, determining the desirable global concentration (and net emission levels) for GHGs and allocating the national responsibilities for mitigation and adaptation to meet these target emission levels are international decisions. However, such decisions must be implemented by individual sovereign nations. Country decision makers

(continued)

BOX 3.6 Application of the Sustainable Energy Development
Methodology in Sri Lanka *(continued)*

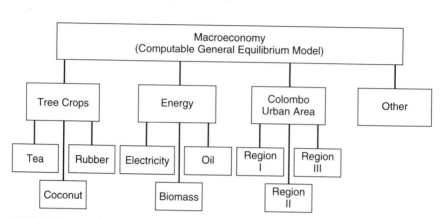

FIGURE B3.4. Modeling framework for Sri Lanka study. "Other" includes submodels
dealing with other key potential environmental issues (e.g., coastal zone). From
Munasinghe et al. (1995).

must decide on appropriate mitigation measures (policies, programs,
and projects) at the national level that will contribute to a coherent
global climate change response strategy.

Developing countries face especially difficult choices. While sharing
worldwide concerns about climate change, they must also address other
urgent issues, like poverty, hunger, and disease, as well as rapid population growth and high expectations concerning future progress. More important, the paucity of resources constrains the ability of developing
countries to undertake costly measures to protect the global commons.
Thus, it would be most effective if climate change measures could be systematically integrated with other national objectives. Furthermore, since
energy production and use constitute the single most important source
of anthropogenic GHG emissions, climate change policies must be incorporated within an SED framework of the type shown in Figure 3.2 (Munasinghe 1991).

The hierarchical SED framework, which identifies the interactions
among different levels of national decision making, can be adapted to
explain how climate change measures are translated from collective decisions at the global level into tangible measures (i.e., policies and projects) at the national, sectoral, and subsectoral (or project) levels. To the
extent that climate change policies require measures in the energy sector, the SED framework can be applied directly. At the same time, it
could be readily adapted to analyze and coordinate similar actions in
other sectors. The application of the framework is illustrated here by the
case study of the Sri Lankan energy sector.

An ongoing study in Sri Lanka focuses on establishing development–
environment linkages (Munasinghe et al. 1995). While in-country envi-

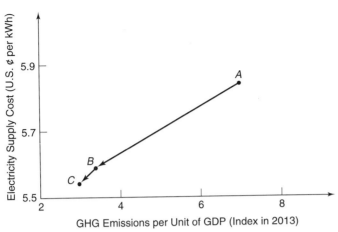

FIGURE B3.5. Effects of electricity pricing policy on GHG emissions. Point *A*: constant (subsidized) price; *B*: 30% price increase by 2005; *C*: 50% price increase by 2005 (price equals long-run marginal cost). From Munasinghe (1996).

ronmental problems have priority, the decision-making framework can be readily adapted to include global environmental issues. As shown in Figure B3.4, the overall set of models include a multisector computable general equilibrium (CGE) model, which links together several sectoral/regional models (including subsector models). Of specific interest is the energy model that helps determine the economic as well as environmental consequences of policies. The specific developmental implications of climate change policies are examined using this model.

To illustrate the approach, we examine how the power sector submodel is used to trace the environmental effects of various electricity supply options. First, future electricity needs are forecast on the basis of projected growth in various economic sectors of the macro model. Next, various supply options for meeting the demand are examined. The existing power system is more than 85% hydroelectric, but large-scale coal-fired generation is envisaged in the future. While economic efficiency requires that power demand be supplied by the cheapest supply sources (i.e., least-cost supply planning), other criteria indicated in the SED framework (see Figure 3.2) – for example, the emission of GHGs – complicate the decision.

One key option tested was the effect of electricity pricing policy (e.g., partially induced by a carbon tax on imported coal). The CGE model traces the effects of the energy price increase throughout the economy: the key outcome is that productive resources shift within the economy (to more energy-efficient sectors) and gross domestic product (GDP) rises. More specifically within the energy sector, Figure B3.5 shows that

(continued)

BOX 3.6 Application of the Sustainable Energy Development
Methodology in Sri Lanka *(continued)*

when power prices are raised by 30% from the present subsidized level,
both the costs of supply and the GHG emissions per unit of economic
output fall sharply (from *A* to *B*). Raising the electricity price to the full
long-run marginal cost provides even further gains (at *C*). Thus, eliminating subsidies to electricity users results in a win–win case – with simultaneous economic and environmental benefits. As indicated in the SED
framework, other objectives such as social/equity considerations must
also be met, primarily by structuring the higher electricity prices to pro-

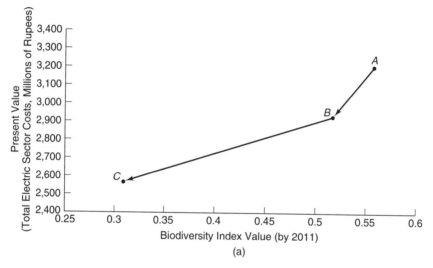

(a)

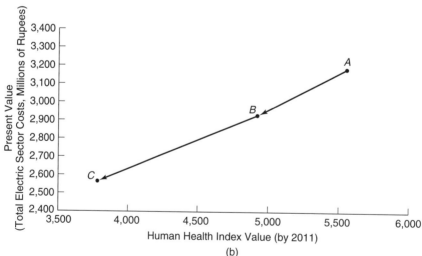

(b)

FIGURE B3.6. Effects of electricity pricing policy on selected in-country environmental indices: (a) biodiversity and (b) human health. Points *A, B,* and *C* are as in
Figure B3.5. From Munasinghe (1996).

vide a safety net in terms of low-cost or "lifeline" rates for poor consumers. To summarize, embedding GHG reduction measures within
national energy pricing policies is an attractive option because it also addresses the main objectives of sustainable development. Energy conservation and demand management policies offer similar advantages.

The interaction of energy policies, local environmental impacts, and
climate change measures were also analyzed. It was very difficult to incorporate environmental externalities into conventional cost–benefit analysis. As explained in Chapter 4, the nature of certain effects, such as the
impact of pollutants from coal-fired generating stations on health, the
potential loss of biodiversity associated with large-scale hydro reservoirs,
and the effects of GHG emissions, were all exceptionally difficult to value
in economic terms. Therefore, multicriteria analysis was used to assess
energy policy alternatives when costs and benefits were not all measurable in terms of a single criterion (e.g., monetary units used in cost–
benefit analysis). For example, Figure B3.6 shows how electricity price increases significantly reduce both the loss of biodiversity and effects on
human health. Furthermore, Chapter 4 describes how multicriteria
analysis was used in the Sri Lanka case to choose among the many technological options available on the basis of the trade-off curve approach
(see Figures 4.4 and 4.5).

The conceptual SED framework facilitates policy making and does not imply rigid centralized planning. Thus, such a process should result in the development of a flexible and constantly updated energy strategy designed to meet
national sustainable development goals. This national energy strategy (of
which the investment program and pricing policy are important elements)
can be implemented through a set of energy supply and demand management policies and programs that make effective use of decentralized market
forces and incentives. In particular, the implementation of climate change
measures within countries could be far more effective if they were incorporated into the strong current trends toward privatization and decentralization
of the energy sector.

3.3 SUMMARY

This chapter began by reviewing the main causal linkages leading from anthropogenic emissions of GHGs to climate change damage and possible policy responses. In this context, we set forth a set of key issues that decision makers must address in dealing with the problem of global climate change. The
framework for decision making includes three main elements. First, the

global optimization step seeks primarily efficient solutions. However, while the ideal approach would be to reduce emissions up to the point where the marginal costs of mitigation were exactly equal to the marginal benefits of avoided climate change damage, uncertainties in knowledge suggest that a more basic precautionary rule be used – that is, the adoption of a safe minimum standard for future GHG emissions and concentrations. Second, the collective decision-making step attempts to achieve both procedural and consequential equity. Response strategies consist of mitigatory, adaptive, other policy, and cooperative approaches. Financial transfers and technology cooperation (especially North to South) are also relevant. Third, the procedure and mechanism step explores how the UNFCCC process might be developed to create innovative financing mechanisms, insurance schemes, and in-country policy application frameworks for implementing the international agreements that might emerge from the negotiations.

The decision-making framework must be both flexible and heuristic, given that a conventional approach based on deterministic optimization in the long term is not possible in the face of high levels of uncertainty (see Chapter 2). At the same time, uncertainty should not be used to justify inaction, since a business-as-usual scenario results in a high level of climate change risk (see Chapter 1). The framework facilitates the making of short-term decisions (such as no-regrets measures) that are sensible and robust but adjustable in the longer term, as more accurate information becomes available.

REFERENCES

Arrow, K. J., J. Parikh, and G. Pillet. 1996. Decision-making frameworks for addressing climate change. In J. P. Bruce, H. Lee, and E. F. Haites (eds.), *Climate change 1995: Economic and social dimensions of climate change.* Contribution of Working Group III to the Second Assessment Report of the Intergovernmental Panel on Climate Change. Cambridge University Press, pp. 53–78.

Ayres, R. U., and P. M. Weaver. 1994. The case for proactive risk management. In *Environmental hazard: Liability or opportunity?* Report of the SANDOZ/INSEAD colloquium, June 29, 1993. Paris: INSEAD, Center for the Management of Environmental Resources, pp. 3–9.

Chichilnisky, G., and G. Heal. 1993. Global environmental risks. *Journal of Economic Perspectives,* **7**(4), 65–86.

Eeckhoudt, L., and C. Gollier. 1995. *Risk evaluation, management and sharing,* translated by V. Lambson. New York: Harvester Wheatsheaf.

IIASA (International Institute for Applied Systems Analysis). 1993. *Processes of international negotiation.* June, pp. 4–12.

Munasinghe, M. 1990. *Energy analysis and policy.* London: Butterworth–Heinemann, 1990.

Munasinghe, M. 1991. Electricity and environment in developing countries. In N. Ferrari et al. (eds.), *Energy and environment in the 21st century.* Cambridge, MA: MIT Press, pp. 361–85.

Munasinghe, M. 1992. Environmental challenge facing the developing countries. *EPA Journal*, **16**(Mar.–Apr.), 52–4.

Munasinghe, M. (ed.). 1996. *Environmental impacts of macroeconomic and sectoral policies.* Washington, DC: International Society of Ecological Economics, World Bank, and United Nations Environment Program.

Munasinghe, M., and C. Clarke. 1995. *Disaster prevention for sustainable development.* Geneva and Washington, DC: International Decade for Natural Disaster Reduction and World Bank.

Munasinghe, M., S. A. Karunaratne, W. Cruz, A. Persson, K. Lvosky, P. Meier, and S. Gupta. 1995. *Understanding macroeconomic–environmental linkages: A policy modeling framework for Sri Lanka.* Washington, DC: World Bank, Environment Department.

Munasinghe, M., and K. King. 1992. Implementing the Montreal Protocol to restore the ozone layer. *Columbia Journal of World Business* (Fall/Winter), 136–43.

Munasinghe, M., P. Meier, M. Hoel, S. W. Hong, and A. Aaheim. 1996. Applicability of techniques of cost–benefit analysis to climate change. In J. P. Bruce, H. Lee, and E. F. Haites (eds.), *Climate change 1995: Economic and social dimensions of climate change.* Contribution of Working Group III to the Second Assessment Report of the Intergovernmental Panel on Climate Change. Cambridge University Press, pp. 145–78.

Munasinghe, M., and S. Munasinghe. 1993. Enhancing South–North cooperation to reduce global warming. Paper presented at the IPCC Conference on Climate Change, Montreal, May.

Parikh, J. 1994. Joint implementation and sharing commitments: A southern perspective. In N. Nakicenovic et al. (eds.), *Public regulation: New perspectives on institutions and policies.* Cambridge, MA: MIT Press, pp. 227–53.

Pearce, D. W., et al. 1996. The social costs of climate change: Greenhouse damage and the benefits of control. In J. P. Bruce, H. Lee, and E. F. Haites (eds.), *Climate change 1995: Economic and social dimensions of climate change.* Contribution of Working Group III to the Second Assessment Report of the Intergovernmental Panel on Climate Change. Cambridge University Press, pp. 179–224.

Smith, K. 1991. Allocating responsibility for global warming: The natural debt index. *Ambio*, **20**(2), 95–6.

UNFCC (United Nations Framework Convention on Climate Change). 1992. Articles. New York: United Nations.

Weilenmann, U. 1994. Insurable risk associated with climate change. In J. Pillet and B. Gassmann (eds.), *Steps towards a decision making framework to address climate change.* Report of the IPCC Working Group III Writing Team II, Montreaux meeting, March 3–6. Würenlingen and Villigen (Switzerland): Paul Scherrer Institute, pp. 11–17.

WCED (World Commission on Environment and Development). 1987. *Our common future.* London: Oxford University Press.

4

DECISION CRITERIA AND TOOLS

When public or collective decisions must be made, it is essential to have a transparent set of decision-making criteria and tools around which a consensus can be developed. Accordingly, this chapter reviews the strengths and weaknesses of the various analytical techniques available for making decisions concerning global climate change. The initial focus is on achieving economically efficient outcomes, while equity aspects are introduced later in the chapter.

4.1 CHOICE THEORY AND DECISION-MAKING CRITERIA

In basic terms, the theory of choice rests on the principle that individuals are free to select from a certain range of actions – each having some corresponding consequences. Generally, an individual will have some notion of the relative desirability of such consequences and will select the option whose expected consequences are preferred to those of alternative actions (Arrow 1951).

In the utility-based approach, the concept of a utility function provides the basis for systematically comparing consequences and thereby ranking alternative actions. Utility basically denotes the level of satisfaction (or welfare) achieved, and the utility function of an individual indicates how a particular choice might affect his or her welfare. A consistent and rational human being is assumed to choose the actions which will maximize his or her utility (or welfare) (Savage 1954). Conceptually, welfare could be measured in abstract units, but it is more practical if the welfare consequences of selected actions can be valued in economic terms. In this case, the individual will seek to choose the outcome that has the greatest economic value. Basically, a value-based utility function can be used to derive consumer demand curves that indicate how individuals value different goods and services in terms of their willingness to pay (see Section 4.2.2).

In the simplest case, projections of the future consequences of current decisions are assumed to be determined with certainty. If there is some element of uncertainty, this complication is usually handled by focusing on the expected value of the outcome. In this approach, the future is typically characterized by several possible outcomes, each having a different risk probability. The expected outcome is the sum of the various outcomes, each weighted by the corresponding probability of occurrence (see Chapter 3). The certain outcome, the probability of which is 100%, can be considered a special case of the more general situation involving uncertainty.

In brief, all utility-based criteria operate broadly under the fundamental concept of maximizing expected utility, subject to constraints on the range of feasible actions – usually determined by the availability of resources. Utility is maximized whether it is deterministic utility (in the certainty case) or expected utility (in the probabilistic case). Although utility-based criteria relate mainly to economic efficiency, they can also be modified to reflect equity considerations (as discussed in Section 3.2.2). When the consequences of actions can be measured in economic terms, the well-known criteria of cost–benefit analysis come into play (see Box 4.1).

BOX 4.1 Criteria for Economic Decisions

In economic cost–benefit analysis, benefits are defined relative to improvements in human well-being, while costs are measured by their opportunity costs, which are the benefits forgone when resources are not used in the best available application (Munasinghe 1990). The economic analysis of an activity seeks to evaluate the overall effect of that activity in relation to the whole economy. Rather than financial prices, shadow prices are used that reflect opportunity cost, including valuation of externalities wherever practical (as described later in this box). The most basic criterion for accepting a decision is a comparison of costs and benefits to ensure that the net present value (NPV) of the benefits of that action is positive,

$$\text{NPV} = \sum_{t=0}^{T} [(B_t - C_t)/(1 + r)^t]$$

where B_t and C_t are the benefits and costs in year t, r is the discount rate (see Chapter 2), and T is the time horizon. Both benefits and costs are defined as the difference between what would occur with and without the decision being implemented. In economic testing B, C, and r are defined in economic terms and appropriately shadow-priced.

If projects are to be compared or ranked, the one with the highest (and positive) NPV is the preferred one; that is, if $\text{NPV}_\text{I} > \text{NPV}_\text{II}$ (where NPV_i is the net present value for decision i), then decision I is preferred

(continued)

BOX 4.1   Criteria for Economic Decisions *(continued)*

to decision II, provided that the scale of the alternatives is roughly the same. More accurately, the scale and scope of each of the decisions under review must be altered so that, at the margin, the last increment of investment yields net benefits that are equal (and greater than zero) for all the projects. Complexities may arise in the analysis of interdependent projects.

The internal rate of return (IRR) is also used as a decision criterion. It can be defined by

$$\sum_{t=0}^{T} [(B_t - C_t)/(1 + \text{IRR})t] = 0$$

Thus, the IRR is the discount rate that reduces the NPV to zero. The decision is acceptable if IRR $> r$, which in most normal cases implies NPV > 0 (ignoring cases where multiple roots could occur, which may happen if the annual net benefit stream changes sign several times). Problems of interpretation occur if alternative projects have widely differing lifetimes, so that the discount rate plays a critical role.

Another frequently used criterion is the benefit–cost ratio (BCR):

$$\text{BCR} = \sum_{t=0}^{T} [B_t/(1 + r)t] / \sum_{t=0}^{T} [C_t/(1 + r)t]$$

If BCR > 1, then NPV > 0 and the decision is acceptable.

Each of these criteria has strengths and weaknesses, but NPV is probably the most useful. As a simple illustration, consider an investment of 1,000 rupees in year 2, which is expected to yield a onetime benefit of 1,405 rupees in year 5. Assuming a discount rate of 10%, the NPV today (in year 0) is given by NPV $= -1,000/(1.10)^2 + 1405/(1.10)^5 = 46$ rupees. The corresponding IRR is about 12%, since $-1,000/(1.12)^2 + 1,405/(1.12)^5 = 0$. Finally, the BCR is given by $[1,405/(1.10)^5]/[1,000/(1.10)^2] = 1.055$.

The NPV test can also be used to derive the least-cost rule. This approach is applicable in cases where the benefits of two alternative decisions or actions are equal (i.e., they serve the same purpose). Then the comparison of alternatives is simplified as follows:

$$\text{NPV}_{\text{I}} - \text{NPV}_{\text{II}} = \sum_{t=0}^{T} [C_{\text{II},t} - C_{\text{I},t}]/(1 + r)^t$$

since the benefit streams cancel out. Therefore, if

$$\sum_{t=0}^{T} C_{\text{II},t}/(1 + r)^t > \sum_{t=0}^{T} C_{\text{I},t}/(1 + r)^t$$

this implies that NPV$_{\text{I}} >$ NPV$_{\text{II}}$. In other words, the decision that has the lower present value of costs is preferred. This is called the least-cost alternative (when benefits are equal). However, even after selecting the least-cost alternative, it would be necessary to ensure that this project provided a positive NPV.

Shadow Pricing
In the economist's idealized world of perfect competition, the interaction of atomistic profit-maximizing producers and utility-maximizing consumers gives rise to a situation called Pareto-optimal. In this state, prices reflect the true marginal social costs, scarce resources are efficiently allocated, and, for a given income distribution, no one person can be made better off without making someone else worse off. However, conditions are likely to be far from ideal in the real world. Distortions due to monopoly practices, external economies and diseconomies (such as environmental effects that are not internalized in the private market), interventions in the market process through taxes, import duties, and subsidies all result in market (or financial) prices for goods and services that may diverge substantially from their shadow prices, or true economic values. Such considerations necessitate the use of appropriate shadow prices (instead of market prices) of resource inputs and outputs to determine the optimal investment decisions and policies – especially in developing countries, where market distortions are more prevalent than in industrialized countries. Since distortions in the prices of environmental resources are the most relevant in valuing climate change effects, this topic is discussed in detail in the main text.

While utility-based criteria are useful for decision making in the context of climate change, they do have some significant limitations. First, since climate change is a global issue, it is necessary to develop a measure of global welfare (see Section 3.2.2). Unfortunately, the addition of individual utility functions to derive a value for global welfare is highly problematic (Arrow, Parikh, and Pillet 1996). Clearly, decisions on global climate change require the analysis of global welfare. The main issue is whether each person's utility should be given equal weight. For example, from a global viewpoint, is the loss of $100 to a poor person equivalent to a gain of $100, or $200, to a rich individual? Is it realistic to ignore this issue by assuming the existence of mechanisms by which those who benefit could transfer at least a part of their gain to compensate the losers?

Second, it is preferable to make choices pertaining to climate change through a collective process because of so-called externalities.[1] Thus, uncoor-

---

[1] An externality is defined as the cost (or benefit) imposed on one individual by the actions of another, for which the latter does not pay (or receive) compensation.

dinated decision making by different stakeholders could prove to be both counterproductive and inequitable if it were based on the maximization of individual welfare pursued at the expense of everyone else's welfare. As indicated earlier, each individual action leading to the emission of GHGs often has unperceived implications for others who are distant in both space and time. Ideally, if markets and prices correctly reflect these (external) consequences, then individual decisions will also reflect what is collectively desirable. Unfortunately, many important issues are not well represented in markets. Among such considerations are poverty and the fragility of ecosystems and species, which are difficult to incorporate within a value-based utility criterion.

Third, decisions cannot be based on fixed or static situations, especially when high levels of risk and uncertainty are involved. The response to climate change will have to evolve dynamically, as more information becomes available. Introducing a probabilistic approach and expected values would partially address this issue. However, this modification applies mainly to risk associated with individual choice and uncertainty associated with economic and technical variability. Climate change entails risk and uncertainty on a much greater scale and may thereby elicit unexpected reactions. For example, it has been observed that the public is often quite averse to the risk of high-impact but rare (i.e., low-probability) events. Using the expected value concept introduced earlier in this section, one could compare (a) a nuclear accident that causes 10,000 deaths and has a probability of occurrence of 0.001 per year; (b) commercial airplane crashes, which cause 100 deaths per average event and occur 0.1 time per year; and (c) road accidents, which cause (on average) only 1 death but happen 10 times per year. In each case, the expected number of fatalities is the same (i.e., $10,000 \times 0.001 = 100 \times 0.1 = 1 \times 10 = 10$ expected deaths per year). The public is likely to react most adversely to a potential nuclear catastrophe, to exhibit considerably less negativity in response to airplane crashes, and to show the least concern toward automobile accidents. The perception that we do not have good information on the risk probability of a nuclear accident may influence public reactions. In this context, probability distributions are rarely available to describe climate change effects. The precautionary approach introduced in Chapter 3 will play a particularly important role in the presence of risk aversion and potentially irreversible and catastrophic outcomes. Therefore, while the expected utility criterion is acceptable as a rough initial guide to decision making on climate change, other approaches such as decision analysis (using risk and hedging strategies) will be required, as described in Section 4.6.

In contrast to the utilitarian approach, which defines collective or social welfare in terms of the sum of individual utilities, the Rawlsian approach to distributive justice is based on the principle that social policy should focus on the welfare of the worst-off individuals (Rawls 1971). Given the existence of inequality among individuals (or generations), the decision rule should seek to improve the position of the poorest. In the intergenerational case, the interests of future generations should be considered in relation to the interests of the current generation. The Rawlsian principle focuses on those who might be the most adversely affected by future climate change (e.g., poor coastal communities and small island states).

4.2 COST–BENEFIT-BASED DECISION TOOLS

Cost–benefit analysis (CBA) provides an analytical framework for comparing the consequences of alternative policy actions on a quantitative rather than qualitative basis. The basic principles are well understood and straightforward. (See Box 4.1 for an explanation of criteria that are utility-based and assume that costs and benefits can be valued.) An action is justified if the costs of the action are lower than the benefits derived from it. If there are several alternatives, one picks that option whose benefits most exceed the costs, and when there is a shortage of capital, investments should be made in projects that yield the highest returns per unit of capital (Munasinghe 1990).

4.2.1 Traditional and Modern CBA

Traditional CBA was developed initially as a means to evaluate discrete project alternatives (that were limited in scale, geographic extent, and time span). Not surprisingly, therefore, this original approach is too narrow to be relevant to the phenomenon of global climate change. The latter exhibits several problematic features (see Chapter 2), which are quite different from those assumed within the conventional context in which CBA is applied. In traditional CBA, costs and benefits arise within a time span of typically no more than 15 to 25 years, corresponding roughly to the physical life of most projects over which benefits are derived. At the same time, the physical changes are usually fairly localized and incremental (i.e., relatively small). Finally, the elements of uncertainty are tractable and can often be translated into risks defined by probability distributions.

All of these characteristics are different in the context of climate change. Various climate change scenarios and the potential response strategies are continuous rather than discrete alternatives. For analytical purposes, discreteness is introduced by constructing specific future scenarios of climate change. Furthermore, the relevant time spans extend to a century or more and involve the entire globe, with features such as large time lags, irreversibilities, and nonlinearities that are extremely difficult to model. The uncertainties in the chain of causality discussed earlier (in Chapters 2 and 3) are also extremely large, and few elements of such uncertainty are amenable to characterization as probability distributions. Moreover, the effects of cascading the uncertainty implied in each link of the chain of causality greatly amplifies the total uncertainty in the final outcome – namely, the extent of damage caused by climate change. The possibility of catastrophic consequences (arising from nonlinear responses in natural systems), which is especially problematic for risk-averse decision makers, is yet another complication that traditional CBA does not handle well.

In this context, modern CBA has evolved into a family of decision techniques that is more useful for climate change analysis, including (a) economic cost–benefit analysis (ECBA), which compares all costs and benefits expressed in terms of a common numeraire, usually monetary units; (b) cost-effectiveness analysis (CEA), which is designed to identify the lowest-cost method of achieving an objective already defined using other criteria; (c) multicriteria analysis (MCA), which attempts to find desirable options when some or all of the costs and benefits are not measurable in monetary terms; and (d) decision analysis (DA), which is useful when choices have to be made under conditions of uncertainty. Integrated assessment is a comprehensive method that makes use of modern CBA and related techniques to analyze complex phenomena like climate change.

This group of techniques can help to provide better responses to several of the key public policy questions posed earlier, including (a) how much emissions of greenhouse gases (GHGs) should be reduced, (b) when emissions should be reduced, and (c) how emissions should be reduced. Another important question, concerning who should reduce emissions, involves equity. Therefore, it is not amenable to resolution by ECBA alone, although modern decision techniques (especially MCA) can elucidate the trade-offs between economic efficiency and equity.

4.2.2 Valuation

The robustness of CBA depends critically on how reliable the values attached to each item are. As indicated earlier, values used in economic CBA should reflect opportunity costs. If the prices of marketed goods and services are distorted, appropriate shadow prices are required. For nonmarketed goods and services, such as many environmental services, values have to be estimated in order to aggregate costs and benefits and obtain an overall evaluation of choice of policy. Estimated prices may depend on the method chosen to create them, and one should therefore interpret the corresponding results with caution.

Costs, Benefits, and Economic Value   An important reason for averting climate change is that it will give rise to significant damage. As explained earlier in the context of Figure 3.1, it is necessary to estimate both the marginal avoided damage (MAD) and marginal abatement cost (MAC) curves in order to determine an economically optimal response. In fact, if MAD > MAC, it is possible to infer that further abatement is desirable, since severe marginal damage (high willingness to pay) could be avoided at low emission curtailment costs, and vice versa. Valuation of environmental effects in CBA may also be helpful in making cost-effective decisions. However, it is difficult to assess these values. Furthermore, the valuation should be based on a reasonably well founded methodology, since speculative assumptions could confuse decision makers. An action that yields negative net benefits may yet be worth accepting, especially if some important beneficial effects could not be explicitly valued in the calculation but were nevertheless well documented. Different types of costs and benefits are estimated in subsequent chapters, and each requires somewhat different approaches to quantification and valuation. Among the broad categories listed below, the second and third give rise to significant valuation problems. More specific types of costs and benefits to be evaluated have already been listed in Chapter 1.

Costs of Abatement or Mitigation Measures Taken before the Actual Effects Occur. These measures primarily involve GHG emission reductions or the removal of GHGs by augmenting sinks such as forests. Serious issues involving the valuation of effects generally arise in this category when cost estimates have to be based on the performance of technologies 50 to 100 years in the future.

Costs of Adaptation after Effects Become Apparent. Techno-engineering preventive measures to minimize the impact of climate change fall into this category. Building dikes to protect coastal areas is a typical example. Based on actual experience (such as in the Netherlands), the costs of such actions are relatively easy to establish. Other ways in which society might adapt to the effects of climate change may be more difficult to value. For example, climate change will affect crop yields and result in poleward shifts in cultivated land. Some areas will gain and some will lose, which makes this an equity issue (between regions and countries) as much as a cost issue. Some of the costs of adaptation will vary, depending on ex ante actions (e.g., greater expenditures on agricultural research and development to develop drought- or saline-resistant crops). Particularly difficult to value will be social effects, such as the cost of forced adaptation and population movements – a problem already encountered, albeit on a lesser scale, in CBA of the impact of building large reservoirs when a significant number of individuals must be relocated involuntarily.

Costs (and Benefits) of Nonadaptation. These are the most difficult to value. Some useful techniques of valuation already being implemented are outlined in this subsection (see also Chapter 5).

The distinction between mitigation and adaptation measures is not always clear-cut (see Chapter 5). Some actions may fall in between, as in the financing of research (before climate change occurs) on crops that are better adapted to growth in saline water, as more areas become inundated by the rise in mean sea levels.

A fundamental concept underlying valuation is that of the total economic value (TEV) of a resource. TEV consists of its use value (UV) and non–use value (NUV) (see, e.g., Munasinghe 1993). Use values can be broken down into direct use value (DUV), indirect use value (IUV), and option value (OV) or potential use value. One must be careful not to double-count both the value of indirect supporting functions and the value of the resulting direct use. A major category of non–use value is existence value (EV). Thus, we may write

$$TEV = UV + NUV$$

or

$$TEV = [DUV + IUV + OV] + [NUV]$$

Figure 4.1 shows this disaggregation of TEV in schematic form. Below each valuation concept is a short description of its meaning and a few typical examples (based on a tropical rain forest) of the environmental resources underlying the perceived value. Option values, existence values, and non–use values, are shaded to caution the analyst about some of the ambiguities associated

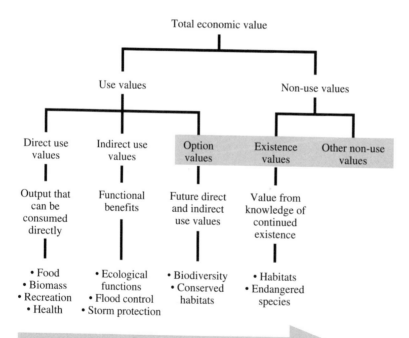

FIGURE 4.1. Categories of economic values attributed to environmental assets (with examples from a tropical rain forest). From Munasinghe (1993).

with defining these concepts; as shown in the examples, they can spring from similar or identical resources, while their estimation could also be interlinked. However, these concepts of value are generally quite distinct. Option value is based on how much individuals are willing to pay today for the option of preserving the asset for future direct and indirect use. This concept is very similar to that of holding a financial asset in the expectation of making some future capital gain. In the presence of uncertainty, quasi–option value is said to define the value of preserving options for future use in the expectation that knowledge about the potential benefits or costs associated with the option will grow over time (see Fisher and Hanneman 1987; Pearce and Turner 1990). The latter approach may be quite relevant, given the great uncertainties associated with climate change, as elaborated in Section 4.6. Existence value is a more altruistic concept, based on the perceived value of the environmental asset unrelated to either current or optional future use – that is, the value arising simply because it exists.

Methods of Economic Valuation   A variety of valuation techniques may be used to quantify the above concepts of value (for methodological details, see Freeman 1993; for some applications, see Munasinghe 1993 and Chapter 5). As shown in Box 4.2, valuation methods can be categorized according to what type of market they rely on and how they make use of actual or potential behavior.

BOX 4.2   Types of Valuation Techniques

TABLE B4.1   Types of Valuation Techniques

| Type of Behavior | Type of Market | | |
	Conventional Market	Implicit Market	Constructed Market
Based on actual behavior	Effect on production Effect on health Defensive or preventive costs	Travel costs Wage differences Property values Proxy-marketed goods	Artificial market
Based on intended behavior	Replacement costs Shadow project		Contingent valuation

Source: Munasinghe (1993).

(continued)

BOX 4.2 Types of Valuation Techniques *(continued)*

Effect on Production

An investment decision often has environmental effects, which in turn affect the quantity, quality, or production costs of a range of productive outputs that may be valued readily in economic terms.

Effect on Health

This approach is based on the impact of pollution and environmental degradation on health. One practical measure related to the effect on production is the value of human output lost due to ill health or premature death. The loss of potential net earnings (called the human capital technique) is one proxy for forgone output, to which the costs of health care or prevention may be added.

Defensive or Preventive Costs

Often, costs may be incurred to reduce the damage caused by an adverse environmental impact. For example, if drinking water is polluted, extra purification may be needed. Then, such additional defensive or preventive expenditures (ex post) could be taken as a minimum estimate of the benefits of mitigation.

Replacement Costs and Shadow Project

If an impaired environmental resource is likely to be replaced in the future by another asset that provides equivalent services, then the costs of replacement may be used as a proxy for the environmental damage – assuming that the benefits of the original resource are at least as valuable as the replacement expenses. A shadow project is usually designed specifically to offset the environmental damage caused by another project. For example, if the original project was a dam that inundated some forestland, the shadow project might involve replanting an equivalent area of forest elsewhere.

Property Values

In areas where relatively competitive markets exist for land, it is possible to decompose real estate prices into components attributable to different characteristics like house and lot size, air and water quality. The marginal willingness to pay (WTP) for improved local environmental quality is reflected in the increased price of housing in cleaner neighborhoods. This method has limited application in developing countries, since it requires a competitive housing market, as well as sophisticated data and tools of statistical analysis.

Wage Differences

As in the case of property values, the wage differential method attempts to relate changes in the wage rate to environmental conditions, after accounting for the effects of all factors other than environment (e.g., age, skill level, and job responsibility) that might influence wages.

Proxy-Marketed Goods

This method is useful when an environmental good or service has no readily determined market value but a close substitute exists that does have a competitively determined price. In such a case, the market price of the substitute may be used as a proxy for the value of the environmental resource.

Artificial Market

Such markets are constructed for experimental purposes to determine consumer WTP for a good or service. For example, a home water purification kit might be marketed at various prices, or access to a game reserve may be offered on the basis of different admission fees, thereby facilitating the estimation of values.

Contingent Valuation

This method puts direct questions to individuals to determine how much they might be willing to pay for an environmental resource, or how much compensation they would be willing to accept if they were deprived of the same resource. The contingent valuation method (CVM) is more effective when the respondents are familiar with the environmental good or service, and have adequate information on which to base their preferences. However, WTP tends to be affected by affordability in the case of the poor. Recent studies indicate that CVM, cautiously and rigorously applied, could provide rough estimates of value that would be helpful in economic decision making, especially if other valuation methods were unavailable.

Source: Munasinghe (1993).

The basic concept of economic valuation underlying all of these methods is individuals' willingness to pay (WTP) for an environmental service or resource. Willingness to pay is itself based on the area under the demand curve, as illustrated here by a simplified case. In Figure 4.2, curve $D(S_0)$ indicates the demand for an environmental resource (e.g., the number of visits made per

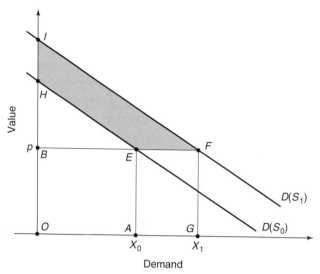

FIGURE 4.2. Increase in benefits due to improvements in the quality of an environmental asset. From Munasinghe (1993).

month to a freshwater source like a river).[2] X_0 is the original demand level at price p (e.g., the cost of making a trip, including the value of time spent traveling). The total WTP or value of the services provided by the environmental resource is measured by the area $OHEA$.[3] This area consists of two main components: the area $OBEA$, or (pX_0), which represents the total cost, and the area BEH, which is called the consumer surplus or net benefit (i.e., the net value over and above actual expenses). Point H represents the "choke price," at which demand falls to zero.

Next, we extend this example to examine what could happen if the quality of the environmental service were improved (e.g., by cleanup of the freshwater source). The normal response would be an increase in demand, represented in the figure by a shift in the demand curve from $D(S_0)$ to $D(S_1)$. The new level of demand is X_1 (assuming the same price p), yielding the corre-

---

[2] The theoretically correct demand function to use in estimating value is the compensated (or Hicksian) one, which indicates how demand varies with price while keeping the user's utility level (or well-being) constant. Problems of measurement may arise because the commonly estimated demand function is the Marshallian one, which indicates how demand varies with the price of the environmental good, while keeping the consumer's income level constant. In practice, it has been shown that the Marshallian and Hicksian estimates of WTP are in good agreement for a variety of conditions. Furthermore, in a few cases the Hicksian function may be derived once the Marshallian demand function has been estimated (Willig 1976).

[3] Equivalently, the change in value of an environmental asset could be defined in terms of the difference between the values of two expenditure or cost functions: $E_2 - E_1$. In this expression, E_1 and E_2 are the respective minimum expenditures required to achieve a given level of well-being for a household (or output for a firm) with and without access to the environmental resource in question, while keeping all other aspects constant.

sponding total WTP equal to area *OGFI* and the new net benefit measured by area *BFI*. Thus, the quality improvement would result in an incremental increase in the value of the environmental resource, given by the striped area *EFIH*.[4]

People's willingness to accept (WTA) compensation for environmental damage is another measure of economic value and is closely related to WTP. Empirical evidence indicates that questions about WTA compensation yield responses with higher values than questions about WTP to retain the same environmental asset. One reason for this observation may depend on *affordability*, because WTP may be constrained by the basic inability of poorer consumers to pay, whereas this consideration does not affect WTA. The result WTA > WTP may also arise from *strategic behavior*, whereby individuals tend to inflate the amounts they would like to receive and deflate the sums they may have to pay out, in the hope of influencing the actual outcome. Another explanation is *status quo preference*, in which people are conservative about changes and prefer no change. There may be an aversion to the risk of losing environmental benefits that already exist, or substitutes might be difficult to find, especially in the case of basic needs like clean drinking water. Another possibility is that people who are used to a clean environment could be quite resistant to the loss of implied property rights posed by the WTA criterion.

In practice, either or both measures are used for valuation, and therefore references to WTP here may be broadly interpreted to include WTA also, unless otherwise stated. Generally, WTP is considered to be a more consistent and credible measure than WTA. However, when significant discrepancies exist between the two measures, then the higher values of WTA may be more appropriate when valuing losses in environmental benefits.

As already mentioned, monetary values placed on environmental goods and services are traditionally lower in poorer areas because ability to pay restricts the expression of WTP. One way of addressing this concern within cost–benefit calculations is to weight monetary values in inverse proportion to income. However, the use of income weights is problematic, especially when data on income or consumption levels of the concerned groups are unavailable. Generally, it is more practical to deal with equity and distributional issues separately rather than attempt to combine them with the economic efficiency–based approach that underlies WTP estimates in CBA (see Chapter 2).

Valuation techniques obviously should be selected with some care, and in particular one must recognize that a particular valuation technique may not

---

[4] From a theoretically strict viewpoint, this measure of value is correct in the present case only if the good in question (i.e., visits to the water source) and the environmental quality attribute (i.e., how clean the water is) are so-called weak complements (Maler 1974).

necessarily capture the entire value. For example, if the replacement cost approach is used to value the loss of a primary forest area being inundated by a dam, it would most likely capture only the use value. The value of biodiversity loss involved in the loss of the primary forest, or any other ecosystem, may not be included.

We note that these valuation techniques have been developed for more conventional environmental impact analysis and would require significant modification and/or careful interpretation when applied to global climate change (e.g., long-term intergenerational impact, biodiversity loss, welfare comparisons across cultures, or situations involving large gaps between gainers and losers). Nevertheless, whatever the difficulties may be, the importance of valuation remains, and the development of better techniques should be viewed as an important item in the climate change research agenda. Certainly, ignoring an impact because it cannot be satisfactorily valued carries high risk and is one of the reasons for the use of MCA (see Section 4.5).

4.3 ECONOMIC COST–BENEFIT ANALYSIS

In light of the complexity of the climate change problem, what can we say about the suitability of CBA? In this section we explore how basic economic cost–benefit analysis (ECBA) can be adapted to the initial process of analyzing climate change issues.

Traditional ECBA evolved as a technique for evaluating and comparing discrete project alternatives. In the early years of its application, there was little concern with externalities, and the analysis took into account only the direct costs and direct benefits of projects. Since it was developed in the industrialized world, market prices provided appropriate guidance on how to evaluate benefits. When the development community began to apply the technique to nonmarket economies, where prices were subject to significant distortions, shadow-pricing techniques provided simple corrections (see Box 4.1). For example, if an oil-importing country keeps the domestic price of oil at artificially low levels, ECBA requires the use of the international or border price (not the domestic price) as a basis for valuing oil.

Discounting is one of the central concepts in ECBA (see Chapter 2), which addresses the fact that costs and benefits may not occur at the same point in time. For example, while costly actions undertaken to avert future climate change may have to be taken in the near term, most of the benefits of such actions will occur far in the future. Discounting enables one to take into account the time value of money. In the case of evaluating simple investment alternatives over shorter time horizons (e.g., <15 years), it is well accepted that the opportunity cost of capital should be the discount rate to be applied to both costs and benefits. However, in the case of global climate change character-ized by large geographic scales and timescales, nonlinearities and cata-strophic consequences, irreversibilities, and other complications, there is sharp disagreement as to what discount rate is appropriate.

A major problem in applying CBA to the climate change problem follows directly from the chain of causality discussed in Chapter 1 (Figure 1.1). Thus, while estimating the costs of emission reduction involves the beginning of the causal chain, estimating the benefits (or the costs of avoided damage) involves the very end of the chain. Since some level of uncertainty is associated with each of the linkages, estimates at the last stage of the chain are subject to com-pound uncertainties, which may be very large indeed. It is also difficult to de-termine discrete alternative scenarios and response strategies for purposes of comparison, because climate change phenomena and effects form a con-tinuum.

4.4 COST-EFFECTIVENESS ANALYSIS AND LEAST-COST SOLUTION

As CBA began to be applied to much broader contexts, and particularly to the comparison of alternative portfolios of projects and broad policy choices, the increasing complexity (resulting especially in problems of comparing greatly varying benefit streams or simply valuing benefits) made it desirable either to keep the level of benefits constant or to measure benefits in terms of a consis-tent physical (nonmonetary) unit. This approach led to cost-effectiveness analysis (CEA), which implicitly assumes that the projects under considera-tion will pass the NPV test (i.e., provided that the economic costs and benefits are available).

If benefits are held constant, CEA reduces to finding the most effective, or least-cost, solution (LCS) that meets the desired level of benefits (see Box 4.1). An additional advantage is that benefits do not necessarily have to be ex-plicitly valued, provided that they are identified in some consistent manner. For example, in power sector planning, sophisticated models are applied to identify the least-cost capacity expansion plan (where the present value of sys-tem costs are minimized), while achieving some constant level of physical benefits, usually represented by an exogenously specified time path of elec-tricity demand at an exogenously specified level of reliability. If benefits do vary among project alternatives but can still be measured in some consistent (nonmonetary) unit, CEA focuses on maximizing the benefits that could be derived from a limited (or fixed) set of resources. Alternatively, it is possible to obtain the maximum benefit per unit of project expenditure.

As we shall see, the approach based on CEA is a variant of CBA that has seen rather widespread application to the climate change problem, in which one seeks to identify the least-cost options available for achieving given GHG emission reductions, without attempting to specify what the benefits of those reductions might be. However, in order to determine the LCS, it is crucial to clarify how the target is defined, because there are several options in the global climate change context. Most recent analyses of mitigation costs have focused on a target based on future emission levels,[5] such as stabilization of the emission of certain GHGs by a given year. However, it might be more relevant to express the targets in terms of concentrations of atmospheric GHGs at some future time (as mentioned in Chapter 1).

To change the target for climate policy from emissions to atmospheric concentrations requires a radically different CEA strategy. As shown in Chapter 1, maintaining CO_2 emissions at present levels is not sufficient to stabilize future atmospheric concentrations at a reasonably acceptable level. At the same time, an identical level of future stable concentrations could be reached via many alternative emission profiles (Figure 1.16b). Richels and Edmonds (1995) have compared the costs of reaching some particular concentration in the year 2100 by following alternative strategies. They showed that immediate stabilization of emissions would be much more costly than a more gradual reduction of emissions. This result follows from their key assumptions that a more gradual reduction would avert the economic shock associated with a sudden stabilization of emissions, while in the future advanced and more economic technologies might become available to a larger extent – thereby reducing sizable abatement costs through postponement.

4.5 MULTICRITERIA ANALYSIS

The most basic requirement for the application of CBA is that both costs and benefits be given economic value. This is typically a two-step process: first, the effects of a decision or action must be quantified in terms of physical measures; and second, those physical effects must be valued in economic terms. Some applications of the valuation techniques described earlier are likely to be controversial. For example, putting a value on the risk to human health and life has been a major problem in the practical application of CBA in the past, even in situations where there is agreement on the increase in morbidity and mortality that might be caused by some policy or project. Indeed, even if

---

[5] The degree of emission abatement is reported in such studies in two different ways. The first is in terms of a reduction from some baseline – itself defined as the trajectory of GHG emissions for some postulated business-as-usual scenario. The second is in terms of reductions from some reference year (e.g., "Reduce greenhouse emissions to 80% of their 1990 levels by 2010"); see Chapter 7.

one could place economic value on human life, not everyone would agree that doing so is even *appropriate*.[6] Furthermore, comparing or aggregating the value of life across communities or countries raises serious problems (see Chapter 5, especially Box 5.1). Placing economic value on the loss of biodiversity has been equally difficult.

4.5.1 Scope of MCA

Recognizing the problem of economic valuation has led to the use of so-called multicriteria analysis (MCA). Such techniques first gained prominence in the 1970s, when the intangible environmental externalities lying outside traditional CBA were increasingly recognized. In modern practice, CBA concepts are often embedded both in MCA that expressly allows more than one objective and in decision analysis that expressly addresses risk and uncertainty (see Section 4.6), thereby providing a mechanism for integerating various types of criteria.

MCA is more comprehensive than traditional CBA, since it is designed to deal with multiple objectives, of which economic efficiency may be only one. Thus, MCA allows for the appraisal of alternatives with differing objectives and varied impacts that are often assessed in differing units of measurement. MCA is a particularly powerful tool for quantifying and displaying trade-offs between conflicting objectives. In this context, it meets an important objective of modern decision makers, who prefer to be presented with a range of feasible alternatives, as opposed to one "best" solution. It is also helpful when there is concern that monetized values themselves may be inaccurately estimated or that such values might not reflect welfare levels. As in any other assessment involving gainers and losers, the question of who is affected, and how they will perceive the impact, requires careful definition in an MCA.

As noted earlier, conventional CBA cannot provide answers about the optimum level of equity in the same way that it provides answers about the optimum level of economic efficiency. But MCA can identify the *trade-offs* between equity objectives and economic efficiency, as indicated in Figure 4.3. Thus, the best equity result (indicated by option 1 – say, an equal per capita sharing of the burden of GHG emission reduction) may have the highest cost, while the worst equity result (indicated by option 5 – say, one based on the present distribution of GHG emissions) may have the lowest economic cost.[7] Nevertheless, even MCA requires a quantification or at least an ordinal ranking of

---

[6] For further discussion of the ethical and epistemological aspects of the climate change problem, see Brown (1992).

[7] But see Chapter 2 for a discussion of the difficulties of making cross-country comparisons of costs.

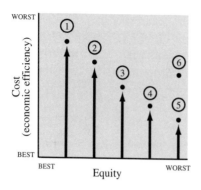

FIGURE 4.3. Multicriteria analysis. From Munasinghe et al. (1996).

the noneconomic efficiency criteria, as suggested in the figure. Even such a noncardinal ranking may prove problematic when a global issue like climate change requires comparisons across countries and cultures.[8]

More generally, as explained in Chapter 2, it is increasingly accepted that the pursuit of sustainable development will require the recognition of goals related to economic efficiency, social equity, and environmental protection (Munasinghe 1993). Economic valuation of the impact of climate change on certain social and environmental aspects (e.g., biodiversity and cultural assets) will be difficult, and MCA-related approaches will be needed to make the trade-offs between otherwise noncompensable costs and benefits.

4.5.2 Selecting Criteria

Even the staunchest advocates of CBA would concede that economic efficiency (or economic value) is not the sole criterion in setting public policy, and that policy makers rightfully need to consider a broader set of objectives. Unfortunately, there is much confusion about what constitutes a coherent set of objectives (see Box 4.3 for an example).

The following criteria have been suggested by the U.S. Environmental Protection Agency (1989) as constituting the basis for selecting public policy:

1. Flexibility
2. Urgency
3. Low cost
4. Irreversibility
5. Consistency
6. Economic efficiency
7. Profitability
8. Political feasibility

---

[8] Some outcomes may be inefficient, i.e., those that lie inside the frontier of efficient points shown in the figure. Such an inefficient point is represented by option 6. This is discussed further in the presentation of MCA in Section 4.5.3.

9. Health and safety
10. Legal and administrative feasibility
11. Equity
12. Environmental quality
13. Private versus public sector
14. Unique or critical resources

The EPA report states that the first four criteria listed would generally be given the highest priority. Note that many of these criteria overlap, and economic efficiency is only one among many.

Criteria 1 to 6 can all be treated by modern CBA; indeed, questions of timing (urgency), flexibility (or robustness), and capital constraints (low cost) are all central elements of the approach. Criterion 13 is really part of 6. (The EPA report amplifies the criterion by questioning whether the strategy minimizes governmental interference with decisions best made by the private sector.) Furthermore, modern valuation techniques permit substantial parts of criteria 9 and 12 to be included in the economic analysis as well. As conceded by the EPA report, "If the principal costs and benefits can be quantified in monetary terms, economic theory provides a rigorous procedure for making trade-offs between present and future costs, and for considering uncertainty, profitability, and most of the other criteria" (1989, p. 393).

There is also a need to separate the basic goals of public policy – such as economic efficiency and equity – which surely have primacy, from implementation issues like legal and administrative feasibility, which are generally secondary.

A premise of modern decision techniques is that one looks first at the primary objectives and then asks how many may have to be sacrificed to achieve practical implementation. This principle has become accepted in many areas of policy making. For example, one key starting point for setting electric utility rates is to calculate the economically efficient tariff (based on marginal costs) and then make adjustments to protect low-income groups (through lifeline rates, special provisions for disconnection in the event of nonpayment, etc.). The essence of the approach is not that noneconomic issues are ignored, but that the trade-offs between economic efficiency and equity (or indeed other objectives not readily monetized) are explicitly quantified and displayed in such a manner that decision makers are made aware of how much of one objective is traded off in the interests of the other.

4.5.3 Making Trade-offs

One of the main advantages of MCA is that it forces political decision makers to look at the trade-offs between their major criteria or objectives, rather than attempting to boil everything down to a single number, particularly when valuation techniques may be controversial.[9] We illustrate the procedure with a basic example involving two criteria. Exactly the same principles may be applied to cases involving more than two criteria.

The first step in the application of MCA methods involves the selection and definition of attributes that describe the decision-making objectives. Suppose that the objectives seek to limit both mitigation efforts and climate change damage as much as possible. Ideally, if both objectives could be valued in monetary terms, then the preferred attribute for both objectives would be economic value. In this case, we have reduced the problem to basic ECBA, which would seek a level of emissions reduction that would minimize total costs consisting of the sum of climate change mitigation and damage costs, as explained earlier in Box 3.1. MCA comes into play if, for example, mitigation efforts can be valued monetarily, whereas damage cannot. In this case, let us select the attribute of economic cost to represent mitigation efforts, and the attribute of cumulative GHG emissions over the study period (say, tons of carbon equivalent emitted during the two centuries from 1991 to 2190) as the proxy for damage due to climate change.

The next step is the quantification of the magnitude of each attribute for each of the policy options under consideration. At this stage of the analysis, trade-off curves (described below) are powerful tools for communicating with decision makers; they are particularly relevant in situations, such as those related to the climate change problem, in which the valuation of avoided damage may be difficult and where decision makers must act largely on the basis of trading off shorter-term mitigation costs against certain levels of GHG emission reduction.

The third step leads to the selection of (a) the best policy alternative or (b) at least a set of superior alternatives (among the many more originally considered). In the first case, it would be necessary to amalgamate the information on the two attributes in some way. For example, we could assign a "weight" (or value) of U.S.$$X$ per ton of carbon equivalent that is emitted (as a proxy for damages) and thereby estimate damage costs. We could then combine the damage costs with the mitigation costs (which we already have), to obtain the total cost. The best policy alternative would be the one that minimized the total cost (see Box 3.1). However, different values for the weight X would generally yield different optimal policy choices. In the second case, the trade-off curve approach could be used to derive a set of superior policy alternatives and the remaining options could be discarded.

---

[9] For example, simply placing a value on human life, or valuing the lives of different individuals unequally, is strenuously opposed by some; thus, one might argue that if the life of a U.S. citizen is worth $1.5 million, then so are the lives of everyone else. Estimating the (preventive) cost of reducing the probability of death by a certain amount (say from 10 per 10,000 people down to 5 per 10,000) is one way of avoiding direct valuation of a life (see Chapter 5).

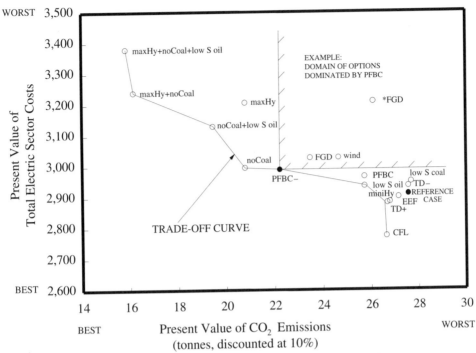

FIGURE 4.4. The trade-off curve. Calculating the present value of tonnes of CO_2 emissions over the life of each measure is equivalent to assuming that the damage due to a tonne of CO_2 remains constant over this period. It is not necessary to know the monetary value of the damage due to CO_2 emissions, since the analysis involves only a comparison of options. For abbreviations, see Table 4.1. From Meier et al. (1995).

Next, the foregoing procedure is illustrated by a practical example. Consider Figure 4.4, taken from a study of options for GHG emission reductions in Sri Lanka (Meier, Munasinghe, and Siyambalapitiya 1993), which shows the two attributes – full system costs of electricity supply and GHG emissions – for the technology policy options identified in Table 4.1.[10] Each point represents a perturbation of the reference case, defined as the official 1993 base-case capacity expansion plan of the Ceylon Electricity Board (CEB), the Sri Lanka power utility.

The trade-off curve is made up of all those options that are not dominated by others (sometimes referred to as the "noninferior set"). These are the options that are "closest" to the origin and therefore represent the "best" set of options, which merit further attention. Several useful concepts arise here.

---

[10] Note that the trade-off analysis will be much more complex as the number of attributes increases. Thus, when more than two attributes are involved, the analysis will involve more than two dimensions and the consideration of trade-off surfaces. However, even with multiple attributes, the simple approach of examining two attributes at a time often provides extremely useful insights.

TABLE 4.1   Technology Interventions for GHG Emission Reductions

Option	Comments	Symbol
Wind energy	Maximum capability used	wind
Mini hydroelectric schemes		miniHy
DSM 1	Involving energy-efficient refrigerators	EEF
DSM 2	Involving compact fluorescent lights	CFL
T&D loss reduction 1	Accelerated program	TD+
T&D loss reduction 2	Delayed program	TD−
Maximum hydroelectric power		maxHy
Clean coal technology 1	Pressurized, fluidized bed combustion – combined cycle units	PFBC
Clean coal technology 2	Revised capital cost assumptions	PFBC−
Clean fuels 1	Use low-sulfur residual oil for diesels	low S oil
Clean fuels 2	Use low-sulfur coal	low S coal
FGD systems 1	Coal plants must have FGD systems, and the model is free to select them, as appropriate	FGD
FGD systems 2	FGD systems forced onto base-case solution	*FGD
No coal	Diesel and hydroelectric plant only	noCoal

Abbreviations: DSM, demand-side management; FGD, flue gas desulfurization; T&D, transmission and distribution.

First, there is the concept of *dominance*.[11] PFBC (a clean coal technology) is said to dominate the options (FGD, wind, and *FGD) within the quadrant. PFBC yields both lower system costs and better (lower) GHG emissions than the other three options. Therefore, it is preferred over the latter options under both criteria. If only these two attributes mattered, there would be no reason to select any of the dominated options in place of PFBC.

Another key concept is illustrated by dividing the solution space of Figure 4.4 into quadrants with respect to the reference case (Figure 4.5). The options that fall into quadrant III are the *"win–win" options*, which are better than the reference case in *both* attributes. In this case, the minihydroelectric power, energy-efficient refrigerators, transmission and distribution system loss reduction, and compact fluorescent options all fall into this quadrant, providing both cost and emission gains.

Finally, MCA leads to implicit valuations whenever two options are compared. For example, in the case of Figure 4.5, a decision maker who prefers

---

[11] Decision analysis distinguishes among several types of dominance, such as strict dominance and significant dominance; see, e.g., Meier and Munasinghe (1995) for an application of these concepts to environmental decision making.

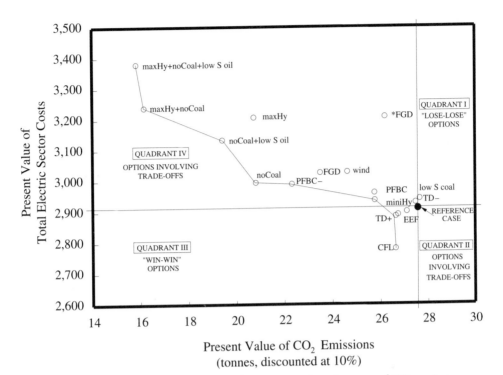

FIGURE 4.5. Win–win options. Calculating the present value of tonnes of CO_2 emissions over the life of each measure is equivalent to assuming that the damage due to a tonne of CO_2 remains constant over this period. It is not necessary to know the monetary value of the damage due to CO_2 emissions, since the analysis involves only a comparison of options. For abbreviations, see Table 4.1. From Meier et al. (1995).

option Y (i.e., "maxHy + noCoal") to option X (i.e., "noCoal + low S oil") makes an implicit valuation of the concomitant reduction of GHG emissions in terms of the increased costs (i.e., equal to the slope of the trade-off curve between X and Y – in this case, about U.S.\$200 per ton of CO_2).[12]

The choice of criteria in an MCA will depend on each country's short- and long-term development plans. Despite a common global objective of stabilizing atmospheric concentrations of GHGs, developing countries may use different criteria because of immediate or urgent needs – for example, to ensure food supplies and service debt requirements. Consequently, different countries may place different weights on the attributes.

To conclude, MCA techniques may be useful in decision making at the global level, especially in the early stages of developing an operational definition of Article 2 of the UNFCCC. For example, more is likely to be known in the next few years about the costs of mitigation than about the costs of climate

---

[12] However, because of the presence of joint products – each represented by a different dimension in the multidimensional trade-off space – such valuations that look only at two dimensions must be interpreted with some caution.

change damage. Therefore, a graphic analysis of mitigation costs versus an index of climate change damage (like cumulative GHG emissions, temperature increase, or sea level rise) for different emission profiles and stabilization cases might provide useful insights into the relative merits of such options – perhaps helping to identify a trade-off curve and eliminating spurious or dominated policy alternatives (see the trade-off curve in Figure 8.1 and the related discussion in Chapter 8).

4.6   DECISION ANALYSIS

While MCA helps one to work around certain shortcomings of conventional CBA (like valuation problems), it does not necessarily deal more effectively with uncertainty than CBA. This complication has led to the development of a further group of decision techniques that are known by the general term *decision analysis* (DA). Here the focus is on making decisions under conditions of uncertainty. These techniques have found prior application in a wide variety of situations, from decision making in the high-risk field of wildcat oil drilling to analysis of financial options. As we shall see, such techniques also provide a more systematic approach to dealing with irreversibility – one of the important potential characteristics of the climate change problem.

At the outset, it is important to understand the difference between risk and uncertainty. Risk occurs when several known outcomes exist in the future, each one having a given probability of occurrence. For example, if a fair coin is tossed into the air, there is a 50% probability that the result will be either a head or a tail. In contrast, uncertainty prevails when little or nothing is known about the probability of future outcomes. For example, the final score of a basketball game will be uncertain, although it may be possible to assign some risk probability to the question of which team might win. In borderline cases between risk and uncertainty, some notional ideas of the relative likelihood of various outcomes may be available. In the extreme case of total ignorance, even the outcomes may be unknown or ill-defined.[13] DA attempts to incorporate uncertainty explicitly into the modern CBA framework and looks for robust solutions that are invariant under the widest possible range of future circumstances. The outcome will depend crucially on whether the decision maker is risk-averse (i.e., hypercautious), risk-neutral, or risk-loving (i.e., as a gambler would be); see Box 4.3.

---

[13] What might happen inside a black hole is an example of this category, although some astronomers may disagree with this judgment.

BOX 4.3 Choice under Conditions of Uncertainty

A simple numerical example is presented here to show how decision analysis might be used under uncertain conditions. If probabilities could be assigned to different outcomes involving climate change, these probabilities might be applied as weights to the corresponding costs and benefits involved, and the best outcome could be selected by maximizing the expected value of net benefits (see also Section 4.1). However, even if such probabilities were not known (which is the case with almost all long-term climate change phenomena), it would be possible to make rough choices based on the systematic analysis of risks.

Consider Table B4.2, which is a matrix of values based on different assumptions about two critical parameters. The rows indicate whether the global climate sensitivity is low, medium, or high. This parameter tells us how sensitively the global climate will respond to changes in the atmospheric GHG concentration, with the high sensitivity denoting the worst state of nature (see Chapter 1). The columns of the matrix show the effects of three different response strategies, corresponding to weak, average, and strong mitigation interventions (e.g., the response strategies that might result in the stabilization cases S1000, S750, and S550, respectively, in Chapter 1). The cells in the matrix contain values (represented by the single numbers) that indicate the total costs imposed on the world economy corresponding to each outcome. These costs are basically the sum of the present values of climate change abatement and damage costs, shown by the respective pairs of numbers in parentheses – up to the time of stabilization of GHG concentrations in the distant future. For

TABLE B4.2 Cost Matrix

	Response Strategy		
Climate Sensitivity	**Weak (W/S1000)**	**Average (A/S750)**	**Strong (S/S550)**
Low (L)	(2 + 16) *18*	(13 + 8) *21*	(30 + 5) *35*
Medium (M)	(2 + 48) *50*	(13 + 20) *33*	(30 + 10) *40*
High (H)	(2 + 98) *100*	(13 + 47) *60*	(30 + 21) *51*
Expected value (assuming equal probability L, M, H)	56	**38**	42

(continued)

BOX 4.3   Choice under Conditions of Uncertainty *(continued)*

a given climate sensitivity, abatement costs would rise and damage costs would fall, as the response strategy became more stringent. Meanwhile, for a given response strategy, damage costs would increase with rising climate sensitivity. These results correspond to curves C and D in Figure B3.1 (Box 3.1). (Thus, we implicitly assume a single mitigation cost curve C and three damage curves [say, D_1, D_2, and D_3] corresponding to the three values of climate sensitivity.)

The costs have been normalized by assigning a value of 100 to the worst outcome, where climate sensitivity is high and the response strategy is weak. All the costs are purely notional and are intended simply to illustrate the decision technique.

As a benchmark, consider the case in which the risk probabilities of the three different states of nature are known. Let us assume the simplest case, in which each level of climate sensitivity is equally probable (i.e., .33). Then, as explained in the main text, the expected value associated with a given response strategy is estimated in the last row of Table B4.2 by summing the product of the costs under that strategy (i.e., the three values in the relevant column of Table B4.2) with their corresponding probabilities (i.e., .33). The lowest value of expected costs will occur with the average response strategy (38 in boldface).

However, in the absence of any knowledge regarding the probability of occurrence of a particular value of climate sensitivity, a rudimentary decision process might proceed as follows. In the first approach, let us choose the maximum cost associated with each response strategy (or column) – that is, the worst outcome for each response. These values are underlined (100, 60, and 51, for W, A, and S, respectively). A risk-averse decision maker might feel justified in trying to achieve the lowest value (i.e., 51 in boldface and underlined) of these worst-case outcomes.

This is called the MINI-MAX criterion (because one seeks to minimize the maximum cost). It results in the choice of the strong response strategy. The main objective of the decision maker in this case is to limit the losses that might occur if the worst state of nature prevailed – a rather pessimistic attitude.

Next, we examine a second approach that offers a striking contrast. Here, one begins by identifying the minimum cost associated with each response strategy (or column), as shown by the values in italics (18, 21, and 35, for W, A, and S, respectively). A risk lover might wish to depend on luck and aim for the lowest value (i.e., 18 in italics and boldface) among these best-case outcomes. This is called the MINI-MIN criterion (where one attempts to minimize the minimum cost) and favors the weak response strategy. The choice reflects an optimistic outlook, in which the decision maker gambles on the best state of nature

TABLE B4.3 Regret Matrix

	Response Strategy		
Climate Sensitivity	**Weak (W/S1000)**	**Average (A/S750)**	**Strong (S/S550)**
Low (L)	(18 − 18) *0*	(21 − 18) *3*	(35 − 18) *17*
Medium (M)	(50 − 33) *17*	(33 − 33) *0*	(40 − 33) *7*
High (H)	(100 − 51) *49*	(60 − 51) *9*	(51 − 51) *0*

prevailing – for example, a political leader who minimizes short-run response costs and hopes for the most desirable long-run outcome.

A third, somewhat more sophisticated approach is also possible, but requires the construction of a regret matrix, as shown in Table B4.3. For each climate sensitivity level or row in Table B4.2, one identifies the lowest total cost. For row L, this value is 18 – when the response strategy is weak. Next, we compare the costs of the other two response options in the same row with the low value of 18. This result is called the measure of regret, or how much one would have lost because of the wrong choice of response strategy. On this basis, regret values of zero (or 18 −18), 3 (or 21 −18), and 17 (or 35 −18) are assigned to the corresponding cells in the top row of Table B4.3. The other two rows of the regret matrix are computed in exactly the same way. Finally, we apply the MINI-MAX rule to the regret matrix. The maximum regret values associated with each response strategy (or column) are underlined (49, 9, and 17 for W, A, and S, respectively). The minimum among this set is 9 (shown in boldface and underlined), which yields the average response strategy. This reflects a more neutral view in which one seeks to limit how badly one would deviate from the ideal outcome.

The results of the analysis are summarized in Table B4.4, which shows that the three rules (each reflecting a different decision-making attitude) will yield three different choices. Nevertheless, while decision the-

TABLE B4.4 Results of Choices

Criterion	Attitude	Preferred Response Strategy	Implied Outcome
MINI-MAX costs	Pessimistic	Strong	S/H
MINI-MIN costs	Optimistic	Weak	W/L
MINI-MAX regret	Neutral	Average	A/M

(continued)

BOX 4.3 Choice under Conditions of Uncertainty *(continued)*

ory cannot eliminate uncertainty and provide an unambiguous choice, it can help decision makers to structure their thoughts systematically, especially by focusing on attitude toward risk and identifying what kinds of information might be helpful in improving future decisions.

4.6.1 Role of Option Values

As discussed earlier, in traditional CBA the usual decision criterion is to undertake a project if the expected benefits exceed the expected costs: $B > C$. Normally, the expected costs are the so-called opportunity costs represented by the returns to the resources used in the project, if they were applied in the best alternative investment. However, there may be another option consisting of committing only a part C_p of the resources available, to obtain a partial benefit B_p, and delaying the rest of the investment $(C - C_p)$ until some future date – in the expectation of realizing a further option benefit B_o.

Assuming that committed investments cannot be recovered (i.e., are irreversible), a more appropriate approach might be to make the original investment only if $B > (B_p + B_o)$. In other words, the original benefits from the full investment should exceed the benefits from the partial investment by an amount at least equal to the value of the future options forgone by undertaking the original (full) project. For example, suppose that an investment option to abate GHG emissions depends on assumptions that are subject to great uncertainty – such as future world oil prices. If one makes an investment decision that is largely irreversible, such as a decision to build a large hydroelectric power plant, then one commits a large amount of resources and loses the flexibility associated with waiting to learn more about the factors that affect oil prices (see also the discussion of the point of bifurcation in Box 7.6, Chapter 7). Preserving that flexibility has some economic value, namely the so-called option value. In financial and commodity markets such options to buy (and sell) are traded, with option prices determined by the market itself (Dixit and Pindyck 1994). Moreover, option value theory is now being applied to other fields involving capital-intensive investments, such as power generation (see, e.g., Crousillat and Martzoukos 1991).

In applying these concepts to the climate change problem, there are, of course, several key differences. First, in one sense the issues are more complicated than those faced, say, by the power sector. In climate change, one can both lose and gain flexibility if one makes short-term investments to reduce GHG emissions. The loss of flexibility (or options) arises because of opportunity costs: investment in GHG emission reductions now is not free, in the sense that resources have to be diverted from other uses and better emission technologies may be available in the future. Thus, to some extent, a commitment now to current technology restricts the option of using better technology later on. The gain in flexibility occurs because the lower emissions resulting from the investments will reduce future effects and damage over time (see Chapter 1) and provide more options for deferring emission reductions in the future. A special consideration arises if a current investment prevents an irreversible change in climate (perhaps due to nonlinear behavior). In this case, choices that avert the irreversible outcome have a corresponding option value, because they leave more possibilities open for the future. Ultimately, the desirability of any investment choice depends on the relative value of the options that are lost and gained. In addition to these static considerations, dynamic aspects may also be important. For example, current investments may spur further research in the private sector, or alternatively, resources not used for investments today might be used to accelerate research and development work.

Second, unlike the situation with private finance and commodities, there is no marketplace to set the value of the option, and therefore public policy analysis is required to establish an appropriate value. However, a marketplace might emerge if a tradable emission quota system were instituted, as discussed in Chapter 7. Chao and Wilson (1993) outline a means of using option values to quantify the flexibility associated with the purchase of a tradable emission permit instead of fixed capital investments in control technology.

4.6.2 Sequential Decision Making and Hedging

A climate change response strategy depends on both uncertainty and the risk of inaction. If there were little or no uncertainty, or if there were plenty of time to take action, the best approach would be to gather more complete information and then determine the optimal (deterministic) strategy. In the presence of uncertainty and the risk of significantly adverse outcomes, a more appropriate approach would be a sequential hedging strategy involving iterative steps alternating between relatively small actions and learning. For example, some actions must be taken before it is known what the full impact of climate change may be in the years to come. Then as information becomes available and the probable future state of the world becomes clearer, further action is taken on the basis of this new information.

The basic premise is that all uncertainty cannot be resolved before decisions are taken, since this would be prohibitively costly and might not lead to a clearer assessment of the situation. Present studies assume that the potential damage due to climate change may not be known until 2010 or 2020, and consequently, appropriate emission levels will also be undetermined until then.

In reality, damage estimates may be quite uncertain even by 2020. In this context, the "act-then-learn" hedging policy consists of making more limited decisions at specific times in the future, whenever fresh information has been collected and analyzed to support the next round of decision making. For instance, energy sector policies in the decades before, say, 2020 may have to be made under conditions of greater uncertainty. Therefore, decisions should be made iteratively, based on an optimal hedging policy for emission levels.

The value of information is especially important to consider in this context, especially in the case of natural system parameters – for example, the benefit derived from knowing the "true" value of a parameter such as the global climate sensitivity (see Chapter 1). The potential yields to more accurate information under uncertain conditions are a topic that has already been analyzed in the private sector, and this experience could be helpful in climate change analysis.[14] The value of research is equally important, especially with respect to variables that are influenced by human actions – for example, the benefit derived from new research discoveries such as more efficient and lower-cost energy technologies. In practice, information gathering and research may be closely intertwined. Nevertheless, in the allocation of resources between pure information gathering to make more accurate estimates of modeling parameters and research aimed at actually improving future technologies, the relative values of information gathering and research should be compared carefully. For example, Peck and Teisberg (1993) estimated that the value of information about future technology (for making predictions) would be about U.S.$10 billion, whereas the value of research that had a 10% probability of actually improving future technology would be around U.S. $100 billion.

Another key element of a sequential hedging strategy is to adopt the simpler and lower cost options in the near term and defer more costly and intensive emission reduction measures to the longer term – when more complete information is likely to be available. For instance, energy conservation measures represent a low (or negative) cost approach that can be adopted with immediate effect, whereas fuel switching represents a more costly strategy suitable for adoption over a longer time horizon. Fuel switching implies greater changes in capital equipment, which must be undertaken over a time frame in which it is optimal for long-lived capital to be switched. This argument for deferring more aggressive mitigation measures must be examined in the context of the option value trade-off concerning the use of current versus

---

[14] For example, in oil exploration, drilling a wildcat well is very expensive. The question arises as to how money should be spent in less expensive work before the drilling – e.g., on low-cost general magnetic surveys or on somewhat more expensive seismic surveys? Neither type of prior survey yields perfect information. For details of how such decision theory models are applied in this field, see, e.g., Dixit and Pindyck (1994).

future technology discussed earlier. Another relevant consideration is the likelihood that more stringent and earlier emission reduction efforts may well spur research and development of more cost-effective technologies.

Some recent examples of the application of hedging approaches provide useful insights. Among the early attempts at applying decision analysis to the climate change problem are those of Manne and Richels (1992), who developed an approach to determining the optimal hedging strategy. The paradigm they use is that of choosing a portfolio of insurance options (see also the subsection on insurance mechanisms in Section 3.2.4, Chapter 3). In the case of climate change, key questions involve the amount of resources to be devoted to a variety of such options, including data gathering to resolve scientific uncertainty, measures for immediate abatement of emissions, and research and development of new supply and conservation technologies to reduce future abatement costs. As mentioned earlier, with perfect information the best course of action can be charted immediately and there is no need to hedge bets. Manne and Richels conclude that there is an inverse relationship between the pressure for precautionary near-term GHG emission reductions and making a sustained commitment to research and development to acquire better climate information (which reduces the need to hedge against an uncertain and potentially hostile future). However, given the inherent predictive uncertainty of climate change (and in particular the reliability of indicators), the limitations of even hedging-type approaches must be recognized.

An obvious practical problem concerns the need for consistency in any long-term climate change response strategy. The flexibility implied by a sequential hedging approach could be used to justify an ad hoc or inconsistent approach. Clearly, there is a mismatch between the long time horizon of the climate change problem and the short time horizon of decision makers who are elected every four to six years in most countries. In particular, decision makers might have a tendency to avoid undertaking major sacrifices during their own limited terms of office.

4.7 INTEGRATED ASSESSMENT

As discussed in Chapter 2, climate change researchers face formidable problems in seeking to understand the interactions of planetary-scale physical, ecological, and socioeconomic systems. The analysis is complicated by additional difficulties such as irreversibility, data scarcity and uncertainty, nonlinear behavior (that could lead to catastrophic outcomes), and multiple objectives (including efficiency and equity). In this context, integrated assessment is not so much a decision tool as a conceptual framework for addressing com-

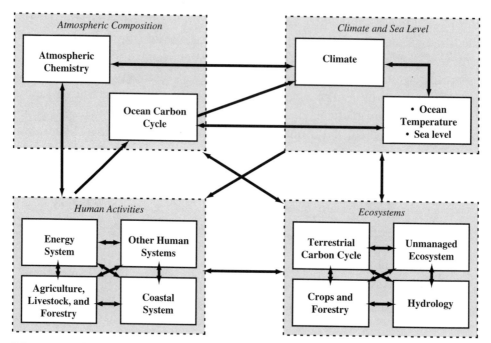

FIGURE 4.6. Schematic representation of the key components of a typical IAM. From Weyant et al. (1996).

plex problems by combining a broad and transdisciplinary range of knowledge in a systematic manner. Many early attempts at integrated assessment sought to make policy recommendations more useful, by using ad hoc (and often qualitative) methods of integrating knowledge and introducing interactions across disciplines that were absent in single discipline analyses.

4.7.1   Integrated Assessment Models

In the context of climate change, formal computerized, integrated assessment models (IAMs) have been developed (see Weyant et al. 1996). Such IAMs link a variety of submodels based on information derived from several disciplines, including meteorology, geophysics and geochemistry, biology and ecology, economics and sociology (see Figure 4.6 for a schematic representation of a typical IAM). The mathematical formalism helps to ensure consistency among the alternative assumptions and inputs from the broad array of sources involved in climate change analysis. However, such rigor also tends to limit the types of data that can be incorporated into IAMs. Nevertheless, these models are being used widely to examine three key questions:

- What is the feasible range of variation for the long-term evolution of human and natural systems?

- What are the implications of different response strategies and policy options?
- How might research and information-gathering efforts be prioritized?

In general, there are two broad categories of IAMs: *policy evaluation models* and *policy optimization models*.

Policy evaluation models basically simulate the socioeconomic and biogeophysical consequences of selected policies. The results of the numerical simulations can then be compared and judged according to different criteria in order to evaluate the relative desirability of those policy alternatives. There are two subcategories of policy evaluation models, depending on whether the future projections are deterministic (i.e., all inputs and outputs have only a single path over time) or stochastic (i.e., some or all of the inputs and outputs could vary probabilistically over time).

By contrast, policy optimization models determine the desirability of policy options as part of the computation itself, by using a built-in mechanism (e.g., the objective function for economic efficiency described in Section 3.2.2, Chapter 3). Thus, the results of such a model yield an optimal future trajectory and a corresponding set of optimal policies that are determined by the assumptions and weights already built into the objective function. There are three subcategories of such models. First, cost–benefit models use the logic described earlier in this chapter to compare economic costs and benefits (e.g., to determine the optimal path of GHG emissions that maximizes global net benefits). Second, target-based models determine optimal policies to achieve a preset goal (such as a predetermined future atmospheric GHG concentration). Third, uncertainty-based models rely rather heavily on the techniques of decision analysis described in the preceding section to deal with the problems of uncertainty.

The results produced by IAMs thus far have given rise to some tentative conclusions:

1. The model results are no more reliable than the underlying scientific and socioeconomic data used as inputs. However, they provide useful insights into the complex linkages involved in climate change analysis.
2. Regionally disaggregate results are beginning to emerge, which will be useful in clarifying issues that remain unresolved in climate change negotiations.
3. The identification of gaps in both scientific and socioeconomic knowledge is improving.
4. Certain parameters (such as the types of objective function, discount rate, and timing of mitigation efforts) appear to affect outcomes and costs more critically than others, especially in the short to medium term.
5. Policies that enhance the flexibility of responses appear to be desirable. Results also tend to favor research and development efforts that facilitate

adaptation to climate change and improve the efficiency with which natural resources are used.

6. Climate may not necessarily be the most important determinant of future changes that affect human beings; agricultural practices and urbanization could have at least as great an impact.

4.7.2 Developing-Country Issues

In the recent past, a number of weaknesses of IAMs have been identified, especially in their treatment of developing countries.[15] Efforts are being made to address some or all of these perceived shortcomings, including the following:

1. All IAMs thus far have been built by developed-country researchers, whereas much of the policy response will have to occur in the developing world (which may not be well represented in such models).
2. Current IAMs have heavy data requirements that may be more readily available in industrialized countries than in the developing world.
3. Model projections tend to assume that developing countries will follow the same technology and consumption patterns as the industrialized world.
4. The market-based and efficiency-oriented neoclassical economics underlying present IAMs might not capture the socioeconomic structures, or either autonomous or policy-induced behavioral responses, in developing countries.
5. Value systems may be quite different in developing countries, and valuation methods based on willingness to pay that have been used in wealthier countries could give misleading or inequitable results when applied to poorer groups in the South.
6. South–North equity concerns must be given greater weight, since IAMs have tended to emphasize the efficiency objective.
7. Policy modeling should pay more attention to incentives or disincentives offered to developing countries to participate in a collective global climate change strategy.
8. Most of the uncertainty in present IAMs is of a scientific nature, whereas developing countries may be more vulnerable to issues arising from socioeconomic uncertainty (e.g., related to poverty and income inequality).
9. Given that developing-country priorities will be focused on growth and the eradication of poverty, more attention must be paid to integrating climate change responses with conventional development policies within coun-

---

[15] See, e.g., the papers presented at the IPCC Asia–Pacific Workshop on Integrated Assessment Models, Tokyo, March 1997.

tries. The effectiveness of policy instruments and measures used in the industrialized world may also be different in developing countries.

Despite these drawbacks, the continuing efforts devoted to integrated assessment modeling worldwide indicate that this is a potentially promising approach to analyzing climate change issues and policies.

4.8 SUMMARY

This chapter began with a review of key decision-making criteria, including utility-based approaches, which favor economic efficiency, and the Rawlsian view, which leans more toward greater equity. The requirements of traditional cost–benefit analysis, including limited scale, geographic extent, and time span, as well as discreteness, make it difficult to apply in the case of climate change decisions. The centuries-long time lags, irreversibilities, nonlinearities (that threaten potentially catastrophic outcomes), and great uncertainty associated with climate change further complicate the analysis. As a response, modern CBA has evolved into a more generic approach that includes a family of decision techniques designed to overcome many of the shortcomings just mentioned.

Modern CBA-based decision techniques relevant to climate change analysis include (a) economic cost–benefit analysis, which compares all costs and benefits expressed in terms of a common numeraire, usually monetary units; (b) cost-effectiveness analysis, which is designed to identify the lowest-cost method of achieving an objective already justified using other criteria; (c) multicriteria analysis, which attempts to find desirable options when some or all of the costs and benefits are not measurable in monetary terms; and (d) decision analysis, which is useful when choices have to be made under conditions of uncertainty.

Such techniques are helpful in resolving climate change decisions involving efficiency, especially to determine by how much, when, and by what means the emission of GHGs should be reduced. Equity-related questions, including who should reduce emissions and who should be compensated for climate change damage, are not so amenable. Even in such cases, techniques like MCA are useful in elucidating the trade-offs between equity and efficiency.

The complications posed by uncertainty merit special attention. In general, because of uncertainty, strictly quantitative application of decision techniques to climate change analysis, and undue reliance on numerical projections into the distant future (e.g., 100 years), are not practical. However, the methods discussed in this chapter could help to structure thinking and provide a more systematic basis for developing qualitative insights, especially concerning the rough magnitudes and directions of change. Quantitative analysis may be more successfully applied to the evaluation of shorter-term alternatives like energy efficiency and fuel switching in the existing portfolio of options. Further (unresolvable) uncertainties must be handled through strategies that include sequential approaches and hedging, insurance (see Chapter 3), portfolios of options, robust solutions, and better information gathering and research.

Finally, integrated assessment is a systematic methodology for testing various policy and strategy options within a consistent framework that pulls together a wide range of data and assumptions from different disciplines and sources. In particular, computerized integrated assessment models provide key insights into the likely future evolution of interlinked human and natural systems, the effectiveness and desirability of policy tools, and the most important gaps in knowledge and data.

REFERENCES

Arrow, K. J. 1951. Alternative approaches to the theory of choice in risk-taking situations. *Econometrica*, **19**, 404–37. Reprinted in K. J. Arrow. 1984. *Collected papers of Kenneth J. Arrow*. Volume 3, *Individual choice under certainty and uncertainty*. Cambridge, MA: Belknap Press of Harvard University Press, pp. 172–208.

Arrow, K. J., J. Parikh, and G. Pillet. 1996. Decision-making frameworks for addressing climate change. In J. P. Bruce, H. Lee, and E. F. Haites (eds.), *Climate change 1995: Economic and social dimensions of climate change*. Contribution of Working Group III to the Second Assessment Report of the Intergovernmental Panel on Climate Change. Cambridge University Press, pp. 53–78.

Brown, P. G. 1992. Climate change and the planetary trust. *Energy Policy*, **20**, 208–22.

Chao, H. P., and R. Wilson. 1993. Option value of emission allowances. *Journal of Regulatory Economics*, **5**, 233–49.

Crousillat, E., and S. Martzoukos. 1991. *Decision-making under uncertainty: An option valuation approach to power planning*. Industry and Energy Department, Energy Series Paper 39. Washington, DC: World Bank.

Dixit, A. V., and R. S. Pindyck. 1994. *Investment under uncertainty*. Princeton, NJ: Princeton University Press.

Fisher, A. C., and M. Hanemann. 1987. Quasi-option value: Some misconceptions dispelled. *Journal of Environmental Economics and Management*, **14**, 183–90.

Freeman, A. M. 1993. *The measurement of environmental and resource values*. Washington, DC: Resources for the Future.

Maler, K. G. 1974. *Environmental economics: A theoretical enquiry*. Baltimore: Johns Hopkins University Press.

Manne, A. S., and R. Richels. 1992. *Buying greenhouse insurance: The economic cost of carbon dioxide emission limits*. Cambridge, MA: MIT Press.

Meier, P. M., and M. Munasinghe. 1995. *Incorporating environmental considerations into power sector decisionmaking*. Washington, DC: World Bank.

Meier, P., M. Munasinghe, and T. Siyambalapitiya. 1995. *Energy sector policy and the environment: A case study of Sri Lanka.* Washington, DC: World Bank, Environment Department. Reprinted in M. Munasinghe (ed.). 1996. *Environmental impacts of macroeconomic and sectoral policies.* Washington, DC: World Bank and UN Environment Program.

Munasinghe, M. 1990. *Electric power economics.* London: Butterworth–Heinemann.

Munasinghe, M. 1993. *Environmental economics and sustainability.* Washington, DC: World Bank.

Munasinghe, M., P. Meier, M. Hoel, S. W. Hong, and A. Aaheim. 1996. Applicability of techniques of cost–benefit analysis. In J. P. Bruce, H. Lee, and E. F. Haites (eds.), *Climate change 1995: Economic and social dimensions of climate change.* Contribution of Working Group III to the Second Assessment Report of the Intergovernmental Panel on Climate Change. Cambridge University Press, pp. 145–78.

Pearce, D. W., and R. K. Turner. 1990. *Economics of natural resources and the environment.* New York: Harvester Wheatsheaf.

Peck, S. C., and T. J. Teisberg. 1993. Global warming uncertainties and the value of information: An analysis using CETA. *Resource and Energy Economics,* **15**(1), 71–97.

Rawls, J. 1971. *A theory of justice.* Cambridge, MA: Harvard University Press.

Richels, R., and J. Edmonds. 1995. The economics of stabilizing atmospheric CO_2 concentrations. *Energy Policy,* **23**(3–4), 373–9.

Savage, H. A. 1954. *The foundations of statistics.* New York: Wiley.

U.S. EPA (U.S. Environmental Protection Agency). 1989. *The potential effects of global climate change on the United States.* Report EPA-230-05-89-050, December. Washington, DC.

Weyant, J., et al. 1996. Integrated assessment of climate change: An overview and comparison of approaches and results. In J. P. Bruce, H. Lee, and E. F. Haites (eds.), *Climate change 1995: Economic and social dimensions of climate change.* Contribution of Working Group III to the Second Assessment Report of the Intergovernmental Panel on Climate Change. Cambridge University Press, pp. 367–96.

Willig, R. 1976. Consumer surplus without apology. *American Economic Review,* **66**(4), 589–97.

5

GREENHOUSE DAMAGE AND ADAPTATION

5.1 INTRODUCTION

5.1.1 The Role of Efficiency

Chapter 3 (Section 3.2) described the main elements of a framework for making decisions regarding the abatement of greenhouse gas (GHG) emissions: global optimization, a collective decision-making process providing "equitable" solutions, and effective answers to procedural questions. With regard to global optimization, it was argued that an economically "efficient" solution is based on the notion that the aggregate *net* benefits associated with different future climate change scenarios are to be maximized. In this and the following two chapters we will elaborate on this efficiency concept, as well as on the role of no-regrets options. This is not to say that we will ignore the equity and procedural issues associated with GHG abatement decision making, or that these elements are not as important as efficiency. Indeed, if one tries to get a sense of the damage avoided through abatement action, one cannot escape problems of valuation that are deeply linked with equity, as we shall see throughout this chapter. Also, if one tries to assess the costs of abatement strategies, one will continuously face complications that need to be weighted, and such weighting relates to equity aspects as well. However, notwithstanding the fact that, fundamentally, equity and efficiency elements cannot be fully disentangled as far as the climate change issue is concerned, the main outlines of Chapters 5, 6, and 7 are based on the economic efficiency concept.

An economically efficient policy for GHG emission reduction (assuming that the cost functions are known) is one that maximizes the net benefits of reduced climate change, that is:

- the benefits in terms of costs of greenhouse damage avoided through GHG abatement measures, as well as secondary benefits, such as reduced air pollution, noise, congestion, and accidents and the preservation of species and biodiversity (which comprise curve *D* in Figure B3.1a, Box 3.1), minus

• the costs associated with GHG abatement efforts (curve *C*).

To put it differently, it is economically efficient to expand the scale of GHG mitigation actions, as long as the costs of greenhouse damage averted by a particular action plus the secondary benefits of that action are expected to surpass the costs of the action itself. Examples of the costs of averting greenhouse damage will be discussed later. Such actions result in a reduction in loss of agricultural productivity; forest loss; sea level rise; species loss; suffering in terms of human health, morbidity, amenity, and life; local pollution; adjustment costs involved with infrastructure, migration, and water supply; and suffering due to hurricanes, flooding, and desertification. Secondary benefits include local or regional air quality improvements, fewer traffic accidents or less congestion, preservation of biodiversity and wildlife, and the creation of employment and safety.

   Since it seems fair to assume a priori that, on the one hand, the increase in damage avoided plus secondary benefits will decline with each additional unit invested in abatement action (in the language of the economist, there will be declining *marginal benefits* because the greater the GHG abatement, the more difficult it will be to reduce damage further) and, on the other hand, that the actual emission reduction per unit of cost incurred in GHG abatement will tend to decline as more abatement efforts are made (in economists' jargon, increasing *marginal costs* of GHG action), the efficiency criterion tells us that an optimal level of policy action can, at least theoretically, be determined. At this point the net benefits of abatement action are maximized, which is at the intersection of the marginal cost curve and the curve representing the marginal benefits of abatement action (whereby the benefits consist of the damages avoided plus secondary benefits). This maximum (or optimum) is obviously at such a level of GHG abatement that the damage avoided plus secondary benefits declining with expanding GHG abatement efforts equal the costs incurred (increasing with expanding abatement efforts). See Figure B3.1b, Box 3.1.

   At this point it may be worth emphasizing what the efficiency criterion does *not* mean. It does not mean that the overall, or total, global costs and benefits of abatement strategies should be confronted in order to determine the optimal strategy. To give a fictitious example, if there were evidence that the overall benefits of greenhouse strategies aiming at a stabilization of GHG emissions would amount to 3% of world GDP, and that the costs involved would also be estimated at 3% of world GDP, efficiency would definitely not imply that no action should be taken. This is because there might very well be some level of action lower than that addressed by these cost estimates where benefits would exceed costs: this is where marginal benefits would exceed marginal costs. Although such a notion is common among economists, it is not always clearly understood by others.

It cannot be emphasized often enough that the above conceptual framework is highly restrictive – for example, because it presupposes a decision-making mechanism that is very much driven by economic efficiency, because it does not deal with the delicate political or equity aspects of international decision making that play such an overriding role in real-life international negotiations on greenhouse policy, and because it assumes that information about the marginal abatement cost and benefit curves is available and/or that both curves have a rather well-defined slope. This having been said, it remains of interest to see what exactly is currently known about the GHG abatement benefits, on the one hand, and GHG abatement costs, on the other. This is particularly the case because of recent increases in empirical information about such costs and benefits and about the efficiency and other implications of the instruments used to implement greenhouse policies. Therefore, Chapters 5 to 7 focus on the damage and abatement cost functions.

5.1.2 Chapter Outline

According to the augmented 1992 IPCC scenarios in conjunction with the "Bern model," which links emissions and concentrations (see Chapter 1, Figure 1.9), it is likely that CO_2 equivalent concentrations will double, approximately sometime in the middle of the next century, probably after 2050, particularly if no serious measures are taken to reduce CO_2 emissions and increase CO_2 absorption. The damage that this benchmark climate change and global warming will cause to the world community, and the extent to which different regions, sectors, and individuals will suffer, is the central theme of this chapter. With respect to this it will be assumed that up until the benchmark year the damage due to global warming will most likely be less than its level in the benchmark year, but after having reached the doubling of CO_2 concentration levels, the damage will be greater than in the benchmark year.

It is anticipated that mitigation measures may only postpone the moment at which the benchmark CO_2 concentration will be reached. (Mitigation response options are discussed in Chapter 6.) Therefore, it is vital to investigate how and to what extent adaptation measures – that is, policies that do not combat the *causes* of climate change but instead try to tackle its *consequences* – can be applied in order to reduce or at least postpone greenhouse damage. This chapter first gives a general idea of what the order of magnitude of greenhouse damage will be if no action is taken (Section 5.2). Next, it focuses on the methodological complications involved in assessing the benefits of adaptation (Section 5.3). Subsequently, the costs and benefits of adaptation policies are discussed in general (Section 5.4) and then with regard to the issue of coastal protection as a special case (Section 5.5). Finally, some information is provided about the secondary benefits that might emerge from abatement strategies (Section 5.6).

5.2 GREENHOUSE DAMAGE: NO ACTION

5.2.1 How to Assess Greenhouse Damage

To assess the degree to which the world might suffer from damage due to climate change, one has to realize that such damage is most likely to increase over time, since CO_2 equivalent atmospheric concentrations tend to increase. The damage is associated mainly with a commonly used benchmark of twice the preindustrial atmospheric CO_2 equivalent concentrations (i.e., the equilibrium temperature change associated with $2 \times CO_2$, or 560 ppmv). Under a "business-as-usual" scenario, say IS92a, this level would be reached sometime in the future, probably after 2050, with damage continuously increasing thereafter. However, leading up to this, annual damage might well build up and manifest itself in rather unexpected and unpredictable forms and patterns (see Chapter 1). This is important, because damage is related not only to climate change itself, but also and particularly to the magnitude of climate change. Under a stabilization scenario, say at 550 ppmv, this damage would occur later.

In order to assess the scope and order of magnitude of taking no actions to avert greenhouse damage, either worldwide or regionally, theoretically one could try to determine the *cumulative damage* over a certain period of time. Figure 5.1 illustrates this. Here it is assumed that, in the business-as-usual scenario, annual damage increases over time, because the main causal factors – increasing emission levels and consequently increasing atmospheric concentrations – are projected to increase. Furthermore, damage in the figure is assumed to be more or less proportional to concentrations. (Indeed, this assumption can be questioned, but it is used here for convenience's sake; it implies that the shape of the curve in Figure 5.1 would correspond roughly to the curve in Figure 1.2a.) At a certain point in the future, a doubling of the atmospheric concentration will be reached. If this point is used as a benchmark, the cumulative damage suffered until then can be determined simply by adding up the damage over the preceding years (the striped area in Figure 5.1). Obviously, this area does not represent the cumulative damage associated with GHG emissions until then, because past emissions will continue to

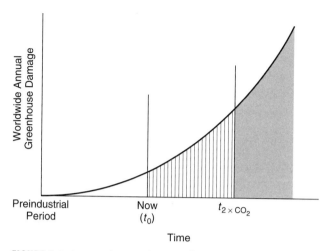

FIGURE 5.1. A greenhouse damage function.

contribute to climate change damage well into the future. (For emissions before the $2 \times CO_2$ benchmark year, this damage is represented in the figure by the shaded area to the right of the benchmark year.)

However, the problem with determining cumulative damage and thus adding annual damage figures is that, precisely because a large portion of the damage will be suffered in the distant future, one has to find a way of making the damage of different years comparable to the present situation. One procedure that might facilitate such an intertemporal comparison is *discounting*, but this leaves us with the delicate question of which discount factor to use (see Chapter 2). For example, if worldwide greenhouse damage were estimated at U.S.\$150 billion for 2020 and U.S.\$250 billion for 2050, how could these amounts be combined or compared? Furthermore, how could cumulative damage be incorporated into current decisions if so little is known about climate change and its potential damage and the way to deal with these issues? (Consider the potential technological breakthroughs of the future that one can now only dream of.)

Therefore, in order to say something meaningful about greenhouse damage sensitivities, one often makes the bold assumption of both the benchmark doubling of atmospheric concentrations of CO_2 and the damage associated with it being imposed on the world economy *today*. Obviously, this assumption underestimates the flexibility of the system to use the available time to develop suitable technologies and to undertake any necessary actions. On the other hand, it may provide a good starting point for getting an idea of the long-term implications of a no-action scenario. It is also very useful because one ought not to rule out the possibility that CO_2 atmospheric concentrations will rise beyond the $2 \times CO_2$ level before the equilibrium is reached. More-

TABLE 5.1 Monetized $2 \times CO_2$ Damage in Different World Regions
(Annual Damage)

Region	Fankhauser (1995)		Tol (1995)	
	U.S.\$ (Billions)	% GDP[a]	U.S.\$ (Billions)	% GDP[a]
European Union	63.6	1.4		
United States	61.0	1.3		
Other OECD	55.9	1.4		
OECD America			74.2	1.5
OECD Europe			56.5	1.3
OECD Pacific			59.0	2.8
Total OECD	180.5	1.3	189.5	1.6
E. Europe/former USSR	18.2[b]	0.7[b]	−7.9	−0.3
Centrally planned Asia	16.7[c]	4.7[c]	18.0	5.2
South and Southeast Asia			53.5	8.6
Africa			30.3	8.7
Latin America			31.0	4.3
Middle East			1.3	4.1
Total non-OECD	89.1	1.6	126.2	2.7
World[d]	269.6	1.4	315.7	1.9

[a] Note that the GDP base may differ among the studies.
[b] Former Soviet Union only.
[c] China only.
[d] Percentage of GDP figures are based on market exchange rate GDP. The order of magnitude of estimates does not change if uncorrected damage categories are purchasing-power-parity-adjusted and expressed as a fraction of PPP-corrected GDP.
Source: Pearce et al. (1996).

over, as indicated in Chapter 1, the damage will continue to increase long after any stabilization of equilibrium atmospheric concentration has taken place.

5.2.2 Some Aggregate Damage Estimates

Some studies have tried to provide rough estimates of global warming damage, both worldwide and in major countries and/or regions, on the basis of the above assumption. Table 5.1 provides a summary of various research outcomes. Note that the sources of many of the estimates are the same, and that to a certain extent some of the outcomes have been derived by extrapolating the more detailed U.S. estimates to other regions. With these limitations in mind, the broad conclusions that can be inferred from the table are as follows:

- the magnitude of $2 \times CO_2$ damage worldwide is projected to be of the order of 1.4 to 1.9% of world GDP; the same range would apply to the OECD region;
- in developing countries, the expected $2 \times CO_2$ damage is most likely to be larger relative to their GDP than in the OECD, and may range in the major developing regions from some 4 to 5% of GDP for Latin America and China to significantly more for South and Southeast Asia and Africa (more than 8%); and
- overall net damage in Central and Eastern Europe will probably be rather minor.

Although the data on which the estimates for the non-OECD area are based are much weaker than the data for the OECD area, they provide a clear indication that climate change will have its worst impact in the developing world.

It should be emphasized that the ranges in Table 5.1 do not represent the range of estimated values, but only the available "best guess" estimates of the impact of climate change on various countries. So the uncertainty of the best guesses per country is much greater than the uncertainty presented in the table. Indeed, many of the data allow for a considerable margin of error. Moreover, it is impossible to objectively put a price tag on many of the damage aspects. (For a more elaborate discussion of this subject, see Chapter 3, as well as Box 5.1 and Section 5.3, in this chapter.) Also, the estimates refer to midrange global warming expectations, whereas the ultimate outcome could be substantially different in the case of upper- or lower-bound warming.[1] Nevertheless, the general picture appears to be that if an average global warming of 1 to 3.5°C based on the present-day economy (as a percentage of GDP) is projected to future world product, the damage may actually range from zero in the best cases to almost 10% of GDP in the worst cases.

BOX 5.1 Valuing a Statistical Human Life

Now that we have listed some rough estimates of nonmarket damage, we can describe in more general terms whether such damage can be valued materially and, if so, how. As mentioned in the main text, opinions differ strongly with respect to valuing damage in terms of irreversible loss of ecosystems or biodiversity. The problem will only increase when one is trying to value a statistical human life, or to put it differently, trying to determine what value to attach to reducing the statistical risk of losing one human life. Not surprisingly, opinions on this subject differ even more.

---

[1] Note that damage is expected to increase more than linearly with increasing warming.

Economists generally tend to accept that a material value can be attributed to a statistical human life; they support their argument by indicating that, when deciding upon expenditures for lifesaving health care, governments in fact implicitly attach values to statistical human lives through their policies. Individuals may similarly attach a value to human life when they decide to take specific action in order to statistically reduce the risk of a deadly casualty. At the same time, there are those who reject the idea of attributing a material value to a human life on moral grounds, even if it is just a statistical concept.

Among those who believe valuing a statistical life is acceptable are those who nevertheless take a prescriptive view when it comes to aggregating risks over different groups of human beings. They emphasize that all human lives should be treated equally. Others, however, adopt a descriptive point of view by arguing that the actual behavior of individuals and governments implicitly expresses the value they attach to a statistical human life and can therefore be aggregated. They are, as a consequence, prepared to accept behavior as a basis for valuing, say, losses due to climate change in terms of the number of potential casualties of climate change events. (Alternatively, if adaptation policies can reduce the risk of casualties, the same approach is used to value these benefits.)

With regard to estimating the value of a statistical life based on willingness to pay, it cannot be emphasized enough that a value is not placed on human life itself; rather, people assign a value to a change in the risk of death among a population. These valuation approaches obviously tend to discriminate against the poor, because willingness to pay depends on a person's income and wealth or the local government's disposal of public resources: poor people simply cannot afford to pay a large amount in order to reduce risks, nor can poor governments. Similar discrimination against the poor, as well as against less-educated, disabled, and elderly people, would apply if one were to accept the so-called human capital approach, according to which the value of a statistical life is based on (lifetime) earnings or the capacity to generate personal output over personal consumption. (For an example of this approach, see Cline 1992a, who valued a U.S. statistical life at $595,000.)

Notwithstanding the fact that risks to lives are marketed implicitly or explicitly by governments, insurance companies, and so on continuously, the issue of how to aggregate the risks of various groups of human beings remains an extremely sensitive topic in international negotiations. This was altogether clear in the proceedings leading up to the acceptance of the summary for policy makers on the basis of the IPCC Second Assessment Report. The controversy over valuing statistical lives inspired the country representatives, convening in order to establish the IPCC Working Group III Summary for Policymakers, to conclude after intense de-

(continued)

BOX 5.1   Valuing a Statistical Human Life *(continued)*

bate, rightly or wrongly, that "there is no consensus about how to value statistical lives or how to aggregate statistical lives across countries" (Bruce, Lee, and Haites 1996b, p. 13). Furthermore, the summary suggests, "Human life is an element outside the market, and societies may want to preserve it in an equal way." This would, of course, "increase the share of the developing countries in the total damage estimate." However, "other aggregation methods can be used to adjust for differences in the wealth or incomes of countries in calculations of monetary damages" (p. 15).

However, researchers differ significantly with respect to their appreciation of the differences in the literature regarding the contributions of the various categories of damage to the overall damage figure. The latter result again illustrates the rather tentative nature of damage estimates. As already indicated, most of the more detailed greenhouse damage studies refer to the United States; these studies are summarized in Table 5.2. As discussed in Chapter 4, one can value greenhouse damage categories only by making rather strong valuation assumptions. One may therefore prefer to present damage in terms of physical quantities – the forest area lost in square kilometers, the number of protected habitats lost, the number of hurrican casualties, and so on. Such an overview is presented in Table 5.3. Again, the table assumes a benchmark average global warming of 2.5°C, which is associated with a doubling of atmospheric CO_2 concentrations.

5.2.3   Complications of Drawing an Overall Picture of Greenhouse Damage

Greenhouse damage will probably manifest in many different ways, as outlined in Chapter 1. This is one reason for the complexity of drawing an overall picture of the potential damage on a global or even regional scale. Some remarks with regard to the effects on agriculture may illustrate this point. Keeping in mind the anticipated altered frequency of extreme weather conditions, average climate change patterns, and the fertilization effect of greater atmospheric concentrations of CO_2, various researchers have tried to translate the anticipated effects of climate change on agricultural production into welfare economic scenarios at a national or regional level. A first major complication thereby involves the translation of physical data from climate change models into an economic analytical framework. After all, a different pattern of agricultural production in the various regions of the world will lead to new market

TABLE 5.2 Monetized $2 \times CO_2$ Damage to Present U.S. Economy
(Base Year 1990; Billions of Dollars of Annual Damage)

Damage Category	Cline (2.5°C)	Fankhauser (2.5°C)	Nordhaus (3°C)[a]	Titus (4°C)	Tol (2.5°C)[b]
Agriculture	17.5	8.4	1.1	1.2	10.0
Forest loss	3.3	0.7	Small	43.6	—
Species loss	4.0 + a[c]	8.4	—[c]	—	5.0
Sea level rise	7.0	9.0	12.2	5.7	8.5
Electricity	11.2	7.9	1.1	5.6	—
Nonelectrical heating	−1.3	—		—	—
Human amenity	+b[c]	—		—	12.0
Human morbidity	+c[c]	—		—	—
Human life	5.8	11.4		9.4	37.4
Migration	0.5	0.6		—	1.0
Hurricanes	0.8	0.2		—	0.3
Construction	±d[c]	—		—	—
Leisure activities	1.7	—	_d_	—	—
Water supply					
Availability	7.0	15.6		11.4	—
Pollution	—	—		32.6	—
Urban infrastructure	0.1	—		—	—
Air pollution					
Tropical O_3	3.5	7.3		27.2	—
Other	+e[c]	—		—	—
Mobile air- conditioning	—	—		2.5	—
Total	61.1	69.5	55.5	139.2	74.2
	+a + b + c ± d + e[c]				
(% of GDP)	(1.1)	(1.3)	(1.0)	(2.5)	(1.5)[b]

Note: Figures represent *best guesses* of the respective authors: Cline (1992a), Fankhauser (1995),
Nordhaus (1991b), Titus (1992), Tol (1995). Although none of the studies reports explicit
confidence intervals, figures should be seen as reflecting orders of magnitude only.
[a] Transformed to 1990 base.
[b] United States and Canada, base year 1988.
[c] Costs that have been identified but not estimated.
[d] Not assessed categories, estimated at 0.75% of GDP.
Source: Pearce et al. (1996).

equilibria, different prices of agricultural products, new patterns of agricultural trade, and so on. Therefore, a link between the physical and economic models ought to be established. The next complication arises with respect to deriving the distributional implications – who will gain and who will suffer – from the economic data. Rather detailed modeling is required to evaluate the impact on the interests of different groups. A final complication is the fact that highly divergent assumptions can be made with respect to possible adap-

TABLE 5.3 $2 \times CO_2$ Damage in Physical Units: Different World Regions (2.5°C Warming)

Type of Damage	Damage Indicator	EU	USA	Ex-USSR	China	Non-OECD	OECD	World
Agriculture	Welfare loss (% GNP)	0.21	0.16	0.24	2.10	0.28	0.17	0.23
Forestry	Forest area lost (km²)	52	282	908	121	334	901	1,235
Fishery	Reduced catch (1,000 t)	558	452	814	464	4,326	2,503	6,829
Energy	Rise in electricity demand (TWh)	54.2	92.0	54.6	17.1	142.7	211.2	353.9
Water	Reduced water availability (km²)	15.3	32.7	24.7	32.2	168.5	62.2	230.7
Coastal protection	Annual capital costs (m$/yr)	133	176	51	24	514	493	1,007
Dryland loss	Area lost (1,000 km²)	1.6	10.7	23.9	0	99.5	40.4	139.9
Wetland loss	Area lost (1,000 km²)	9.9	11.1	9.8	11.9	219.1	33.9	253.0
Ecosystem loss	Number of protected habitats lost, assuming 2% loss	16	8	NA	4	53	53	106
Health/mortality	Number of deaths (1,000)	8.8	6.6	7.7	29.4	114.8	22.9	137.7
Air pollution	Equivalent increase in emissions							
Tropical O_3	(1,000 t NO_x)	566	1,073	1,584	227	2,602	1,943	4,545
SO_2	(1,000 t sulfur)	285	422	1,100	258	1,864	873	2,737
Migration	Additional immigrants (1,000)	229	100	153	583	2,279	455	2,734
Hurricanes								
Casualties	Number of deaths	0	72	44	779	7,687	313	8,000
	m$	0	115	1	13	124	506	630

Source: Pearce et al. (1996), after Fankhauser (1995).

tations made by both producers and consumers (see also Section 5.4). Therefore, caution is required in drawing any conclusions from the related modeling exercises (see, e.g., Nordhaus 1991a; Cline 1992a; CRU/ERL 1992; Kane,
Reilly, and Toby 1992; Adams et al. 1993; Mendelsohn, Nordhaus, and Shaw
1993; Rosenzweig and Willel 1993).

5.3 GREENHOUSE DAMAGE AND ADAPTATION: METHODOLOGY

5.3.1 Adaptation Policies as a Welfare Economic Measure

Some issues complicate the process of determining the social costs of climate
change, the assessment of greenhouse damage, and the benefits of control
through adaptation. (Mitigation response options are discussed in the next
chapter.) This chapter focuses on adaptation policies by considering them as
a welfare economic measure. This means that the costs as well as the benefits
of adaptation measures have to be determined. It is only by comparing costs
and benefits that one can determine the optimal policies in an economically
meaningful way. (See Chapters 3 and 4 for details of cost–benefit analysis.)

Generally, the costs of adaptation actions can be determined relatively easily, because one has only to acquire information on adaptation investment
costs – for example, the investment in coastal protection measures or the introduction of new drought-resistant species in agriculture. In some cases, however, the costs of adaptation are far more difficult to quantify. For example, if
coastal protection entails the loss of wetlands or if changes in farming practice
require technological developments and information exchange, the costs are
rather difficult to establish. Serious complications arise when one is trying to
assess the benefits that can be derived from investments in adaptation. (See
Chapter 4 for a discussion of valuation of environmental damage.)

Some of the benefits are very tangible and relatively easy to quantify and
value. Clear examples are cases in which high-productivity agricultural land
threatened by flooding due to the rise in sea level can be retained through
coastal protection or in which progressively increasing droughts are offset by
the introduction of irrigation systems. Other benefits, however, are much less
tangible and are difficult to express in monetary terms. Examples include a reduction in the number of people injured or killed by increased flooding and
the preservation of vulnerable ecosystems that are threatened by climate
change. The issue of whether and how these nonmarket benefits can be assessed has become the subject of intense discussion among experts. This is the
first serious complication of greenhouse damage assessment.

A second complication related to the assessment of the benefits of adaptation results from the rather long delay between the payment of the costs of

adaptation and the reaping of the benefits (although this delay is generally much longer with mitigation policies; see Section 5.6). Because the benefits of adaptation often accrue – at least partly – to future generations, whereas costs are incurred by the present generation, the issue is how and to what extent we should take benefits for future generations into account in making decisions about adaptations by the present generation. In other words, the fundamental issue here is intergenerational equity. This topic has also led to heated debates, as described in Chapter 2.

A final complication is that greenhouse policies usually have secondary effects, mainly positive. For instance, coastal protection can be used to create new wetland systems (although it can also cause wetland loss in itself), generate new jobs, and improve the potential of exploiting wave energy. Improved weather information systems for detecting approaching hurricanes at an early stage can also provide other useful climate information. Assessing and valuing these secondary effects can prove to be complicated. However, such benefits help to explain why the overall benefits of adaptation (as well as of GHG abatement) may surpass damage due to climate change if no policy measures are taken, even if the implications of dynamic aspects of global warming and potential irreversibilities are disregarded.

In the following sections the above-mentioned complications are dealt with in more detail.

5.3.2 Complication 1: How to Assess Nonmarket Damage

Before one can actually assess damage due to climate change or damage avoided by means of greenhouse abatement action, an answer must be found to the question of how to value nonmarket costs, such as damage to nature and reduced quality of life or even the loss of human life. Some general remarks on this subject were made in Chapter 4 and will not be repeated here. Instead, we will focus on the main differences in perspective regarding the evaluation of nonmarket damage. Basically two fundamentally different starting points are used. These two approaches can lead to entirely different policy conclusions. We shall examine these, keeping in mind that most policy makers will take an intermediate position.

One approach is based on the acceptance of a so-called cost–benefit framework. It assumes that benefits and costs are ultimately defined in terms of human preferences – for example, the willingness to pay (WTP) for a benefit and the willingness to accept compensation for certain losses. Once this fundamental starting point has been accepted, several techniques are available that express the financial value of various "intangible" assets in order to compare the costs and benefits of actions in an economically useful manner.

The other approach, which is based on what is sometimes called the strong sustainability framework, assumes that it may be impossible to attach a material value to certain intangible or nonmarket matters. Consequently, certain nonmarket values simply cannot be compared with values that one can put a price on, since they are assumed to have different dimensions. This implies, for instance, that one does not accept that some nonmarket harm knowingly inflicted on future generations by the present generation (e.g., loss of biodiversity) will be compensated for by possible improvements in the material well-being of future generations (e.g., a higher standard of living traditionally measured). As a consequence, the cost of avoiding certain nonmarket damage is considered economically irrelevant (unless it is totally unacceptable); damage must be prevented for its own sake (see the discussion on absolute standards in Chapter 3, especially Figure 3.1a).

It must be clear that these two points of view may have fundamentally different consequences for environmental policies. Let us take an example. Suppose a number of firms in a developing country were to exploit a "global common" in the same country – in this case a richly endowed piece of tropical forest – in an unsustainable way. As a result, a rapid process of net deforestation would take place, leading to the gradual disappearance of the tropical forest and, along with it, a large amount of genetic material. Some would argue that the global community could not hold the developing country responsible for this environmental degradation if this community were not prepared to offer the developing country some compensation for returning to sustainable forest management (assuming returning to sustainable management would imply an economic sacrifice for the developing country).

If we continue along this line of thinking, on the basis of the cost–benefit framework the world community's WTP will influence whether or not unsustainable management practices leading to deforestation can be halted. This WTP will be based on consumers' wish to preserve the forest as a place to visit in the future (option value); on their willingness to pay for halting deforestation to conserve the forests for their children (bequest value); or simply on their appreciation of the current existence of the forest (existence value). If, however, it turns out that the overall WTP based on these functions – objectively measured in one way or another and incorporating the present generation's evaluation of the effects of deforestation on future generations – is insufficient to save the forest,[2] from the cost–benefit point of view one will have to accept the gradual disappearance of the forest. Implicitly, in this case human preferences indicate that people are not prepared – for whatever reasons – to make a sacrifice large enough to save the forest.

---

[2] Assuming that a reasonable compensation for the developing country has been established.

According to the strong sustainability approach, this outcome would probably be unacceptable. First, individuals' preferences are not considered to be an acceptable basis for evaluation, not only because people may have an incentive to put their own interests first and only then to consider general interests, but also because it is assumed that individuals are ill-informed, so that human preferences may not – or insufficiently – capture intrinsic values of nature. Second, it is unacceptable to exclude future generations' interests from decisions that have irreversible ecological consequences and are undertaken by the present generation (e.g., a decision to destroy the forest for the sake of current economic welfare). Moreover, according to this point of view, it is unacceptable that today's actions may give rise to quantitatively unknown but potentially great risks. This latter principle is commonly referred to as the precautionary principle (see Chapter 3).

Another, related complication is the question of whether and how nonmarket damage can be *aggregated*. This issue arises even if one accepts that nonmarket damage can somehow be valued as a statistical entity and in a specific context. To illustrate this, let us assume that one may accept – in line with the cost–benefit approach just described – that reducing the statistical risk of a loss of a human life in a particular country can be valued at a certain amount of dollars. In terms of aggregation, this does not mean, however, that once such values of statistical lives have been derived for a set of individual countries, they can then be aggregated, just as if the context in which the valuation had taken place were no longer relevant. This issue of international aggregation of nonmarket damage has fueled an intense debate, as argued on pp. 166–8.

5.3.3 Complication 2: How to Discount Damage

In assessing climate damage, it may help to assess future damage as if it occurred today – that is, the present value of damage. This is somewhat similar to determining the present value of an asset on the basis of a continuous flow of future returns, although in this case the procedure is far more complicated – for example, due to uncertainties concerning the development of technologies that may prove highly effective in reducing future damage. The most straightforward procedure for comparing a series of future flows with the present situation is *discounting* (see Chapter 2). Although we acknowledge the limitations of this approach for the present purpose, we will pursue it nevertheless.

The principle of discounting (irrespective of the value of the discount rate considered most appropriate) is that the value of a return in year $t + 1$ or year $t + 2$, and so on, can be expressed as a corresponding value in the year t (the present situation) if the discount factor is known. The latter is the degree to which a certain return will be valued less if it is acquired one or more years into the future. Obviously such a discount factor is nothing but an expression for an individual's time preference – that is, the degree to which one prefers the return (or for that matter the possibility of consuming) now rather than in the future. (For the sake of convenience we disregard inflation, which, if present, would also have to be compensated for.) This preference can vary among individuals: those who are keen on spending their return rapidly will have less appreciation for a future than those who are indifferent as to when their return will be available. However, as soon as an agreement has been reached on the value of the discount factor considered most appropriate for the given purposes, a present value can be determined on the basis of the expected future flow of returns through a simple discounting procedure (for an illustration, see Box 2.4).

The same principle applies to damage caused by climate change. Suppose such damage manifests itself through a continuous annual flow according to a certain function expressing the time pattern of damage (for an example, see Figure 5.1). Reducing total damage to one figure, as if it all occurs currently, one may discount future damage in order to determine its present value.

The question now arises as to whether there is a discount factor that would be appropriate here. This issue has already been dealt with extensively in Chapter 2. Therefore, we will mention only a few generalizations.

The bottom line of damage assessment, namely that damage is experienced mainly by future generations – people who do not yet exist and therefore cannot defend their interests or who are too young to defend their interests – but has been caused by the present generation, is that the present generation does not tend to be concerned about the interests of future generations except on the basis of the moral argument that their children and grandchildren will be a part of these future generations. Therefore, the discussion about what discount factor is suitable for long-term damage assessment boils down to the *moral* issue of intergenerational equity, or the extent to which the present generation ought to respect the interests of future generations. The fundamental conclusion will therefore probably be as follows. Just as there is no generally accepted principle of economic efficiency that, when used in a cost–benefit approach, will tell us exactly whether we should proceed with mitigation or what the optimal program would be, there will also be no generally accepted principle of equity that will provide us with unique answers to the same problems.

With these remarks in mind, again, two fundamentally different approaches can be taken in dealing with the issue of discounting: the cost–benefit type of approach, which includes the market returns to investment and the social rate of time preference as relevant criteria (as discussed in Section 2.4.4), and the so-called sustainability approach. The first approach is based on the notion that concern for future generations can be expressed semiobjectively in a social discount rate (this is essentially the approach taken in Section 2.4.4). The sustainability approach, however, rejects this idea and uses a normative discount rate to protect the interests of future generations.

Because the discount factor as derived on the basis of the cost–benefit approach is likely to be higher than that derived by means of the sustainability approach, a moral approach will generally favor future generations' interests more than a (semi)objective approach. A similar time profile of damage can easily produce a smaller present value of damage in the cost–benefit approach than in the sustainability approach. It is therefore conceivable that the type of approach selected may affect the decision as to whether to take action now or at some later time.

5.4 COSTS AND BENEFITS OF ADAPTATION

5.4.1 The Need to Deal with Adaptation Measures

As argued earlier, the augmented IS92 scenarios suggest that benchmark $2 \times CO_2$ warming is likely to occur in the next century, although it can be delayed somewhat by abatement policies. Society should realize that damage due to global warming will take place, irrespective of current actions. What climate change patterns will look like is still highly uncertain, even though our knowledge about this is constantly improving (Houghton et al. 1995). Many developments seem possible, even much slower scenarios or low-probability surprise scenarios in which the climate will change to a much greater extent and much faster than general consensus indicates, due, for example, to melting of the West Antarctic ice sheet and structural changes in ocean currents.

For this reason, one cannot rely on abatement alone, but must also prepare for adaptation. This explains why the important role of adaptation policies is being increasingly recognized, and why the search for more information about the costs and benefits of such policies is accelerating. This is especially so because of growing evidence that damage due to climate change (e.g., in agriculture) is highly sensitive to the availability of low-cost adaptation measures. Indeed, the IPCC Second Assessment Report of Working Group II indicates that the availability of such measures may even result in a net benefit for agriculture rather than net damage (Watson, Zinyowera, and Moss 1996). However, there are no comprehensive surveys of the adaptation options and their costs, probably because adaptation covers such a broad range of potential actions and also because of the large uncertainty surrounding these options. The literature on this subject is limited but growing.[3] In any case, it is obvious that society already encounters large costs in adapting to climate extremes; climate changes will only increase these costs. When talking about adaptation, one must keep several key questions in mind: (a) what effects to adapt to, (b) how to adapt, and (c) when to adapt. In this section only the first two questions are considered extensively; moreover, no attention will be paid to the issue of insurance, which can be considered an adaptation option in its own right (see Chapter 3). The third question refers to such issues as implementing no-regrets adaptation options now (such as developing drought-resistant crops and techniques) and of weighing the implementation of mitigation options now against adaptation options in the future. Section 5.6 contains some remarks on this subject.

5.4.2 Adaptation to What?

Adaptation to different degrees and in different forms may be necessary to cope with ecosystem changes that are closely related to human (economic, social, political, legal, and cultural) activities, depending on sequence, severity, and importance of impact (see Watson et al. 1996 on the effects of such changes). Changes in temperature and related amounts of rainfall may lead to droughts or heavy rain in some places, thus affecting worldwide surface-water and groundwater availability, which in turn will affect agronomic practices and yields in agriculture. (This may also be influenced by a CO_2-enriched atmosphere.) Fisheries and forestry will be affected by changes in temperature and the availability and quality of water (e.g., salinity). Livestock population and output may change due to, among other things, heat stresses caused by temperature increases. The populations of other animal life, including parasites and insects, may grow, which will affect animal and human health.

Climate change may cause an accelerated rise in the sea level, possibly accompanied by floods or hurricanes, changes in regional temperatures, an increase in storminess, and changes in the runoff of river water due to changing precipitation (in terms of mean value and variability). Global research on sea level rise is becoming increasingly common (see, e.g., Fankhauser 1994). Testimony to the "early" awareness of the risks of sea level rise is the fact that in 1993 the World Coast Conference pointed out the need to integrate responses

---

[3] For a survey of the costs of greenhouse damage, see Chapter 3 of this book and Watson, Zinyowera, and Moss (1996). Modeling of adaptation to climate change has been carried out for sea level rise, storminess, and changes in river discharges. A method for assessing damage can be found in Howe et al. (1991) and Green et al. (1994). Penning-Rowsell and Fordham (1994) present a general method for adaptation. Models for flood hazard assessment and management can be found in Klaus et al. (1994). With respect to river zone management, including the institutional setting, Correia et al. (1994) present a framework for analysis.

TABLE 5.4 Agricultural Yield Changes with a Doubling of CO_2 Concentration (% of Gross Agricultural Product)

	Model[a]					
	UKMO Scenarios[b]			GISS Scenario[b]	GFDL Scenarios[b]	
Region[c]	1	2	3	1	2	3
OECD-A	−20.0	−5.0	−5.0	10.0	10.0	10.0
OECD-E	5.0	5.0	5.0	10.0	10.0	10.0
OECD-P	7.5	7.5	7.5	7.5	7.5	7.5
CEE & SU	−7.5	−7.5	−7.5	22.5	22.5	22.5
ME	−22.5	−22.5	−7.5	−7.5	−7.5	7.5
LA	−22.5	−22.5	−8.5	−15.0	−15.0	−1.0
S & SEA	−20.0	−20.0	−10.0	−10.0	−10.0	0.0
CPA	−7.5	7.5	7.5	7.5	22.5	22.5
AFR	−20.0	−20.0	−20.0	−7.5	−7.5	27.5

[a] The climate change scenarios used are the equilibrium $2 \times CO_2$ experiments according to the central circulation models of the UK Meteorological Office (UKMO), the Goddard Institute for Space Studies (GISS), and Geophysical Fluid Dynamics Laboratory (GFDL).
[b] The scenarios involve no adaptation (1), minor shifts (2), and major shifts (3) in behavior.
[c] OECD-A: OECD America; OECD-E: OECD Europe; OECD-P: OECD Pacific; CEE & SU: Central and Eastern Europe and former Soviet Union; ME: Middle East; LA: Latin America; S & SEA: South and Southeast Asia; CPA: Centrally Planned Asia; AFR: Africa.
Source: Tol (1994), p. 5, after Rosenzweig and Hillel (1993); see also Fisher and Hahnemann (1993), Rosenzweig and Parry (1994), and Reilly et al. (1994).

to long-term threats such as climate change and related sea level rise with existing planning and management efforts in order to realize integrated coastal zone management (IPCC 1994). However, research on the possible consequences of climate change still differs considerably as to the assumptions of scenario building: IPCC (1994) works with a 1-m rise in sea level every 100 years,[4] whereas others work with a scenario of 50 cm. In addition, the appropriate or adequate response options to be called for differ significantly with respect to all of the above-mentioned effects.

As far as the impact of changes in river discharges is concerned, only limited and scattered information is available. For example, the discharge of the Rhine River in the Netherlands is expected to fall by 10 to 15% due to an anticipated rise in temperature of 4°C in the Alpine part of the basin (see Kwadijk 1991, as cited in Penning-Rowsell and Fordham 1994, p. 26). However, there has been only a small amount of research on the correlation of the effects of precipitation changes with the effects of temperature rise. A major

[4] A sea level rise of 1 m would affect the supply of rice to more than 200 million people in Asia (IPCC 1994).

conclusion of the IPCC (1994) is in fact that it is very difficult to differentiate between sea level rise and non-climate-related factors, such as subsidence and excessive groundwater abstraction, which may be equally important determinants in *relative* sea level rise.

Obviously, countries where sea level rise could become prominent might therefore face challenges beyond those that would be triggered by a climate change only: agriculture might, for instance, be affected by regional changes in temperature and sea level rise. Likewise, the impact on agriculture depends on the full range of possible effects of climate change rather than on temperature only (for some estimates see Table 5.4).

5.4.3 How to Adapt

To answer this question, let us first focus on the issue of how to adapt to sea level rise. Generally, such adaptation will take the form of retreat, accommodation, or protection. The pattern of adapting to changing river water discharges is similar to the pattern of adapting to a rise in the sea level. Here, no global costs are available.

Retreat will usually cause loss of dryland as well as loss of wetland. IPCC (1994) computed that a 1-m sea level rise could threaten 170,000 km² (or 56%) of global coastal wetland. Loss of dry land will mean losses in agriculture, forestry, species, and physical assets and will imply human migration.[5] Some sources present estimates of land protection costs insofar as land loss is prevented on economic grounds by means of coastal infrastructure and other measures. Table 5.5 presents a survey of these types of costs (on an annual basis).

Not all estimates include the side effects of resettling people who lived on the lost land. These costs include the costs of receiving refugees and the costs incurred by (the trouble caused to) those people. Combining various sources of information, Tol has estimated the global annual costs of relocation due to sea level rise at some U.S.$14 billion. These costs vary between 0.01 and 0.03% of GDP for the OECD and countries in transition, and between 0.3 and 0.8% for developing regions (Tol 1993, p. 52).

Accommodation to sea level rise means not only adapting existing structures to a higher sea level but also, for example, eliminating subsidized insurance for building new structures along the seashores of industrialized countries. During the transition, countries may have to deal with inundation, which may cost lives and cause damage to physical assets, agriculture, and the environment (Penning-Rowsell and Fordham 1994).

---

[5] Attempts to express these losses in monetary terms can be found in Ayres and Walter (1991), Rÿsberman (1991), Cline (1992a,b), Suliman (1990), Nordhaus (1991b, 1993), Fankhauser (1994), and Tol (1994).

TABLE 5.5 Annual Damage Costs of Sea Level Rise

Region[a]	Wetlands (U.S.$ Millions)	Drylands (U.S.$ Millions)	Protection (U.S.$ Millions)	Total (U.S.$ Millions)	Total (% of GDP)
OECD-A	5,000	2,000	1,500	8,500	0.15
OECD-E	4,000	500	1,700	6,200	0.14
OECD-P	4,500	4,000	1,800	10,300	0.45
USSR & EE	1,250	1,250	500	3,000	0.07
ME	0	0	0	0	0
LA	1,500	500	1,000	3,000	0.35
S & SEA	1,500	1,000	2,000	4,500	0.65
CPA	500	0	500	1,000	0.29
AFR	500	500	500	1,500	0.43

Note: All figures are estimates ±50%.

[a] EE denotes Eastern Europe. For all other abbreviations, see Table 5.4.

Source: Tol (1993), p. 50.

Estimates of the costs of *protection* against sea level rise differ considerably.[6] Tol (1994), assuming a 50-cm rise, estimates global coastal defense to be U.S.$[1988]9.5 billion on an annual basis; IPCC (1994), assuming a 1-m rise, estimates costs at U.S.$[1994]10.0 billion per year; while Ayres and Walter (1991) come up with U.S.$[1981]50 to 100 billion. Protection costs for an increased intensity of storms are not available on a global scale.

Some Observations Regarding How to Adapt Adaptation to changing temperatures involves adjustments in health care, heating and cooling, and household activities, improvements in infrastructure, and adaptation of agriculture and fisheries. Improving infrastructure involves irrigation / water storage systems, including (small) dams, drainage and sewer systems, dikes and locks, as well as urban construction. In agriculture, various technical responses are available: changes in crop variety; changes in irrigation, fertilizers, and drainage; changes in crop management; and changes in farming strategies. Some salt-tolerant crops, for example, can be grown successfully along the shoreline of coastal deserts irrigated with ocean water. Global estimates of the potential impact of adaptation in agriculture, including shifts per region, can be inferred from Table 5.4. Given our limited understanding of climate change, extending the range of adaptation strategies seems worthwhile. In this respect, special attention should be given to the following:

1. Capacity building – in both industrialized and developing countries – to educate people in developing countries about the effects of their activities

---

[6] A further factor contributing to the complexity of these cost estimates is wetland loss often caused by coastal protection.

on carbon-trapping biota, and people in industrialized countries about possible responses to the effects of natural climatic variability and potential climate changes in the future.

2. Changes in land-use allocation, including the development of tropical (plant) species. Since most of the world's plant food comes from just 20 species, the potential of the majority of species is still to be developed.

3. Food security policies and reduction of postharvest losses. Postharvest losses, due to deficient systems of storage and transport, amount to 50% or more in many developing countries, which means that there is ample room for improvement.

4. Conversion to "controlled-environment agriculture." The widespread introduction of integrated controlled-environment agriculture in developing countries may require an investment of several tens of billions of dollars, or billions of dollars per annum if introduced over several decades.

5. Aquaculture. Climate change affects ocean circulation in the upper layers, upwelling, and ice extent, all of which affect marine biological production and thus marine fisheries. One way to adapt is to intensify efforts to develop aquaculture. Integrating aquaculture and controlled-environment agriculture has great potential; recently, dramatic progress has been made in marine biotechnology. The almost sterile, nutrient-rich bottom water from ocean thermal energy conversion systems is very promising as a cultivation medium for kelp, abalone, oysters, and a variety of fish species.

It must be mentioned here that for marginal groups the risks of damage due to climate change will grow larger as the land distribution system becomes more unequal. Change in land tenure may, as a side effect, reduce these risks and can be considered an indirect adaptation option in itself. Finally, scarcity and surplus will change across regions and over time, presenting new opportunities for trade between nations in order to stabilize supply.

5.4.4 Adaptation Measures in Developing Countries

The quest for adaptation options in developing countries existed long before the global debate on climate change started. Countries tried to find long- and short-term responses to recurrent droughts in arid and semiarid zones. Countries in heavy-rainfall regions and countries affected by storms and cyclones in coastal areas tried to deal with recurrent floods (non)structurally. Generally, whether adaptation must be applied immediately or can be delayed depends on the type and intensity of the effects of climate change. Current bilateral and multilateral activities are expected to lead to a more systematic assessment of adaptation options and their costs. Our knowledge of adaptation potentials and their costs as far as developing countries are concerned is, however, still rather limited. And in the case of Africa, no real adaptation studies have yet been carried out.

Short-term responses to recurrent droughts have included improvement in drought anticipation and have focused mainly on drought relief and recovery activities. With respect to drought relief, supplementary food programs and programs to protect and replenish the national livestock have been initiated. Drought recovery entails such activities as the provision of seed and land preparation supplements to farmers after a period of drought. However, even for these short-term responses, no systematic studies have been carried out.

Long-term measures have included regional and national research on drought-resistant crops and the breeding of hardy livestock. Consequently, the incorporation of drought and salt resistance in crop varieties is a major item on the research agenda of some developing countries. Further activities, particularly those that strengthen research capacity, including increased flows of financial resources, are necessary and will probably prove to be cost-effective. In areas where water resource management will become crucial because of large changes in rainfall regimes (e.g., increased frequency and severity of floods or droughts), an improved and more environmentally sound infrastructure will be necessary and policies encouraging water conservation (e.g., pricing mechanisms in which prices reflect social scarcity) will have to be introduced.

The electricity generation sector, which will also be seriously affected by changes in climatic patterns, has received support from sources not directly involved in current climate change activities. Facing massive river and dam silting and below-average precipitation to replenish hydroelectric installations, nations have sought to develop alternative base-load systems, such as coal thermal. A more systematic assessment of these responses will prove crucial, particularly in the light of the importance of decarbonizing the fuel base to reduce emissions.

5.4.5 Adaptation Policies and Damage

As we have already extensively argued, individuals and society can often reduce the unfavorable consequences of global warming by adapting to climate change. People can search for more drought-resistant crops, alter planting patterns, defend vulnerable coastlines, improve draining capacities of rivers by migrating from risky areas, improve information systems on approaching extreme weather conditions, make adjustments in the construction of houses and buildings, and so on.

However, adaptation measures do not consist only of physical measures; they also comprise a mixture of economic and societal features. For instance, personal savings and, in some cases, insurance may cover the damage expected by individuals, and the government may provide safety nets, charities, and other forms of support to the victims of damage caused by climate change. Their effectiveness will depend to a great extent on the flexibility and speed of the different feedback mechanisms.

Given the fact that many forms of adaptation will be applied, the question arises as to how this will affect the assessment of damage. If one recognizes the merits of adaptation, the final damage costs are twofold: the net costs associated with adaptation and the remaining, or residual, damage due to climate change.

Let us begin by considering the first component: adaptation costs. The empirical information about such costs shows a rather scattered pattern. In agriculture, for instance, there is evidence – based on a simulation of cereal production in 18 countries by 2060 (assuming $2 \times CO_2$ equivalent warming by that year) – that the projected global welfare loss (between U.S.\$0 and \$61 billion) could be reduced to values ranging between a U.S.\$7 billion net gain and a loss of about U.S.\$38 billion if moderate adaptations were carried out at the farm level (see Rosenzweig and Willel 1993; Reilly, Hohmann, and Kane 1994). This suggests considerable scope for damage reduction in agriculture through adaptation measures.

A similar conclusion was reached in an exercise focusing on agricultural production change in certain U.S. regions in 2030. The (probably relatively optimistic) projection for this year, indicating an agricultural production loss in the area of about 17% as compared with the present situation (or, if the positive impact on agricultural production of carbon fertilization were included, some 8%) could be reduced to about 12 and 3%, respectively, due to farmer adaptations. Thus, according to this model, (low-cost) adaptation measures could succeed in reducing agricultural damage by 30% or more (Easterling et al. 1993).

However, much of the literature dealing with the details of climate change damage focuses on the costs and benefits of coastal protection. This seems logical, because coastal protection is one of the most evident areas of cost-effective adaptation. The following section, therefore, focuses on this topic as a special case study on damage assessment.

5.5   COSTS AND BENEFITS OF COASTAL PROTECTION

In the area of coastal protection, adaptation measures appear to be obvious. In the 1995 review of IPCC Working Group I, a central estimate was adopted for sea level rise by 2100: 45 cm (the assessment for 1990 was 66 cm) (Houghton et al. 1996). In addition, this report argues that, due to increasing ocean–land temperature differentials, storm damage and the risk of (unpredictable) flooding may be higher than projected earlier. This obviously calls for increased investments in coastal protection, especially since one assumes that it will be much longer (some centuries) before the equilibrium effects of sea level rise will occur as compared with the point estimates corresponding to the first year in which the equilibrium warming is realized (which explains why some researchers use a benchmark of a 1-m sea level rise). A particular cause of uncertain damage to coastal and other areas is tropical cyclones, the intensity and frequency of which may increase if sea surface temperature becomes higher as a result of global warming. On average, annually about 70 to 80 tropical cyclones are recorded worldwide, causing annual damage of some billions of dollars (Fankhauser [1995] recently mentioned worldwide hurricane damage of about U.S.$2.7 billion)[7] and casualties numbering in the tens of thousands.

5.5.1 Costs

The investments involved can be considerable. The U.S. Environmental Protection Agency (U.S. EPA), for instance, estimated that in the United States a 1-m rise in sea level by the year 2100 would require $73 to $111 billion in cumulative capital to protect developed areas by building bulkheads and levees, pumping sand, and barricading barrier islands (Smith and Tirpak 1989). This would amount to capital construction costs of about $1.2 billion per annum if spread over a 100-year period (Cline 1992a; note, however, that Gleick and Maurer [1990] estimated annual capital construction and maintenance costs for protecting the San Francisco Bay alone against a 1-m rise at $200 million). Similarly, Fankhauser (1995) estimated the annual costs of coastal protection against a 50-cm rise at about U.S.$1 billion worldwide, with about half of it being spent within the OECD.

Would the major investments in coastal protection that have been proposed in the literature be justified? The answer is, it all depends. The reason, and also why it is difficult to determine the optimal level of investment in coastal protection, is similar to the reason for the difficulties of applying cost–benefit analysis to climate change in general: investment costs, which may be uncertain in the future, have to be weighed against the investment benefits of protection, which are equally difficult to quantify. These benefits are basically twofold. On the one hand, land loss is prevented, and thus the capacity to produce food, to retain infrastructure or space for housing, and so forth, is preserved. These benefits can more or less be valued objectively insofar as they

---

[7] Note that in 1990 European storms resulted in a total loss of DM 25.3 billion (Munich Re 1993); in 1992 Hurricane Andrew alone caused $30 billion in economic damage, about half of which was covered by insurance (Dlugolecki et al. 1994).

accrue now. On the other hand, there are uncertain future benefits, as well as hard-to-value nonmarket benefits or damage avoided in terms of less human insecurity and suffering, as well as less loss of biodiversity due to the conservation of coastal ecosystems.

The problem, therefore, is not only to try to develop a full and sound quantitative understanding of the various benefits, but also to attach material values to these benefits in order to make an economically meaningful assessment of coastal protection investment. In this context we are again faced with the delicate issue of discounting and of how to value certain nonmarket assets. This applies particularly to human insecurity and suffering. Suppose that the number of ecological refugees from various coastal areas could be reduced by investing in coastal protection or improved weather information systems. How could one determine the value of the reduced insecurity and risk of suffering? Or, in the extreme case of casualties of coastal flooding that might have been prevented, how could one determine the value of a statistical human life in the different regions of the world?

5.5.2 Material Benefits

Keeping in mind the immense valuation problems, let us turn to the more or less readily quantifiable coastal protection benefits. Information about such benefits is gradually expanding, mainly at the national and regional levels. It is impossible to provide a complete survey of all the information that has recently become available. The following data, therefore, will merely serve as an illustration of what is known about the damage that might occur if the sea level rises significantly above the current level.

United States In 1989 the U.S. EPA already estimated that a 1-m sea level rise would cause a loss of unprotected dryland of 6,650 mi^2 in the United States. In addition, the U.S. wetland area (about 13,000 mi^2) would be reduced by half (Titus 1992). Similar results were found for the OECD (excluding Canada, Australia, and New Zealand) by Rijsberman (1991), who argued that with a 1-m sea level rise the loss of coastal wetland would probably be between 48,000 and 64,000 km^2.

Attempts to value land loss due to sea level rise have been rather crude so far. For instance, the assumed costs of U.S. wetland preservation programs vary between $10,000 and $30,000 per acre. Cline (1992a) estimates annual U.S. losses with a 1-m rise at $4.1 billion for wetlands and $1.7 billion for drylands by assuming $10,000 capital costs per acre of wetland, a coastal dryland value of $4,000, and a rental opportunity cost of 10%. A worldwide estimate of annual forgone land services with a 50-cm sea level rise can be found in Fankhauser (1995). He estimated this cost at $45.6 billion by assuming that a third of all remaining wetland would disappear. Note that more than 85% of coastal wetland loss is expected to take place in developing countries.

Japan The Japanese economy is affected by rising sea level because the three cities that account for more than half of Japan's industrial production, Tokyo, Osaka, and Nagoya, are located in coastal areas. About 860 km² of coastal land is already below mean high-water level, an area supporting 2 million people and physical assets of U.S.$450 billion. A 1-m sea level rise would expand this area by a factor of 2.7 to embrace 4.1 million people and assets worth U.S.$900 billion; the flood-prone area would expand from 6,270 to 8,900 km², with an additional 3 million people at risk (Nishioka et al. 1993). On the basis of all this, the cost to Japan of adjusting its existing protection measures has been estimated at about U.S.$80 billion. Moreover, one has to keep in mind that pressure on natural shorelines may increase if extensive coastal protection were to be introduced.

Other Areas Although a rising sea level will affect the OECD countries – as just illustrated for the United States and Japan – the developing countries and the lower-situated island states in particular seem to be the most vulnerable areas. According to the IPCC (1992), 49 of the 50 countries and territories identified as having annual shore protection costs of over 0.5% of GDP as a consequence of a 1-m rise in sea level are developing countries (see Box 5.2), the exception being New Zealand.

BOX 5.2 The Vulnerability of Developing Countries to Greenhouse Damage

Some additional damage estimates will illustrate the particularly vulnerable position of developing countries.

Small Island States. Various small island states are among the countries most vulnerable to sea level rise. A profound impact, including nonexistence in the worst case, would be experienced by the Tokelau Islands, Marshall Islands, Tuvalu, Line Islands, and Kiribati; a severe impact resulting in major population displacement would be experienced by Micronesia, Palau, Nauru, French Polynesia, the Cook Islands, Niue, and Tonga. A moderate to severe impact would be felt by Fiji, American Samoa, New Caledonia, the Northern Marianas, and the Solomon Islands. And local severe to catastrophic events would be experienced by Vanuatu, Wallis and Futuna, Papua New Guinea, Guam, and Western Samoa. One has to keep in mind that protection measures are both limited and expensive. For the Marshall Islands, for example, it has been estimated that protecting the Majuro Atoll alone would cost 1.5 to 3 times the current GDP (IPCC 1992).

Egypt. It is predicted that a 1-m sea level rise could destroy up to 25% of the Nile delta agricultural land and displace about 8 million people (El-Raey 1990).

Nigeria. According to the IPCC (1992, 1994), in the absence of protection, a 1-m rise in sea level could flood more than 18,000 km^2 of Nigeria's land, damaging much of the country's oil industry. Also, over 3 million people would have to be relocated. The costs of protecting the highly developed areas are estimated at U.S.$550 to $700 million.

Senegal. A case study of Senegal – where two-thirds of the population and 90% of the industry are located in the coastal zone – estimates that in the absence of protection a 1-m sea level rise would cause a loss of 6,000 km^2 of land. Protecting these areas would cost U.S.$250 to $850 million, about three-quarters of which would go toward beach nourishment (IPCC 1992, 1994).

Bangladesh. A 45-cm sea level rise would flood some 15,700 km^2 of land in Bangladesh (about 11% of the total land area, including some 75% of the Sundarban mangrove forests). According to Asaduzzaman (1994), a 1-m rise would affect about 21% of the land area (nearly 30,000 km^2) and would cause all Sundarban forests to disappear. As a consequence, in 2070 several ports would be affected, damage would result from saline intrusions, rice output would decline, among other effects, leading to an estimated overall macroeconomic impact of the order of U.S.$4.8 billion in terms of lost output, or some 30 and 5% of current Bangladesh GNP in the coastal zone and overall GNP, respectively.

India. The Asian Development Bank has reported estimates with respect to the impact of a 1-m sea level rise on India (Asthana 1993). According to this source, in the absence of protection, approximately 7 million people would be displaced. Some 5,700 km^2 would be lost, contributing to an annual cost of some 1% of GDP. (For some figures with respect to specifically vulnerable regions only, see IPCC 1992; the same source indicates that protecting the Orissa and West Bengal region would require the construction of an additional 4,000 km of dikes and seawalls).

Indonesia. Indonesia, with a shoreline of more than 80,000 km, is also vulnerable to sea level rise, particularly as far as local rice and maize pro-

(continued)

BOX 5.2 The Vulnerability of Developing Countries to Greenhouse Damage *(continued)*

duction in the coastal areas is concerned. Without protection a total area of 3.4 million ha could be inundated.

Thailand. For the Suratthani Province in southern Thailand, a case study calculated that 37% of the area would be affected by a 1-m rise in sea level, with losses of more than 4,200 ha of agricultural land and many shrimp ponds (Parry et al. 1991).

Argentina. In Argentina a 1-m sea level rise would inundate an area of about 3,400 km² (0.1% of total area). Erosion would claim assets and land worth U.S.$5 billion.

Venezuela. Venezuela could lose about 5,700 km², or 0.6% of its area due to a 1-m rise (assuming no protection). Particularly at risk would be the country's low-lying coastal plains and deltas.

Uruguay. Although only a small area would be at risk in Uruguay, the coastlines affected would be highly valuable tourist regions. Uruguay's tourist industry brings in over U.S.$200 million in revenue per annum and attracts more than 1 million visitors each summer. Protecting the beaches would be expensive. The capital costs of protecting developed areas (mainly through beach nourishment) were estimated at U.S.$2.9 to $8.6 billion, more than five times the costs expected for Venezuela. If spread over 50 years, this would correspond to annual investments of 6 to 19% compared with 1987 gross investments (Nicholls et al. 1992).

An increased sea level will not only cause present land to disappear, but will also increase the expected damage due to flooding. Some calculations do indeed show that this damage will increase substantially as the sea level rises. Some estimates for the United States indicate that a 1-ft sea level rise by the year 2100 will probably increase the expected annual flood damage by about 50%; however, a sea level rise of 3 ft might cause a doubling or even a tripling of annual average damage (a comparable outcome with respect to the European Union is suggested in CRU/ERL 1992). The corresponding estimates of the U.S. areas inundated by a 100-year flood are 19,500 mi² in case of no sea level rise, and 23,000 and 27,000 mi² for a 1-ft and 3-ft sea level rise, respectively.

5.5.3 Nonmarket Benefits

There is no clear dividing line between material and nonmarket costs or benefits. A lower level of material welfare may provide the same feeling of disillusionment as, say, a specific degradation of the environment. However, those goods and services that are traded on a regular market have a price that can be accepted as a basis for determining their value expressed in material terms; some, mainly nonmarket, goods and services are not or cannot be traded on a market, however. If this is the case, one can only try to attach a material value to them indirectly, if at all.

Coastal protection can provide some of the latter type of benefits. As has already been mentioned, a sea level rise may force many inhabitants of coastal areas to move, sometimes, in the case of unexpected flooding, abruptly. For example, in Bangladesh about 5% of the population would have to be displaced in the case of a 45-cm sea level rise, and 13.5% in the case of a 1-m rise, in terms of projected 2070 populations of 12 million and 32 million people, respectively (Asaduzzaman 1994).[8] Another example is Indonesia, where a 15- to 90-cm sea level rise by 2070 is predicted to result in the migration of about 3.3 million people (projected costs, U.S.$8 billion), assuming that the Indonesian population stabilizes between 2030 and 2045 (ADB 1994).

A single event in poor countries (small islands are particularly vulnerable) can have disastrous consequences: in 1970 one cyclone caused more than 500,000 deaths in the area we now refer to as Bangladesh; a similar disaster in the same region in 1995 killed another 100,000 people. Most economic studies indicate that the adverse nontangible consequences of sea level rise and the increasing intensity and frequency of hurricanes are likely to be more severe in developing countries than in industrialized countries, not only in an absolute sense but also relative to their wealth. Summary studies for Bangladesh, Brazil, China, Egypt, Malaysia, Nigeria, and Senegal (e.g., Nicholls et al. 1990) suggest that a 1-m sea level rise could displace nearly 100 million people in these countries alone (China accounting for 75%). In the future this could lead to massive human deprivation and suffering.

Not only can increasing sea levels combined with flooding cause nontangible damage like human suffering, they can also cause irreversible damage to ecosystems, some of the clearest examples being wetlands and mangrove forests.

All in all, it has become clear by now that an integral assessment of both material and nontangible benefits of adaptation through coastal protection investment is very hard to achieve. Precisely because such information has to be available before an optimal adaptation strategy can be determined, it is fair to say that a clear picture of the economically most sensible investment decisions on coastal protection simply does not yet exist, or at least is a controversial subject.

---

[8] In practice, migration from the areas most likely to be affected is already taking place, partly for weather-related reasons.

5.6 ADAPTATION VERSUS MITIGATION

5.6.1 Main Characteristics

Having dealt with the issue of what future damage to expect as a result of climate change given the complications of damage assessment, and having examined some information about adaptation costs, particularly with regard to coastal protection, we will now turn to an important question: What more can be said about the benefits of greenhouse policies in terms of damage avoided? The reader should keep in mind that it is sometimes hard to distinguish between mitigation and adaptation.[9]

The next chapter will elaborate on technical response options. In this section we attempt to find an answer to the question just posed and therefore must distinguish fundamentally between two different types of technical policy options available to counter greenhouse warming and their possible feedbacks: mitigation options and adaptation options (the latter being the focus of this section).[10] *Mitigation options* are those that reduce the net emission of GHGs into the atmosphere, either by reducing GHG emissions or by increasing the sinks for GHGs. *Adaptation options* focus on reducing the expected damage caused by rapid climate change by combating or averting the detrimental effects. The benefits in terms of damage avoided through mitigation options, therefore, are fundamentally different from those derived from adaptation options.

5.6.2 The Free-Rider Problem

Whereas mitigation policies can have only a long-term global impact on greenhouse damage – that is, by reducing carbon emissions or increasing carbon uptake, altering the concentration of GHGs in the atmosphere, and thus possibly creating certain favorable global climate consequences and reducing global damage in the long term – adaptation policies generally have a positive direct effect in the short term – for those who implement these policies. To give an example, if the government of the Maldive Islands were to discourage the use of energy-absorbing speedboats to carry tourists to atoll resorts, the impact on sea level rise would probably be long term, indirect, and rather modest. If, instead, the government decided to protect the islands with the help of stone walls and the like, the damage avoided would probably be far

[9] In some exceptional cases, however, the distinction between adaptation and mitigation options can become blurred, e.g., when research and development is promoted in order to create plant species that are more drought-resistant and require less fertilizer (see the example of saline-resistant crops in Section 4.2.2, Chapter 4).

[10] The next chapter will also distinguish between direct and indirect policy options.

more short term, direct, and potentially significant. Because of the differences between mitigation and adaptation policies, decision making with respect to the former can be expected to be more complicated than that with respect to the latter. Mitigation policies will generally be successful only if the countries involved have reached a consensus about the international division of responsibilities for greenhouse abatement action. In contrast, many adaptation policies will be carried out, either privately or officially, by those who consider themselves potential victims of the likely consequences of climate change. The bottom line is that the free-rider problem associated with mitigation policies may encompass the whole world, whereas the incidence and scope of free riding will very likely be much easier to handle in the case of adaptation policies.

In assessing the benefits of greenhouse abatement action, not only does one face complications with regard to damage assessment, but it is very hard to compare the benefits of mitigation and adaptation policies, simply because the free-rider problem is clearly a more serious issue with mitigation policies than with adaptation policies (see Box 5.3). Nevertheless, some general remarks about the trade-off between mitigation and adaptation strategies are in order.

BOX 5.3 The Free-Rider Problem

The free-rider problem can be associated with public goods. A public good has two properties: if an additional individual uses the good, the overall cost of using it will not increase (in other words, users are not each other's rivals: nonrivalry); it is virtually impossible to prevent individuals from enjoying the good (nonexcludability). The earth's atmosphere clearly has both properties, and global greenhouse strategies can therefore be considered a public good. Because most, if not all, countries may benefit from such strategies, they are considered an international public good.

If human behavior alters the characteristics of the atmosphere with possible adverse global implications, one could argue that the atmosphere is (and has been) collectively abused. However, precisely because greenhouse strategies are a public good, it is hard to organize collective action if this will involve costs for individual countries and if there is no international system to force countries to comply with international norms. The reason is that countries can use the good – that is, take advantage of the greenhouse strategies of other countries – without an increase in the overall costs of such strategies. In addition, it is virtually impossible to prevent individual countries from enjoying the benefits of greenhouse strategies carried out by others. Under these circumstances,

BOX 5.3   The Free-Rider Problem *(continued)*

individual countries may be tempted to behave as free riders. This means
that they will take advantage of the strategies of others without committing themselves to similar strategies. It seems logical to assume that because of this kind of behavior, the overall effort in terms of abatement
strategies will be less than optimal. In Chapter 7 we will discuss the free-
rider problem further.

5.6.3   The Trade-off between Mitigation and Adaptation

In deciding on the perfect blend of adaptation and mitigation strategies, one
should theoretically compare the costs and benefits of both strategies. In
other words, if adaptation to climate change (e.g., through the adjustment of
agricultural crops and cultivation patterns or through coastal protection or
improved weather forecasts) could be carried out at almost negligible cost,
adaptation might be more cost-effective, at least in the short term, than the alternative mitigation policy. If the same applied to other sectors, one might
well conclude that, from an overall cost–benefit perspective, the balance between adaptation and mitigation should be shifted toward adaptation for the
time being. Conversely, if adaptation costs prove to rise steeply if applied on a
large scale and damage due to global warming is expected to rise exponentially beyond certain atmospheric CO_2 concentrations, the decision of allocating resources to adaptation or mitigation policies could easily favor mitigation. It is fair to say that the overall information about the costs and benefits of
adaptation on the one hand and mitigation on the other hand is as yet insufficient to assess the ideal blend. Still, one should recognize the potential
economic trade-off between these two strategies.

A final complication in assessing the benefits of damage avoided through
greenhouse mitigation measures is that many greenhouse abatement actions
may be the result of policies that are not aimed at protecting the climate; that
is, positive mitigation effects may be an intended or unintended side effect.
For example, if a car producer streamlines his prototype because this meets
the demands of his customers, the reduction in gasoline used per kilometer
may simply be a by-product that was not aimed for initially. Considering the
producer's investment to improve his prototype, what share is to be considered an investment in the mitigation policy? It will be clear from this example
that it is not easy to answer this question. Yet an answer is required in order to
assess the efficiency of this option in terms of GHG abatement per monetary
unit invested for this purpose (e.g., with a view to subsidies or credits).

In short, considering the remarks made in this chapter, we may conclude
that the theoretical model – which can be used as a basis for comparing the

costs and benefits of certain greenhouse policies in order to determine the economically optimal (or most efficient) level of greenhouse abatement action – is a good starting point and framework for analysis. However, when it comes to describing the precise benefits of particular policies, we are faced with many complications that cannot be solved overnight. The relation between adaptation policies and benefits or possible damage avoided is already extremely complex, as shown in this chapter. The issue is complicated even further when mitigation policies, or policies with a mitigating effect, are considered, let alone when one is trying to identify the trade-off between adaptation and mitigation policies. Having established this, we will focus in the next chapter on the costs of greenhouse policies.

5.7 SECONDARY BENEFITS

In assessing the benefits of GHG abatement, one should not make the mistake of focusing on the greenhouse damage avoided alone. This is because most GHG abatement strategies also have net positive secondary benefits – for instance, a decrease in local pollution or the preservation of biodiversity. Such secondary benefits not only can be significant, but quite often are felt immediately or at least much sooner than the avoidance of damage associated with the greenhouse effect itself. Some results from a number of studies all roughly aiming at stabilizing CO_2 emissions at about 1990 levels can be found in Table 5.6. The results clearly show how widely the secondary benefits vary across countries. Nevertheless, it is clear that in some cases secondary benefits can offset a significant portion of the initial abatement costs not only in industrialized countries but particularly in countries in transition and in developing countries – according to some studies sometimes more than 100%![11]

It should be mentioned that the term "secondary benefits" is sometimes confusing. It suggests that an abatement strategy's main aim is to reduce carbon emissions, whereas in reality environmental policies are often focused on combating local pollution. From the perspective of policy makers, it is then the positive greenhouse abatement implications of their policies that they will consider to be secondary benefits (if they are taken into account at all). In other words, one should not confuse the role of secondary benefits in determining the optimal greenhouse abatement strategies, on the one hand, and the policy makers' intentions when they are actually deciding on environmental policies, on the other.

---

[11] See, e.g., Amano (1994), who has calculated secondary benefits for several Asian regions. His results, however, may have been biased by the approach taken, namely using Norwegian benefit–abatement ratios derived by Alfsen, Brendemsen, and Glomrod (1992) (air quality benefits, as a percentage of GDP, per percent decline in air pollution).

TABLE 5.6 Reduced Air Emissions Due to CO_2 Abatement (% Reduction from Baseline)

Country	Year	Policy/Scenario	CO_2	SO_x	NO_x	CO	TSP[a]	VOC[b]	Secondary Benefits ($/tC)	Source
Regional studies										
World	2000	World CO_2 emissions stabilized at 1990 level	9	14	10	—	—	—	—	Complainville and Martins (1994)
U.S.			8	13	8	—	—	—	—	
Japan			4	4	3	—	—	—	—	
EU			5	7	4	—	—	—	—	
Other OECD			10	14	11	—	—	—	—	
China			18[c]	19	19	—	—	—	—	
Ex-USSR			17[c]	21	18	—	—	—	—	
India			17[c]	17	16	—	—	—	—	
E. Europe			11[c]	11	11	—	—	—	—	
OECD	2000	CO_2 emissions in OECD stabilized at 1990 level	—	—	—	—	—	—	—	Complainville and Martins (1994)
U.S.			18	28	25	—	—	—	—	
Japan			14	12	15	—	—	—	—	
EU			14	18	15	—	—	—	—	
Other OECD			21	29	32	—	—	—	—	
Europe[d]	2000	EU carbon/energy tax								Alfsen et al. (1993)
Current structure			9.4	7.4[e]	6.2	—	—	—	6.1	
Cost-efficient regime			9.7	9.3[e]	6.4	—	—	—	6.6	

Country studies

	Year									
Norway	2000	Emission stabilization (at 1989 level)	15.0	20.8[e]	10.8	24.1	4.3	—	40–140[f]	Alfsen et al. (1992)
UK	2005	EU carbon/energy tax	12.1	38.3[e]	10.6	9.6	30.3	1.1	40–1,040	Barker (1993)
U.S.	2000	Emission stabilization								
		Through carbon tax	8.6	1.9	6.6	1.5	1.0/1.8[g]	1.4	2.0–20	Scheraga and Leary (1994)
		Through Btu tax	8.6	2.2	6.6	3.4	1.6/2.2[g]	2.7	3.5–28	

[a] Total suspended particles.

[b] Volatile organic compounds.

[c] The study uses the hypothetical scenario of a global carbon tax. Note that the UN Framework Convention on Climate Change does not contain any obligations for developing countries to reduce their CO_2 emissions. Economies in transition are granted a "certain degree of flexibility."

[d] Western and Eastern Europe (UN ECE region). Tax in six EU countries (France, Germany, United Kingdom, Italy, Netherlands, Denmark) and three Nordic countries (Norway, Finland, Sweden) only.

[e] SO_2.

[f] Including road traffic benefits (reduced congestion, noise, accident, and road damage costs).

[g] PM_{10}/TSP (PM refers to fine particles less than 10^{-6} m in diameter).

Source: Pearce et al. (1996).

5.8 SUMMARY

Greenhouse abatement strategies provide direct benefits in the form of damage avoided, as well as secondary benefits. Policy makers often consider the secondary benefits to be more important because of their local and mostly short-term positive impact. From the perspective of greenhouse abatement, however, these effects are of lesser importance, because the primary aim of abatement strategies is to reduce the greenhouse effect.

Various complications arise in the valuation of damage avoided. This is particularly so when the types of damage are nonmarket and are related to nontangible assets, such as health, biodiversity, or statistical human lives. A further difficulty concerns discounting when the avoidance of damage will take place in the (distant) future and, in addition, there is uncertainty. Moreover, even if one agreed upon a procedure for valuing the nonmarket aspects of (future) damage, one would still face the issue of aggregating nontangible damage over different countries. The clearest case in point involves comparing the value of statistical human lives – if valuing them is acceptable at all – between industrialized and developing countries.

This raises a more general point, namely that equity issues play a role in determining the scope of abatement benefits, so that economists' main goal – determining the optimal abatement strategy on the basis of an efficiency criterion (e.g., marginal abatement benefits equal to marginal abatement costs) – cannot be met without taking equity considerations into account.

One strategy for dealing with the greenhouse issue is adaptation. In fact, human beings continuously adapt themselves to changing circumstances. If, for example, climate change results in a rise in sea level, individuals can adapt by investing in coastal protection. Information about the various forms of adaptation and their costs is now rapidly increasing.

Governments may face a dilemma as to what extent they would focus on mitigation or adaptation strategies. Mitigation strategies basically try to deal with the causes of climate change; adaptation strategies deal with the consequences only. The dilemma is now twofold. First, the impact of mitigation will be felt only in the long term, at least if the secondary benefits, which usually have a more immediate effect, are disregarded. The impact of adaptation is generally more immediate, or at least is felt in the relatively short term. Second, the impact of mitigation will be global and may therefore accrue mainly to parts of the world other than those where the mitigation action is implemented. In contrast, the effects of adaptation will commonly directly benefit those who have invested in the adaptation activities. Both aspects of the dilemma lead to a free-rider problem associated with mitigation strategies. Unless this problem is solved, it may well create a serious bias in favor of adaptation.

REFERENCES

Adams, R. M., R. A. Fleming, C. C. Chang, B. A. McCarl, and C. Rosenzweig. 1993. *A reassessment of the economic effects of global climate change on U.S. agriculture.* Report prepared for the U.S. Environmental Protection Agency, Washington, DC, September.

ADB (Asian Development Bank). 1994. *Climate change in Asia: Bangladesh.* Country Studies, vol. 8. Manila: Asian Development Bank.

Alfsen, K. H., H. Birkelund, and M. Aaserud. 1993. *Secondary benefits of the EC carbon/energy tax.* Research Department Discussion Paper no. 104. Oslo: Norwegian Central Bureau of Statistics.

Alfsen, K. H., A. Brendemoen, and S. Glomrod. 1992. *Benefits of climate policies: Some tentative calculations.* Discussion Paper no. 69. Oslo: Norwegian Central Bureau of Statistics.

Amano, A. 1994. *Estimating secondary benefits of limiting CO_2 emissions in the Asian region.* Mimeo, Kobe University, School of Business Administration.

Asaduzzaman, M. 1994. *Economic and social impacts of climate change: A case study of Bangladesh coastal zone.* Mimeo, Bangladesh Institute of Development Studies, Dhaka.

Asthana, V. 1993. *Report on impact of sea level rise on the islands and coasts of India.* Report to the Ministry of Environment and Forests, Government of India, New Delhi.

Ayres, R. U., and J. Walter. 1991. The greenhouse effect: Damages, costs and abatement. *Environmental and Resource Economics,* **1,** 237–70.

Barker, T. 1993. *Secondary benefits of greenhouse gas abatement: The effects of a UK carbon/energy tax on air pollution.* Energy Environment Economic Modelling Discussion Paper no. 4. Cambridge University, Department of Applied Economics.

Bruce, J. P., H. Lee, and E. F. Haites (eds.). 1996. Summary for policymakers. *Climate change 1995: Economic and social dimensions of climate change.* Contribution of Working Group III to the Second Assessment Report of the Intergovernmental Panel on Climate Change. Cambridge University Press, pp. 1–16.

Cline, W. R. 1992a. *The economics of global warming.* Washington, DC: Institute for International Economics.

Cline, W. R. 1992b. *Global warming: The benefits of emission abatement.* Paris: OECD.

Complainville, C., and J. O. Martins. 1994. *NO_x and SO_x emissions and carbon abatement.* Economic Department Working Paper no. 151. Paris: OECD.

Correia, F. N., et al. 1994. The planning of flood alleviation measures: Interface with the public. In E. C. Penning-Rowsell and M. H. Fordham (eds.), *Floods across Europe: Flood hazard assessment, modelling and management.* London: Middlesex University Press, pp. 167–93.

CRU/ERL (Climate Research Unit / Environmental Resources Limited). 1992. *Development of a framework for the evaluation of policy options to deal with the greenhouse effect: Economic evaluation of impacts and adaptive measures in the European Community.* Report for the Commission of European Communities, Climate Research Unit, University of East Anglia, and Environmental Resources Limited, London.

Dlugolecki, A., et al. 1994. *The impact of changing weather patterns on property insurance.* London: Chartered Insurance Institute.

Easterling, W. E., III, P. R. Crosson, N. J. Rosenberg, M. S. McKenney, L. A. Katz, and K. M. Lemon. 1993. Agriculture impacts of and responses to climate change in the Missouri–Iowa–Nebraska–Kansas (MINK) region. *Climatic Change,* **24**(1/2), 23–62.

El-Raey, M. 1990. Responses to the impacts of greenhouse-induced sea level rise on the northern coastal regions of Egypt. In J. G. Titus (ed.), *Changing climate and the coast,* vol. 2. Washington, DC: U.S. Environmental Protection Agency.

Fankhauser, S. 1994. *Global warming damage costs: Some monetary estimates,* rev. ed. London: Centre for Social and Economic Research on the Global Environment.

Fankhauser, S. 1995. *Valuing climate change: The economics of the greenhouse effect.* London: Earthscan.

Fisher, A. C., and W. M. Hahnemann. 1993. Assessing climate change risks: Valuation of effects. In J. Darmstadter and M. Toman (eds.), *Assessing surprises and nonlinearities in greenhouse warming.* Washington, DC: Resources for the Future.

Gleick, P. H., and E. P. Maurer. 1990. *Assessing the costs of adapting to sea level rise: A case study of San Francisco Bay.* Mimeo, Pacific Institute for Studies in Development, Environment and Security, Berkeley, CA, and Stockholm Environment Institute, Stockholm.

Green, C. H., A. van der Veen, E. Wierstra, and E. C. Penning-Rowsell. 1994. Vulnerability refined: Analysing full flood impacts. In E. C. Penning-Rowsell and M. H. Fordham (eds.), *Floods across Europe: Flood hazard assessment, modelling and management.* London: Middlesex University Press, pp. 32–68.

Houghton, J. T., et al. (eds.). 1995. *Climate change 1994: Radiative forcing of climate change and an evaluation of the IS92 emission scenarios.* Cambridge University Press.

Houghton, J. T., et al. (eds.). 1996. *Climate change 1995: The science of climate change.* Contribution of Working Group I to the Second Assessment Report of the Intergovernmental Panel on Climate Change. Cambridge University Press.

Howe, C. H., H. C. Cochrane, J. E. Bunin, and R. W. King. 1991. *Natural hazard damage handbook.* Boulder: University of Colorado Press.

IPCC (Intergovernmental Panel on Climate Change). 1992. Global climate change and the rising challenge of the sea. Coastal Zone Management Subgroup of the IPCC Response Strategies Working Group, WMO and UMEP, Geneva.

IPCC (Intergovernmental Panel on Climate Change). 1994. Preparing to meet the coastal challenges of the 21st century. Report presented at the World Coast Conference 1993, National Institute for Coastal and Marine Management, The Hague.

Kane, S., J. Reilly, and J. Tobey. 1992. An empirical study of the economic effects of climate change on world agriculture. *Climatic Change,* **21,** 17–35.

Klaus, J., W. Pflügner, R. Schmidtke, H. Wind, and C. H. Green. 1994. Models for flood hazard assessment and management. In E. C. Penning-Rowsell and M. H. Fordham (eds.), *Floods across Europe: Flood hazard assessment, modelling and management.* London: Middlesex University Press, pp. 69–106.

Kwadijk, J. C. J. 1991. Sensitivity of the river Rhine discharge to environmental change: A first tentative assessment. *Earth Surface Process and Land Reforms,* **16,** 627–37.

Mendelsohn, R., W. Nordhaus, and D. Shaw. 1993. The impact of climate on agriculture: A Ricardian approach. In Y. Kaya, N. Nakicenovic, W. D. Nordhaus, and F. Toth (eds.), *Costs, impacts and benefits of CO_2 mitigation.* IIASA Collaborative Paper Series, CP93-2, Laxenburg, Austria.

Munich Re. 1993. *Winter storms in Europe: Analysis of 1990 losses and future loss potential.* Munich: Munich Re.

Nicholls, R., K. Dennis, C. Volonte, and S. Leatherman. 1992. Methods and problems in assessing the impacts of accelerated sea level rise. In the proceedings of the workshop "The World at Risk: Natural Hazards and Climate Change," American Institute for Physics, New York.

Nishioka, S., H. Harasawa, H. Hashimoto, T. Ookita, K. Masuda, and T. Morita. 1993. *The potential effects of climate change in Japan.* Tsukuba: Centre for Global Environmental Research and National Institute for Environmental Studies.

Nordhaus, W. D. 1991a. Economic approaches to greenhouse warming. In R. Dornbusch and J. M. Poterba (eds.), *Global warming: Economic policy responses*. Cambridge, MA: MIT Press, pp. 33–69.

Nordhaus, W. D. 1991b. To slow or not to slow: The economics of the greenhouse effect. *Energy Journal*, **101**(July), 920–37.

Nordhaus, W. D. 1993. *Managing the global commons: The economics of climate change*. Cambridge, MA: MIT Press.

Parry, M. L., A. L. Chong, M. Blanton de Rozari, and S. Panich (eds.). 1991. *The potential socio-economic effects of climate change in South-East Asia*. Thailand: UNEP Regional Office for Asia and the Pacific.

Pearce, D. W., et al. 1996. The social costs of climate change: Greenhouse damage and the benefits of control. In J. P. Bruce, H. Lee, and E. F. Haites (eds.), *Climate change 1995: Economic and social dimensions of climate change*. Contribution of Working Group III to the Second Assessment Report of the Intergovernmental Panel on Climate Change. Cambridge University Press, pp. 179–224.

Penning-Rowsell, E. C., and M. H. Fordham (eds.). 1994. *Floods across Europe: Flood hazard assessment, modelling and management*. London: Middlesex University Press.

Reilly, J., N. Hohmann, and S. Kane. 1994. Climate change and agricultural trade: Who benefits, who loses? *Global Environmental Change*, **4**(1), 24–36.

Rÿsberman, F. 1991. Potential costs of adapting to sea level rise in OECD countries. In *Responding to climate change: Selected economic issues*. Paris: OECD.

Rosenzweig, C., and D. Hillel. 1993. Agriculture in a greenhouse world. *National Geographic Research and Exploration*, **9**(2), 208–21.

Rosenzweig, C., and M. L. Parry. 1994. Potential impact of climate change on world food supply. *Nature*, **367**, 133–8.

Scheraga, J. D., and N. A. Leary. 1994. Cost and side benefits of using energy taxes to mitigate global climate change. In *Proceedings of the 86th Annual Conference*, National Tax Association, Washington, DC.

Smith, J. B., and D. Tirpak (eds.). 1989. *The potential effects of global climate change on the United States*. Washington, DC: U.S. Environmental Protection Agency.

Suliman, M. 1990. Introduction: Africa in the IPCC report. In M. Suliman (ed.), *Greenhouse effect and its impact on Africa*. London: Institute for African Alternatives.

Titus, J. G. 1992. The cost of climate change to the United States. In S. K. Majumdar, L. S. Kalkstein, B. Yarnal, E. W. Miller, and L. M. Rosenfeld (eds.), *Global climate change: Implications, challenges and mitigation measures*. Easton: Pennsylvania Academy of Science.

Tol, R. S. J. 1993. *The climate fund: Survey of literature on costs and benefits*. Amsterdam: Free University, Institute for Environmental Studies.

Tol, R. S. J. 1994. The damage costs of climate change: A note on tangibles and intangibles applied to DICE. *Energy Policy*, **22**(5), 436–8.

Tol, R. S. J. 1995. The damage costs of climate change: Towards more comprehensive calculations. *Environment and Resource Economics*, **5**, 353–74.

Watson, R. T., M. C. Zinyowera, and R. H. Moss (eds.). 1996. *Climate change 1995: Impacts, adaptations and mitigation of climate change – Scientific-technical analyses*. Contribution of Working Group II to the Second Assessment Report of the Intergovernmental Panel on Climate Change. Cambridge University Press.

6

ASSESSMENT OF TECHNOLOGY OPTIONS

In this chapter we describe various technical response options – mainly mitigatory – that have been developed and that may help to solve the climate change problem. Several options are discussed from the perspective of our knowledge (or lack of knowledge) of their costs, feasibility, applicability, and potential. Such information is required to make meaningful decisions about future investments in different technologies. Only by weighing the options against each other can one make sensible decisions about the ideal mixture of options and their applications. We will show that despite the considerable technological progress made in designing and improving response options, a great deal of uncertainty still exists about the economic viability and political acceptance of these options in different parts of the world, as well as about the technological potential of options that can be implemented in the near future.

The response options can be classified in many ways – for example, on the basis of technology, sector, impact, and strategic approach. In Section 6.1, we set up a framework for putting the options into perspective. We distinguish between mitigation, adaptation, and indirect policy options. In Section 6.2, we discuss the criteria for assessing the options and the degree to which these criteria can lead to different choices in terms of the optimal use of the options. In Section 6.3, we evaluate the mitigation options in terms of technical and practical applicability, cost effectiveness, and social acceptability, special attention being paid to developing countries and countries in transition.[1]

[1] The (macro)economic features of the options will not be assessed here.

6.1 A CONCEPTUAL FRAMEWORK

Figure 1.1 shows the policy options available for countering greenhouse warming and their possible feedbacks.[2] The diagram distinguishes between the following options:

1. Mitigation options (submodule E1) reduce the net emissions of greenhouse gases (GHGs) into the atmosphere, either by reducing GHG emissions (source-oriented measures) or by increasing the sinks for GHGs (effect-oriented measures).
2. Adaptation options (submodule E2) focus on reducing the expected damage due to rapid climate change by combating or averting their detrimental effects.
3. Indirect policy options (submodule E3) are not directly related to the emission or capture of GHGs but can have a considerable indirect effect on GHG emissions or uptake.

6.1.1 Mitigation Options

In the literature on greenhouse policy options, mitigation options receive by far the most attention. Most commonly, the options are discussed individually and from an engineering perspective. Information about their cost effectiveness – for example, in terms of dollars per ton of carbon not released into the atmosphere – is rapidly increasing. The marginal cost effectiveness of these options is probably highly dependent on the scale of application, the sector, the country or region of application, and whether or not additional options are applied. Moreover, learning curves, and therefore cumulative application and time, almost invariably play a dominant role in determining the options' economic viability. All these factors point in the same direction – namely, the cost functions of mitigation options may change over time, sometimes quite rapidly. The same applies to the social and political acceptability of these options. Conclusions about their economic, social, and political viability are therefore highly scale-, time-, and location-specific. Chapter 7 will focus on these aspects.

In discussing the potential of mitigation options, a distinction has been made between measures concerning CO_2 and measures concerning other GHGs, because the former are in actual practice largely associated with en-

---

[2] This chapter deals only with the broad principles underlying the response options. Their actual application would depend on a host of factors that are highly country-specific and include many economic, social, political, and legal considerations (see Chapter 7).

ergy-related activities (i.e., both energy production and consumption), whereas the latter are also associated with other types of activities.

Thus, except for some "exotic," mainly effect-oriented options such as geo-engineering, orbital shades, iron fertilization, creating algal blooms, and weathering rocks, mitigation options can generally be divided into those affecting CO_2 and those affecting other GHGs. Measures concerning CO_2 include the following:[3]

Source-oriented measures:
1. energy conservation and efficiency improvement
2. fossil fuel switching
3. renewable energy
4. nuclear energy

Sink-enhancement measures:
5. capture and disposal of CO_2
6. enhancement of forest sinks

Measures concerning other GHGs include the phasing out of hydrofluoro-carbons (in addition to hydrochlorofluorocarbons, as per the Montreal Proto-col), as well as a variety of measures for reducing emissions of methane (CH_4), nitrous oxide (N_2O), and other GHGs.

Since the energy sector (in terms of both energy production and consumption) is the single largest source of carbon, much of the CO_2 mitigation efforts

---

[3] It may seem that, although conceptually these categories are distinct, in real life they are not mutually exclusive; that is, some measures can be classified into more than one category. An example would be the planting of forests or biomass to be transformed for energy purposes. Such a measure would seem to fall in both categories 3, renewable energy, and 6, enhancement of forests sinks. Strictly speaking, this cannot be the case, because the enhancement of carbon sinks means a net annual uptake of carbon effected by the measure itself, whereas the main greenhouse effect of measures in the renewable category is the replacement of fossil fuels. In a particular project, however, both aspects can be combined, so that it becomes hard to draw the line betweem them.

In fact, in the case of forests, there are roughly three types of measures for sequestering carbon: (a) the afforestation of new land to let the forest simply mature; (b) the planting of forest and sequestering of the timber derived from it; and (c) as in (b), but in addition use of the wood for energy purposes on a sustainable basis, thus avoiding the alternative use of fossil fuels. Measures (a) and (b) are discussed in Section 4.6 on forestry options, whereas (c) belongs to the renewable/biomass category. With respect to the GHG implications of (c), sometimes a significant amount of additional energy may be required to turn biomass into energy. This is the case, for instance, in the production of ethanol from corn, where additional energy requirements amount to the energy content of the produced ethanol itself (Swisher, Wilson, and Schrattenholzer 1993, p. 442)!

Another example of multiple categorization would be the classification of an integrated gasification combined cycle (IGCC) or the hydrocarb process into categories 1, energy conservation and efficiency improvement, and 2, fossil fuel switching. Ultimately no clear distinction can be made between the two categories.

are concentrated there. Each of the four source-oriented options addresses elements of the energy conversion process, from primary energy production to end-use services.

Both energy conservation and energy efficiency (option 1) aim to reduce total energy use without changing the current fuel mix or the fundamental structure of the energy conversion process. The term "energy conservation" is used here to mean a reduction in energy needs resulting from a change in the nature or level of energy services (e.g., lighting areas only when they are occupied rather than during specified periods). "Energy efficiency" means providing the same type and level of energy service with less total energy (e.g., using more efficient lamps to provide the desired lighting level). Since energy conservation is strongly linked to the preferences and behavior of various economic agents (such as households, firms, and governments), policies aimed at achieving it are more likely than other policies to lead to ambiguous conclusions.[4] Consequently, most studies focus on energy efficiency.

A fossil fuel switch (option 2) is intended to alter the mix of fossil fuels in favor of less carbon-intensive ones, such as natural gas (and perhaps oil), and away from coal. Nuclear energy (option 4) substitutes for fossil fuels as primary energy.

Renewable energy (option 3) is characterized by an extensive natural supply, which is vast compared with current levels of commercial energy use, and by a large long-term potential because of its regeneration capability. Mobilization of this natural supply can in some cases result in severe environmental and societal consequences.

Removal technologies (option 5) extract carbon in one form or another from an energy conversion process before it enters the atmosphere. Subsequently, the carbon has to be utilized, stored, or disposed of.

Option 6 is in essence outside the energy area. It aims at binding carbon after it is combusted and dispersed throughout the atmosphere by combating deforestation or by re- and afforestation. It may also refer to activities designed to preserve or enhance carbon uptake by soils.

6.1.2 Adaptation Options

Adaptation options have been dealt with in Chapter 5, and therefore only some general remarks will be made here.

---

[4] For an example involving energy conservation, see Rubin et al. (1992). Here, 25% of employer-provided parking places and parking places with meters are eliminated to reduce solo commuting in the United States by 15 to 20%. Net costs are estimated to be $22/tC (i.e., a savings).

Adaptation options have two purposes: to reduce the damage from climate change and to increase the resilience of societies and ecosystems in response to the effects of climate change that cannot be avoided. Clearly, adaptation measures are linked with mitigation measures, although they cannot be easily compared (see Chapter 5). Generally, the more one succeeds in limiting climate change, the easier it will be to adapt to it, notwithstanding the fact that there can be reasons for supporting adaptation measures in their own right.

Three types of adaptation measures are commonly distinguished: those involving protection, retreat, and accommodation. As far as their costs are concerned, one can focus either on the opportunity cost – in other words, one can assess the welfare implications of no-action scenarios – or on the net investment cost.

6.1.3 Indirect Policy Options

Potential climate change is perceived as a problem, mainly because such change would interfere (and would continue to do so in the future) with the world's economic, social, and ecological systems, and eventually with its political system. Just as the precise scope and risks of climate change are uncertain, so are the development of technology, resources, and the organization and structure of economic, social, and political systems (see Section 7.3). However, changes in the global climate and structural changes in economic, social, and political systems differ significantly in at least one respect: the speed or time lag of changes to be expected. Whereas severe global climate change might take approximately 50 to 100 years, economic, social, and political systems may change several times within a similar period.

This difference poses a fundamental problem (already touched on in Chapter 2 and to be discussed in more detail in Chapter 7) in assessing response options to climate change: the slowly changing climate system has to be superimposed on rapidly changing economic, social, and political systems that are in constant flux due to numerous factors, (potential) climate change being only one of them. This significantly complicates the assessment process, and even more the process of formulating policy options based on such assessment.[5] Indeed, climate and ecological changes are by no means the only factors that may pose serious problems to society. The pursuit of sustainable development (see Chapter 2), other evolutionary trends, and structural adjustment processes – driven by such forces as population growth, urbanization, information technologies and their dissemination, the international mobility of labor and capital, competition for natural resources, and the pattern

---

[5] However, recent history has shown that if there is a strong political consensus about the need to take action, such action can be undertaken vigorously (e.g., the Montreal Protocol of 1987; the Basel Convention of 1989; and the Framework Convention on Climate Change of 1994).

and speed of technological progress (e.g., in waste management and in re-designing products) – may also play an important and unpredictable role in shaping the economic, political, and social systems of tomorrow, especially if the policy makers' time horizon is at most a few decades, if not much shorter. To illustrate, Western nations may well face a combination of problems, such as urban decay, unemployment, massive migration, and changing patterns of economic competitiveness, that may easily attract more public and political attention than the climate change issue. Similarly, developing countries will continue to be preoccupied with poverty, malnutrition, and other existing problems. However, all of the policies dealing with these problems will most likely have some implications for GHG emissions. (See Box 3.6 for a case study of the impact of economy-wide policies on GHG emissions.)

In short, all the above-mentioned problems already call for response options – for instance, in the spheres of consumption and lifestyle, population and migration, technology and environmental, structural and sectoral adjustments or trade, redistribution, and many other forms of macroeconomic policy. Virtually all of these policies will also greatly affect, albeit indirectly, energy use and thus the global climate. An effective climate change response strategy should therefore address the possibilities of joining climate change response options with responses to other sustainable development issues, and thus increase the probability of actual implementation. Small-scale examples of the potential of this approach in developing countries can be found in applications of the integrated systems approach, as illustrated in Box 6.1.

BOX 6.1 Examples of the Integrated Systems Approach

In many developing countries, crop agriculture is highly dependent on energy use, both directly and indirectly, and farmers rely on outside sources for much of their energy supply. In addition, many of these agricultural systems are based on monocultures (e.g., high-yielding varieties of wheat and rice, which increase soil exhaustion and are vulnerable to massive infestations of pests and disease). Alternatives, like low-external-input sustainable agriculture, reportedly lower the need for external and energy-intensive inputs and increase productivity in an ecologically robust way, while at the same time reducing concern for national food security (Reijntjes, Haverkort, and Waters-Bayer 1992).

Yet another example of a "multifunctional system" is wave energy. In this case, the production of energy is combined with other functions, such as coastal protection or water desalination. However, this technology may also have adverse environmental side effects. All of these systems are particularly promising if applied on a relatively small scale in developing countries.

6.2   CRITERIA FOR ASSESSMENT

In assessing response options, the possibilities of applying the options themselves are evaluated rather than the overall policies in which they might be integrated. For the assessment of these policies, see Chapter 8.

From a methodological point of view, one can distinguish between two fundamentally different approaches to the assessment of response options.[6] On the one hand, the financial costs of individual technologies can be expressed in terms of the amount of dollars to be invested in order to achieve one unit of CO_2 emission reduction or absorption. This could be called the *engineering efficiency* approach. On the other hand, options can be assessed in the tradition of welfare economics. According to this line of thinking, determining the costs and benefits of applying any particular technology involves an assessment of the opportunities forgone by the allocation of the resources. This could be called the *welfare economic* approach.

An example involving afforestation will illustrate the differences between these approaches. What investment has to be made to achieve a predetermined target of net CO_2 absorption during some time interval? Using the engineering efficiency approach, one would try to determine the discounted value of the costs of land acquisition, tree planting, maintenance, security, and other needs. Any future (sustainable) harvesting returns would be equally discounted, so that the *net* costs could be determined in dollars. On the basis of this information, and by comparing the cost efficiency of this option with that of other options, one could decide whether to proceed.

However, if the welfare economic approach is taken, the assessment may be quite different. If the land is used for afforestation purposes, the possibility of using the same land for agricultural purposes is forgone. It therefore matters a great deal whether the area has agricultural potential. If so, the local population may well be forced to migrate or else to suffer income losses. Moreover, the afforestation program, if applied on a large scale, may have additional effects, either positive or negative (e.g., through its impact on local climate and soil fertility, social and cultural life, infrastructure, and tourism).[7] In the assessment, attention should also be paid to the distortion by government measures, such as subsidies and taxes, of the efficiency of the forestry option. If all the direct and indirect welfare consequences of the envisaged afforestation program are to be assessed, an extensive and complicated social cost–benefit type of analysis may well be called for, because not all aspects can be quan-

---

[6] This distinction should not be confused with the distinction between top-down and bottom-up modeling, which is being paid a great deal of attention (see Box 7.2, Chapter 7).

[7] Effective monitoring and extension services established at a local level may actually increase the chances that these effects will be beneficial.

tified or monetized (see Chapter 4). A priori, there is no reason why the outcomes of the engineering efficiency and welfare economic assessments of the same project would coincide. The costs of the land in monetary terms – input in the engineering efficiency approach – may not fully reflect the land-use opportunity costs in welfare terms – the corresponding input in the welfare economic approach – because in the former no full account is taken of indirect effects, nonmaterial consequences, distributional impact, or externalities. In short, the major distinction between the cost assessment methodology in both approaches is that the engineering efficiency approach basically starts with the evaluation of a project from the narrow perspective determined by the project boundaries, whereas the welfare economic approach attempts to account fully for the various interests and consequences inside and outside the project concerned, including the external effects and the social and political acceptability of the option (with regard to the distinction, see Box 6.2). In any case, all of this clearly shows that an assessment based on the engineering efficiency approach alone, but meant to provide an input for policy decision making, may lead to a biased view. A more ideal assessment for policy purposes should recognize different priorities within countries, the impact of externalities, the political acceptability at various levels, and a variety of distributional aspects.

The general impression also arises that optimism about the potential of technology is greater in the engineering efficiency approach than in the welfare economic approach; in the latter the emphasis is more on the obstacles in society to absorbing and applying new technologies than on the technological potential alone (which is the main object of the former approach).

BOX 6.2 The Engineering Efficiency Approach versus the Welfare Economic Approach to Assessing Response Options: A Comparison on the Basis of Economies of Scale

The distinction between the engineering efficiency and welfare economic approaches can be related to various aspects of the economies-of-scale concept, notably:

1. Average costs may decrease at a larger scale of application (internal economies of scale).
2. Costs of a given option may decrease when other options are applied on a larger scale because of positive external effects (external economies of scale).
3. Costs may decrease as the application time progresses (learning effects).

(continued)

BOX 6.2   The Engineering Efficiency Approach versus the Welfare Economic Approach to Assessing Response Options: A Comparison on the Basis of Economies of Scale *(continued)*

4. Costs may increase at a larger scale of application due to increasing resistance and bottlenecks related to social, political, and environmental concerns and to increasing opportunity costs; afforestation projects often provide a clear example.
5. Costs may increase because achieving the required rate of diffusion of technologies, public education, and lifestyle changes may become increasingly difficult on a larger scale; this problem may be particularly relevant if response technologies require a high level of technical expertise.

Focusing on items 1 through 3 leads to optimism about the options' economic potential. This is the perspective taken by the engineering efficiency approach. However, focusing on items 4 and 5 might easily lead to a much more pessimistic view, associated with the welfare economic perspective.

In addition to comparing the engineering efficiency approach and the welfare economic approach to assessing response options, a distinction could be made between the gross and the net abatement achieved through response options. This distinction is based on the question of what the energy costs and benefits of options are. This is further explained in Box 6.3.

BOX 6.3   *Gross versus Net Abatement Costs*

In comparing the feasibility of the various response options, studies differ in the extent to which they take the energy costs and benefits of the options into account (i.e., focus on the gross versus net abatement). This can easily cause some confusion about the relative attractiveness of the options. To illustrate, the application of some options, such as capture and disposal, requires significant energy input, which are often denoted as energy penalties; therefore, gross abatement will be much larger than net abatement. Other options, such as nuclear or renewable energy, may require a much smaller amount of energy input per ton of carbon abated; then the difference between gross and net abatement will be much smaller. Focusing on the gross rather than on the net impact in terms of GHG abatement therefore implies a bias in favor of energy-intensive response options.

Figure B6.1, illustrating cost functions for the European Union, shows that whether one deals with gross or net abatement costs can make quite a difference. Differences between gross and net costs turned out to be notably relevant to the options of energy saving, renewables, nuclear energy, and energy farming.

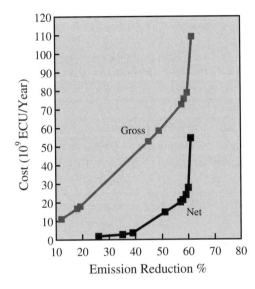

FIGURE B6.1. Options for CO_2 emission reduction in the European Union, net and gross costs, and effectiveness. From TNO (1992).

Notwithstanding the current progress being made in gathering information on the feasibility of different greenhouse response options and in greenhouse-related modeling, there are fewer studies for economies in transition or developing-country economies than there are for industrialized economies. In view of the structural changes that are under way in these regions, it is imperative to improve our understanding of the potential for reducing or absorbing GHG emissions in these economies, with particular regard for their sensitivities to other important considerations such as economic and technological development. Indeed, a welfare economic assessment of climate change response options will encounter several large practical obstacles, particularly in developing countries or countries in transition. First, the policy priorities in those countries, especially with respect to the greenhouse issue, often differ from those in industrialized countries. Second, information about externalities at a local level may not fully reach the public sector because of limitations in data collection, processing, and communication; on the other hand, policies dealing with externalities may fail to reach part of the local population. Third, most developing countries and many countries in transition face a severe lack of institutional and human capacity to deal with these issues (see Box 6.4).

6.3   MITIGATION OPTIONS: ENERGY CONSERVATION AND EFFICIENCY IMPROVEMENT

To put the energy efficiency option into perspective, the Kaya identity (Kaya 1989) may provide a useful starting point:

$$CO_2 = (CO_2/E) \times (E/GDP) \times (GDP/P) \times P$$

where E is energy consumption, GDP is gross domestic product, and P is population. If population growth is given and the future levels of GDP per capita are predetermined, a given CO_2 emission reduction target can be achieved only by a reduction in carbon intensity (CO_2/E) and/or energy intensity (E/GDP). The need to reduce carbon and energy intensities becomes stronger the higher the rate of population growth is and the more ambitious the targets are with respect to GDP increase. This insight obviously reinforces the need to pay specific attention to developing countries.

Historically, carbon and energy intensities in most countries have tended to decline due to ongoing technological change and evolution, as well as structural changes in economies. Energy intensity per unit of value added has been decreasing at a rate of about 1% per year since the 1860s and at about 2% per year in most Western countries in the 1970s and much of the 1980s (2.6% in International Energy Agency [IEA] member countries during 1980–4; Nakicenovic et al. 1993). However, the differences among the countries are enormous, in terms of both levels of energy intensity and its direction over time. Moreover, the carbon and energy intensities in a number of large, rapidly growing developing countries today are much higher than in virtually all presently industrialized countries at a similar stage of technological development (Nakicenovic et al. 1993). Also, in contrast to the postwar trend observed in industrialized countries, some developing countries have not succeeded in reducing energy intensities.

Indeed, within each group, countries vary in terms of the capacity, whether potential or realized, to restrict carbon emissions through energy efficiency. Moreover, within a given country, not all sectors have a similar energy efficiency. During 1973–88, for example, the estimated energy intensity in Japan fell by more than 35% (Ogawa 1992), with the energy intensity of electric refrigerators falling by nearly 67% between 1973 and 1987 and the efficiency of automobiles increasing from around 9.4 to about 13 km/liter (49%). During the same period, the United States, the former West Germany, and France lowered their energy intensities by 27, 22, and 17%, respectively, and IEA member countries by 25% (IEA/OECD 1991). In most cases, changes have been most apparent in the industrial sector. However, low oil prices and economic recession caused a slowdown in energy intensity reduction in the late 1980s and early 1990s.

Over the 1980s, various developing countries managed to lower their industrial energy intensity: China by approximately 30% (Huang 1993), Taiwan (between 1970 and 1985) by some 40% (Li, Shrestha, and Foell 1990), and the Republic of Korea by 44% (Park 1992). However, in other countries, such as Nigeria (Nakicenovic et al. 1993), Egypt (Abdel-Khalek 1988), and Mexico (Guzman, Yunez-Naude, Wionczek 1987), energy intensity actually increased. In addition, Imran and Barnes (1990) have reported energy intensity increases in Brazil (20%), Pakistan (26%), India (25%), and Malaysia (48%) for the period 1970–88.

Reported changes in aggregate energy intensity must be viewed with caution, however, because they depend on how energy use and economic output are measured. In Brazil, for example, official figures show overall energy intensity remaining roughly constant during 1973–88. However, if hydropower is counted on the basis of its direct energy content and GDP is corrected to reflect purchasing power parity with the dollar, then overall energy intensity declined 21% during 1973–88 (Geller and Zylbersztajn 1991).

Carbon intensity, the other variable in the Kaya identity, also shows a declining trend globally. From 1860 to the present, carbon emissions per unit of primary energy consumed have fallen by about 0.3% per year, or from more than 0.8 to somewhat more than 0.5 tC/kWyr (Nakicenovic et al. 1993). Clearly, decarbonization can be achieved by a variety of options, such as fossil fuel switching and the use of nuclear and renewable energy as fossil fuel substitutes.[8] However, projections for developing countries indicate that, without the implementation of serious policies and with changing trends, not only will total emissions increase rapidly, but carbon emissions may increase faster than GDP because demand for energy services is switching from regenerating biofuels to fossil fuels (e.g., for India, see Mongia et al. 1991).

The two factors that underlie reduced energy intensities are improvements in the energy efficiency of individual production processes and structural changes in the economy (in particular, the increasing economic predominance of less energy-intensive sectors, such as many of the service sectors, and the energy efficiency of spatial planning). Only a few studies explicitly incorporate the impact of structural changes. Most focus on energy efficiency measures, which are generally considered to be most relevant. To illustrate, it was estimated that energy efficiency improvements were responsible for about three-quarters of the 26% reduction in U.S. energy intensity during 1973–86

---

[8] If combined, the options can have a considerable impact. For example, a set of three studies for Poland, Hungary, and the former USSR indicate that a combination of energy efficiency improvements, fuel substitution, and structural change (Chandler 1990) could reduce carbon emissions by 40 to 60% from base-case projected levels by 2030: Poland, from 260 Mt in 2030 to 117 Mt (Sitnicki et al. 1990); and the former USSR, −40% (Makarov and Bashmakov 1990).

(Schipper, Howarth, and Geller 1990), while the rest was due to structural shifts in the economy.

Disregarding the impact of structural shifts on energy intensity in an inter-country comparison may easily create a biased view, because the industrialized economies have generally shifted away from the highly energy-intensive secondary toward the less energy-intensive tertiary sectors (a process known as "dematerialization"), whereas developing countries in general are increasingly entering the secondary sector.

One basic reason why the energy efficiency improvement potential is considered substantial is that the ratio of useful energy (i.e., the amount of energy that provides useful services) to overall primary energy (i.e., the amount of energy recovered or gathered directly from natural sources) is estimated at only 34% globally. It is lowest, at 22%, in developing countries and highest, at 42%, in countries in transition (Nakicenovic and Grübler 1993). This ratio, in turn, is the product of two other ratios:

- the ratio of final energy (energy delivered to the point of consumption) to primary energy (with a global average of 74%, a maximum of 80% in developing countries, and a minimum of 69% in countries in transition); and
- the ratio of useful energy to final energy (with an average of 46% globally, 28% in developing countries, 53% in industrialized countries, and 60% in countries in transition)

These numbers suggest that improving energy efficiency is particularly promising with regard to increasing the ratio of useful to final energy. Efficiencies are lowered further if seen from an "exergy" point of view, that is, if the actual services (work) supplied by the energy source are related to the corresponding inputs minimally required: the exergy efficiency of primary inputs in the market economies is only a few percentage points (i.e., of the order of 2.5 to 5%) if energy service is fully taken into account.

Indeed, a back-of-the-envelope calculation shows that if energy efficiencies of the current structure of the OECD technologies were disseminated throughout the world, global primary energy requirements would come down by 17% from 12 to 10 terawatt-years (TWyr) annually. If, instead, the best available technologies instantaneously replaced the current ones, without any altering of the energy system structure, global annual primary energy requirements would decline to 7.2 TWyr (Nakicenovic and Grübler 1993). A similar exercise assuming that Japanese industrial efficiency levels would diffuse globally shows an estimated industrial carbon reduction potential of some 730 MtC worldwide, mainly in the steel, chemical, and cement industries (Matsuo 1991).

Clearly, energy end use is the least efficient part of energy systems, and it is in this area that improvement would bring the greatest benefits. Most studies suggest that a large potential for reducing energy consumption exists in many sectors and regions. A review of 12 studies of long-term energy efficiency potential revealed that in many regions of the world full adoption of cost-effective energy efficiency measures could reduce carbon emissions by 40% or more over the medium to long term, compared with business-as-usual trends (Geller 1994). An example related to the OECD area is IEA/OECD (1991), as shown in Table 6.1. However, at the same time, most studies also acknowledge that institutional, economic, and social barriers may delay or inhibit the achievement of full efficiency potentials in the near future. Moreover, several policy and regulatory reforms have recently begun to address some of these barriers. In the United States, for example, more than 30 states have adopted or experimented with regulatory reforms since 1989 to promote demand-side management (DSM) and to encourage integrated resource planning (IRP).

Other studies focus on energy efficiency improvement potential by analyzing major energy end use (e.g., Blok et al. 1991; U.S. COSEPUP 1991; Congress, OTA 1991; Goldemberg et al. 1988; ESCAP 1991; Gupta and Khanna 1991; Kaya et al. 1991) or focus on specific sectors. To illustrate, recent estimates for the United States show energy-saving potentials of 45% in buildings, 30% in industries, and 30% in cars (Rubin et al. 1992; DeCicco and Ross 1993). In rural areas in developing countries, to give another example, the efficiency of wood and charcoal-fueled cook stoves can be increased from a range of 10–20% to 25–35% using improved stove designs at a capital cost of less than U.S.$10 per stove. Cooking efficiency can be further increased to the 40–65% range by shifting from biomass-based fuels to kerosene, liquefied petroleum gas, or electricity, but at a significantly higher capital cost (U.S. Congress, OTA 1992).

Energy efficiency gains may be particularly promising in the following sectors: power production, transportation, steel and cement production, and residential. However, the relative ranking of sectors in terms of potential for energy efficiency improvement is highly dependent on whether both direct and indirect requirements of energy are taken into account, in other words, whether interindustry demands are included in sectoral comparisons. A comparative study of India (Parikh and Gokarn 1993) shows, for instance, that if direct carbon emission due to fossil fuel use is considered, then electricity generation tops the list of total emissions (one-third of the total). However, if direct and indirect emissions are taken into consideration, the construction sector emerges as the largest carbon-emitting sector in India (22% of the total).

The issue of energy conservation and efficiency in developing countries differs in some respects from that in industrialized countries. First, a substantial portion of the demand for energy is often met by renewable energy sources like biomass. This is likely to remain so in the short to the medium term, and there are estimates to show that the scope for conservation of biomass is enormous in these countries. One reason is that cooking with traditional biomass

TABLE 6.1 Energy Efficiency Potential: Summary of Opportunities and Barriers

	(A) Estimated Share of Total Final Consumption (%)	(B) Estimated Share of Total CO_2 Emissions (%)	(C) Total Energy Savings Possible[a] (%)	(D) Existing Market/Inst. Barriers[b] (%)	(E) Potential Energy Savings Not Likely to Be Achieved[c] (%)
Residential space heating and conditioning	11.4	11	10–50	Some/many	Mixed
Residential water heating	3.4	3.6	Mixed	Some/many	Mixed
Residential refrigeration	1.1	2.1	30–50	Many	10–30
Residential lighting	0.6	1.2	More than 50	Many	30–50
Commercial space heating and conditioning	6.1	6.8	Mixed	Some/many	Mixed
Commercial lighting	1.5	3.4	10–30	Some/many	Mixed
Industrial motors	4.5	9.0	10–30	Few/some	0–10
Steel[d]	4.1	4.6	15–25	Few/some	0–15
Chemicals[d]	8.4	5.9	10–25	Few/some	0–20
Pulp and paper[d]	2.9	1.2	10–30	Few/some	0–10
Cement[d]	0.1	0.9	10–40	Few/some	0–10

Passengers cars	15.2	13.7	30–50	Many	20–30
Goods vehicles	10.1	9.1	20–40	Some	10–20

How to read this table: For example, for residential lighting, more than 50% per unit savings would result if the best available technology were used to replace the average lighting stock in use today over the next 10 to 20 years. Some of these savings would take place under existing market and policy conditions. But due to the many market and institutional barriers, there would remain a 30 to 50% potential for savings that would not be achieved.

[a] Based on a comparison of the average efficiency of existing capital stocks with the efficiency of the best available new technology. This estimate includes the savings likely to be achieved in response to current market forces and government policies as well as those potential savings (indicated in Column E) not likely to be achieved by current efforts.

[b] Extent of existing market and institutional barriers to efficiency investments.

[c] Potential savings (reductions per unit) not likely to be achieved in response to current market forces and government policies (part of total indicated in Column C).

[d] Energy use only.

Source: IEA/OECD (1991).

TABLE 6.2 Regional Potentials for Reducing Industrial Carbon Emissions by Cost Categories (MtC)

	Cost Saving or Moderate Cost	Cost (<100 U.S.$/tC)	Cost (>100 U.S.$/tC)	Sum[a]
Market economies				
Efficiency improvement	15	45	84	144
Structural change/recycling	95	NA	25	120
Fuel substitution	6	NA	NA	>>6
Process technology process	0	2	98	100
	116	47	207	370
Reforming economies				
Efficiency improvement	48	53	NA	>101
Structural change/recycling	165	50	NA	>215
Fuel substitution	10	NA	NA	>>10
Process technology process	0	10	46	56
	223	113	>46	382
Developing countries				
Efficiency improvement	12	41	NA	>53
Structural change/recycling	19	29	NA	>48
Fuel substitution	3	NA	NA	>>3
Process technology process	0	8	56	64
	34	78	>56	168
World				
Efficiency improvement	75	139	84	>298
Structural change/recycling	279	>79	>25	>383
Fuel substitution	19	NA	NA	>>19
Process technology process	0	20	200	220
	372	238	309	920

Note: NA denotes not assessed.

[a]Total reduction potential could be higher because not all measures have been assessed.

Source: Grübler et al. (1993a).

fuels is technically very inefficient, although not necessarily inappropriate from a socioeconomic perspective (U.S. Congress, OTA 1992). Second, energy efficiency in industrial activities in developing countries generally showed little or no improvement (Imran and Barnes 1990). Third, the demand for electricity is growing at a rate that is often hard to keep up with. Some developing countries have allocated a quarter to a third of public investment to the generation of power, and even this is sometimes inadequate to meet the growing demand (World Bank 1993; Munasinghe 1995). However, due to the presently low level of energy efficiency in developing countries and the consequently large scope for improvement, the potentials for energy saving in these countries are considered somewhat similar in magnitude to those in industrialized countries at present, notwithstanding such adverse factors as the fast growth of commercial energy use and the increasing weight of the industrial sectors (Ewing 1985; Levine et al. 1991; U.S. Congress, OTA 1992). Finally, the energy market in developing countries is often distorted by energy pricing policies.

By contrast, energy conservation may be achieved somewhat more easily in industrialized countries, insofar as a trend toward lower material and energy consumption appears to be under way. Various indicators, such as the increasing service orientation of industrial economies, seem to point in this direction. There appears to be ample opportunity for increasing energy conservation in these countries through the introduction of stricter standards with respect to the use of energy and materials and, most of all, through adjustments in lifestyle. This explains why the assessment of energy conservation is often rather qualitative and subjective.

Various studies have been carried out on both the potential for carbon emission reduction via energy efficiency improvement and the net costs involved. An overview of the potential for emission reduction in the industrial sector is presented by Grübler et al. (1993a) in Table 6.2. They argue that a potential reduction of 920 MtC (over 40% of current emissions) could be achieved overall. Of this, 372 MtC could be achieved at net negative or modest positive costs (with about two-thirds of this amount coming from the countries in transition). These estimates disregard the potential for fuel switching and decarbonizing the electricity supply and assume an annuity rate of 10% throughout the lifetime of the investment.

6.4 RESPONSE OPTIONS AND SOME REMARKS ON METHODOLOGY

The preceding survey shows that almost all studies on energy efficiency indicate the usefulness of improving energy efficiency. Moreover, in many cases it is considered to be directly beneficial irrespective of whether or not greenhouse warming will take place; so emission reduction is realized at negative net costs. The question arises as to what extent this impression is correct and/or should be commented on. This section deals with two issues that must be taken into account in the assessment of the energy efficiency potential of any project: the aspect of "no regrets" and, again, the discount factor employed.

6.4.1   Energy Efficiency and the Concept of No Regrets

During the past few years many policy discussions on GHG abatement have increasingly focused on the extent to which improved energy efficiency and conservation can be economically viable in the present while at the same time saving energy and reducing CO_2 emissions. Options that satisfy both requirements are sometimes referred to as *no-regrets*[9] or *win–win* options (see also Box 3.3). However, other, more extensive definitions of no-regrets options are also employed in the literature. Thus, we will describe the fundamentally different concepts of no regrets in the literature. Before doing so, we would like to emphasize the fact that the concept of no regrets is potentially related to all response options and thus could have been discussed elsewhere in this volume. We have chosen to introduce the concept here because much of the discussion on no regrets relates to energy efficiency improvement. What basically distinguishes the various no-regrets concepts is the degree to which externalities are taken into account.

If applied, emission reduction strategies may sometimes generate negative externalities, but will usually generate net positive side effects, through either additional positive effects on the economy or additional environmental benefits. From a welfare economic point of view, when one is assessing emission reduction strategies, these net positive side effects (in economic terms, positive externalities) must be taken into account in order to determine whether this strategy is to be referred to as a no-regrets measure. Thus, on the basis of this welfare economic perspective, all strategies in which the net positive externalities surpass the gross costs of implementation ought to be called no-regrets strategies. These strategies are worth undertaking irrespective of climate-related reasons.

---

[9] An element of confusion is the very term "no regrets." Some have argued that the term might suggest that carrying out only no-regrets options involves some optimal level of abatement policy measures in order to deal with the climate change problem. Obviously, this is not the case. Therefore, some have suggested replacing "no regrets" with the term "worth doing anyway." However, because the former term is generally accepted, we have chosen to use it here, despite the conceptual confusion.

To make matters clearer, the potential positive side effects of abatement strategies can be divided into three categories:

- *The negative cost potential,* that is, the abatement realized by implementing technologies the costs of which per unit of energy are lower than those of the technologies currently in use. For example, replacing a coal-fired electricity-generating plant by a gas boiler plant often results in efficiency gains – in addition to the reduction in GHG emissions – so that the same amount of electricity can be produced not only in a cleaner way but also at a lower price.

- *The economic double dividend.* Due to the abatement measure, both carbon emission is reduced and other economic benefits are achieved. For example, a certain country may introduce carbon taxes in order to fulfill its GHG abatement commitments. This policy could be carried out in such a manner that the payroll (labor) taxes could be lowered – for example, by carefully recycling the carbon tax revenues. If payroll (labor) taxes caused distortions in the labor market (high tax rates might cause firms to hire fewer workers), lowering payroll taxes might have a positive impact on the economy. Thus, additional positive effects on economic growth or employment can be realized. Another example is the overall stimulus to technological development with economic growth as a side effect of (voluntary) agreements between governments and the business community aimed at developing and installing environmentally sound technologies.

- *The environmental double dividend.* This refers to the contribution of GHG abatement policies to the mitigation of nongreenhouse environmental problems, like air pollution, urban congestion, and the degradation of land and natural resources. In other words, this concept refers to what is sometimes called the secondary environmental benefits. (For an example of how secondary benefits can be incorporated into the standard marginal costs and damage curves, see Elkins 1996.) With respect to the example of a coal-fired plant being replaced by a gas boiler plant, a positive side effect could be less local air pollution. Another example would be a situation in which the degradation of tropical forests was halted in some areas as a positive side effect of an afforestation project, even if this policy were implemented primarily for carbon sequestration purposes.

According to the welfare economic point of view, an abatement strategy is a no-regrets strategy if the sum of these three positive side effects outweighs the gross costs; from an overall welfare perspective the net costs are then negative. However, in many current economic models the term "no regrets" is reserved for strategies that account for only the first and/or second category of these positive side effects. There are several reasons for using this more limited definition. First, abatement technologies are often implemented by private agents. Since many of the net positive externalities are irrelevant to the private agent in determining the commercial feasibility of the technologies – simply because externalities accrue to agents other than the private agent itself – she will consider only the direct negative cost potential in order to determine for herself whether the project is of a no-regrets nature.

Second, economic and environmental net positive benefits are often hard to quantify, if they can be quantified at all. For instance, if an abatement strategy has a positive impact on the economic process other than a direct contribution to the alleviation of the greenhouse problem, this externality can usually be quantified only with some degree of precision – for example, if a fairly elaborate economic model connecting the energy sector and other economic sectors is available and operational for such an assessment. Needless to say, this condition is seldom met. Therefore, assessing the *economic* double dividend will usually involve a great deal of guesswork. However, Chapter 7 will discuss in greater detail the outcomes of certain advanced studies that have tried to provide information on the costs of greenhouse abatement technologies by explicitly taking into account the relation between the specific sector in which the technology was applied and the overall economy.

Third, with respect to determining the size of the *environmental* double dividend, as in the case of the economic double dividend, unless highly advanced modeling tools are available, one cannot determine precisely the monetary value of additional environmental benefits. Moreover, one is faced with the problem of pricing nontangibles, a problem mentioned in the preceding chapter.

We have thus explained why different no-regrets concepts are employed in actual practice: some in the welfare economic tradition and some in the field of commercial viability. Much of the debate on the scope and potential of no-regrets strategies can be traced back to this conceptual confusion. We would argue that theoretically the welfare economic perspective is the more correct.

We can now answer the obvious question concerning why no-regrets options are not carried out by the private sector. That is, why is there such a thing as a no-regrets potential? If the narrow no-regrets concept of including only the negative cost potential is applied, the answer is that the market situation is apparently characterized by distorted prices, various institutional barriers, or implicit discount rates that deviate from commercial rates, or a combination of these factors. If the broader no-regrets concept is applied – that is, including all of the externalities – another possible explanation for the existence of no-regrets options is that in deciding whether to introduce a particular technology private agents will not take externalities into account, whereas in a welfare economic assessment they will.

This explains the variety of studies suggesting a considerable scope for no-regrets options, especially in the household and tertiary sector – for example,

Mills, Wilson, and Johansson (1991), Springmann (1991), Rubin et al. (1992), Jackson (1991), Blok et al. (1993), Robinson et al. (1993), and UNEP (1993).

6.4.2 Energy Efficiency and the Discount Rate

As explained in detail throughout this volume (especially Chapter 2, Box 2.4, and Chapter 5), the choice of a financial discount rate is an important factor in evaluating the cost-effective energy efficiency potential in a particular sector or region. One can argue that in the economic behavior of persons or business firms, a discount rate is used implicitly in the decision-making process preceding the introduction of a technology designed to improve energy efficiency. Studies that have tried to assess the implicit consumer discount rates of household investments in energy efficiency reveal ranges that vary (depending on income class and other factors) from only a few percentage points to well over 50%. Train (1985) found a range of 10 to 32% for improvements in the thermal integrity of buildings, 4 to 36% for space heating and fuel type, 3 to 29% for air-conditioning, 39 to 100% for refrigerators, and 18 to 67% for other home appliances.

Thus, it is clear that the estimated scope of no-regrets options depends crucially on the discount factor employed. If one used an interest rate (whether based on market or normative considerations) that was considerably lower than the one applied by the actual investor or consumer, a no-regrets option would not materialize, even if there were no serious bottlenecks in access to information or human and financial resources. However, the practical situation, especially at the grassroots level in developing countries and countries in transition, is such that even the latter conditions are seldom fulfilled (see Box 6.4).

BOX 6.4 High Discount Rates and Institutional Obstacles to Energy Efficiency in Developing Countries

To illustrate how institutional barriers lead to high discount rates in evaluating energy efficiency options, we will consider the introduction of energy-efficient technologies in the consumption of woodfuels in developing countries. Here institutional measures and proper distribution (keeping in mind local societal and cultural factors) are probably quite important (see, e.g., Munasinghe 1996). Popularizing energy-efficient cooking stoves among hundreds of thousands of households may provide a cost-effective contribution to the solution to the climate change problem in many developing countries. However, this would require efforts at many levels. Precisely because of the complicated measures needed for

(continued)

BOX 6.4   High Discount Rates and Institutional Obstacles to Energy
Efficiency in Developing Countries *(continued)*

the successful large-scale introduction of such technology (consider the
difficulty of making people accept the stoves), there is a fair chance that
such projects will not easily get off the ground if there is insufficient institutional support. The alternative – leaving the large-scale adoption of
the cooking stoves to private initiative – is unlikely to work either. Even if
investing in a stove turns out to be efficient from a household perspective, the investment that must be made to install the stove may be so large
that part of the returns might be beyond the time horizon of the household. Moreover, there may not be a financial institution willing to provide a loan to finance the stove. This is another way of saying that institutional obstacles and high (implicit) discount rates limit these options
(even if they are no-regrets measures).

A high implicit discount rate does not mean that substantial energy
efficiency improvements and consequent benefits for the economy are not
possible. Rather, it suggests that significant policy intervention will be required to achieve such improvements. For example, in spite of a high implicit
discount rate, the average energy efficiency of new refrigerators sold in the
United States nearly tripled between 1972 and 1993. This large and steady improvement was due primarily to the adoption of minimum *efficiency standards,*
first at the state and then the national level (Geller and Nadel 1994).

Nevertheless, it is altogether clear that the choice of a discount rate will
have a crucial effect on the overall magnitude of energy efficiency improvements that are considered economical. Meier (1991) argued that by assuming
an annual discount rate of 10% more than a quarter of U.S. electricity demand for refrigerators could be reduced by cost-efficient measures; using a
30% rate results in positive costs for all these measures. Similarly, the Committee on Science, Engineering, and Public Policy (COSEPUP 1991) has shown
how the percent savings in electricity, at the point where the costs of conserved electricity equal the typical operating costs of an existing U.S. power
plant, vary according to the discount rate: at a 3% rate the electricity saving
potential is almost 45%; at a 10% rate, it is about 30%; and at a 30% rate, it is
about 20%.

6.4.3   Fossil Fuel Switching

According to most studies, the present dominance of fossil fuels in global (primary and noncommercial) energy consumption will continue to exist in the
decades to come. According to some authoritative World Energy Council sce-

TABLE 6.3 Energy Mix: Annual Past and Future Global Fuel Use (Gt Oil Equivalent)

Fuel	1960	1990	2020 A	B1	B	C
Coal	1.4	2.3	4.9	3.8	3.0	2.1
Oil	1.0	2.8	4.6	4.5	3.8	2.9
Natural gas	0.4	1.7	3.6	3.6	3.0	2.5
Nuclear	—	0.4	1.0	1.0	0.8	0.
Large hydroelectric	0.15	0.5	1.0	1.0	0.9	0.7
"Traditional"	0.5	0.9	1.3	1.3	1.3	1.1
"New" renewables	—	0.2	0.8	0.8	0.6	1.3
Total	3.3	8.8	17.2	16.0	13.4	11.3

Source: WEC Commission (1993).

narios[10] (WEC Commission 1993; summarized in Table 6.3), fossil fuels will account for between 66% (scenario C, in which renewables are fully explored) and 76% (scenario A, in which fossil fuels remain dominant) of world energy consumption in 2020, compared with 77% in 1990.

All of the scenarios listed in the table show that

- fossil energy will remain dominant;
- the share of natural gas, environmentally the least damaging of the fossil fuels, will increase from the present quarter to one-third at most;
- the share of nuclear power will remain modest; and
- the relative potential of the presently modest "new" renewables is not insignificant, as opposed to the limited size of the projected shifts for large hydroelectric and "traditional" energy sources (for a different point of view, see Kassler 1994).

The continuing dominance of fossil fuels is due to their large resource base,[11] the strongly vested position of current technologies, and price distortions

---

[10] The WEC distinguishes between four scenarios for the energy mix in 2020: scenario A assumes high annual world economic growth (especially in developing countries), high annual energy intensity reduction, and a very high total energy demand; scenario B1 assumes moderate annual world economic growth rates, moderate annual energy intensity reduction, and a high total energy demand; scenario B, the reference scenario, assumes high annual energy intensity reduction; scenario C assumes moderate annual economic growth, very high energy intensity reductions, and a relatively low total energy demand in 2020.

[11] Rogner, Nakicenovic, and Grübler (1993) use a set of definitions to distinguish between different levels of geological certainty and economical and technical feasibility. The *resource base* is described as consisting of (proven) reserves and resources. *Reserves* are those occurrences that are identified, measured, and known to be economically and technically recoverable at current prices, using current technologies. *Resources* refer to the remainder of occurrences with less certain geological and economic characteristics. Additional quantities with unknown certainty of occurrence or with unknown or no economic significance at present are referred to simply as *occurrences*.

(subsidies) that externalize the environmental costs. According to estimates, total identified fossil fuel reserves will suffice to provide for current (1990) levels of energy consumption for the next 130 years.[12] This time span may become considerably shorter, because energy use in the developing countries will increase rapidly.

Of the three fossil fuels, natural gas is the least and coal the most carbon-intensive.[13] Natural gas also produces minimal sulfur emissions and virtually no airborne particulates (WRI 1994). Therefore, a switch from coal and/or oil to natural gas is seen as a response option with multiple benefits. Current estimates of the natural gas resource base, which will most likely be revised upward in the future, allow for a massive switchover for the next century or so to come. If so, the resulting transition of the current energy technology would, as an additional beneficial side effect, pave the way for a broad diffusion of gas from biomass or coal gasification, or of hydrogen,[14] a potentially massive renewable energy source for later in the next century.

The costs of this fuel switching stem from retrofitting or replacing current energy technology and, in some cases, building additional transport grids to connect more remote urban areas with gas fields. Estimates of the costs of switching, even without extending existing networks, depend to a large extent on the type of measure. For example, switching building heating from electric to natural gas (improving overall efficiency by 60 to 70%) would, according to Rubin et al. (1992), yield a net benefit of $90/tCO_2$ in constant 1989 U.S. dollars (assuming a 6% real discount rate). According to the same source, however, switching coal consumption in industrial plants to natural gas or oil consumption, where technically feasible, would involve net direct implementation costs of about $60/tCO_2$ in constant 1989 U.S. dollars.

Ettinger, Jansen, and Jepma (1991) have estimated the investment costs of exploration and extraction for a fuel-switching scenario involving a natural gas supply growth rate of 3.3% per year between 1988 and 2005 plus the costs of extending the existing supply network into a global gas distribution system (based on 1989 data from the Dutch Gas Union and an average transport distance of 2,500 km). They calculate that total costs would be of the order of U.S.$70 billion gross per year, corresponding to $70/tC on average.

---

[12] Total global energy consumption amounted to 10 TWyr in 1990, whereas identified fossil energy reserves are estimated to be 1,280 TWyr (Rogner et al. 1993, p. 463). Obviously, the use of an aggregate figure for fossil fuels (which are mostly coal reserves) should not obscure the fact that the time spans for individual fossil fuels differ widely. The ratio of proven reserves to annual production is estimated to be about 55 years for natural gas, 45 years for oil, and 235 years for coal.

[13] IPCC carbon emission rates are 15.3, 20.0, and 25.8 kg of carbon per gigajoule for natural gas, crude oil, and (bituminous) coal respectively (IPCC 1995).

[14] This is because the types of transport and combustion technologies are roughly the same for natural gas and hydrogen.

However, two caveats should be mentioned. First, many of the advantages of natural gas as a less carbon-intensive fossil fuel are lost if a sizable fraction evaporates into the air by leakage during production and distribution. This is due to the substantially higher global warming potential (GWP) of CH_4, which is about 24.5 times that of CO_2.[15] Estimates of common current leakage rates range from 0.3 to 4% for distribution and from 0.13 to 6% for production (Simpson and Anastasi 1993).[16] The break-even point – that is, the rate at which the reduced total GWP is just offset by leakage of CH_4 – occurs at 7%[17] for switching from coal to gas and at 4% for switching from oil to gas (adopting a GWP index for CH_4 of 24.5 for a 100-year time horizon). These figures point to the need for strict control of leakage rates.[18] Additional questions revolve around what happens to leakage rates in the case of a large-scale fuel switching and whether leakage rates of newly built and/or additional grids (i.e., marginal leakages) can be reduced.

Second, the costs of the fuel-switching option can also be approached on the basis of the opportunity cost concept. For countries such as China and India that dispose of massive coal reserves and that will contribute increasingly in an absolute sense to the global greenhouse problem, the opportunity costs of fossil fuel switching may be considered large, especially if the environmental costs of coal are not taken into account. Note, however, that there can also be significant leakage of CH_4 from coal mines; according to Barnes and Edmonds (1990), the impact of CH_4 leakage from typical bituminous coal mines may well amount to about one-third of the CO_2 emission factor of the use of coal. This type of leakage should therefore also be taken into account in assessing the opportunity costs of fuel switching.

6.4.4 Renewable Energy Technologies

Many technologies have been developed to provide energy on a sustainable basis, in the sense that they harness energy resources that are practically unlimited and require relatively little additional energy input. Moreover, exploitation of renewable energy resources with appropriate technologies has

---

[15] Adopting a 100-year time horizon; for the various ways the GWP measure for CH_4 could be calculated, see, e.g., Reilly and Richards (1993), pp. 41–61.

[16] This would imply that 3 to 41% (for distribution) and 1 to 63% of the carbon reduction of a 100% coal to natural gas fuel switch would be offset by the detrimental effects of leakage.

[17] BEP = $A/[(MER \times GWP) + A]$), where BEP is the break-even point, $A = (26.6 - 13.8) \times 3.67$, MER is the mass/energy ratio for CH_4 = 22 Tg CH_4/EJ, GWP for CH_4 = 21, mass ratio of CO_2 to C = 3.67, and assuming a zero-leakage rate of CH_4 in coal production; for other figures see note 16. Similarly, a 100% oil-to-gas switch shows a break-even point at about 4% leakage.

[18] See, e.g., Jackson (1991) for an analysis of cost effectiveness in the United Kingdom, explicitly incorporating CH_4 leakage.

the advantage of releasing relatively little carbon in net terms.[19] Consequently, a switch from fossil fuels to renewables will result in reduced absolute GHG emissions.

However, renewable technologies are not always sustainable in the sense of being socially and environmentally benign. Particularly in the case of large-scale applications in developing countries, notably of hydropower and biomass, adverse effects may arise for the local population. Moreover, adverse environmental side effects may occur, such as smog from the use of traditional biomass fuels (fuelwood, dung, and crop residues) or changes in biological habitats and local climate.

The following classes of renewable energy resources are commonly distinguished: solar, wind, hydropower, geothermal, ocean, and traditional and modern biomass.[20] To understand the main factors that underlie the costs and energy potential of renewables as a group, a detailed treatment of their diversity is required.[21] Most of these resources, with the exception of biomass energy, vary in supply, and some of them (especially traditional biomass, wind and solar) are relatively more cost-competitive with fossil sources when they are produced on a small scale and near the place of consumption. These aspects make renewables potentially attractive options in remote and underdeveloped areas.

Furthermore, large differences exist in the technical and economical readiness of these options. Hydroelectric, wind, and traditional biomass are relatively well developed, whereas some ocean technologies are still at a demonstration stage, although tidal and wave technologies may soon become more practical economically. Solar, modern biomass, and geothermal lie in between, and photovoltaics may become competitive with fossil fuel power plants within a decade or so (Mills et al. 1991).

---

[19] In this respect, electricity and hydrogen are rather ideal intermittent energy carriers from a technological point of view.

[20] See, e.g., IPCC (1991), Johansson et al. (1993), WEC Commission (1993), and WEC (1994). With respect to the most common classification of renewables, classifying geothermal energy as a renewable resource is technically incorrect, as the center of the earth will slowly but surely cool.

[21] Solar can be broadly subdivided into solar thermal, solar architecture, solar thermal-electric, photovoltaic systems, and thermochemical and photochemical systems. Wind and hydropower are relatively homogeneous energy technologies, the largest differences stemming from scale of operation. Here, a distinction is made between small-/medium-scale and large-scale conversion systems. In contrast, biomass appears to be the most complex of all technologies. A wide range of conversion technologies exist, depending on the type of feedstock used and the form of energy output required. Geothermal consists of hydrothermal, hot dry rock, geopressured, and magma resources technologies. Current ocean technologies encompass tidal, wave, biomass, and salt and thermal gradient technologies.

TABLE 6.4 Current Use and Practicable Potentials of Renewable Energy Technologies (TWh/yr)

Technology	Current Use	Practicable Potential Estimate			
		Johansson et al. (1993) 2020	Swisher et al. (1993) 2030	WEC (1993) 2020	Read (1994b) Average to 2050
Solar	54		1,395	489–1,592	
Wind	3.2		4,931	20,148	
Hydroelectric	2,281.2[a]	6,000–9,000	7,077	8,295[b]	
Geothermal	37–57	>53	1,499	178–405	
Ocean	0.6		247	48–240	
Traditional biomass	4,170		8,003	7,031–7,269	
Modern biomass[c]	543				About 35,000[d]

[a] Includes 81.7 TWh/yr for small hydro.
[b] Includes 211–308 TWh/yr for small hydro.
[c] Modern biomass refers to the use of biomass (e.g., timber or sugarcane) for the production of electricity, liquid fuels, and heat using modern technology.
[d] Assumes 740 million ha become available for biofuel production by 2050 (proportionately less according to technical progress with biofuel productivity per hectare) with a slow start and more rapid buildup after 2010. The 35,000 TWh would yield about 18,000 TWh of electricity given advanced generating technology expected to be in use next century.
Source: Jepma et al. (1996).

Some estimates of "practicable"[22] potentials (relative to current use) are given in Table 6.4. The table clearly shows how small current use is when related to various estimates of practicable potential, whatever discrepancies may exist in estimates of that concept. Notable exceptions are large hydroelectric

---

[22] There are different ways to approach the concept of "practicable," i.e., realizable, potential. Most common distinctions are physical, technical, and economical, the ensuing potentials each being a subset of the one mentioned earlier. *Physical potential* denotes the maximum potential that is constrained by geological, geophysical, and meteorological factors only. *Technical potential* refers to that part of physical potential that can be exploited given the state of technology available. Finally, the technical potential that remains after excluding what is considered infeasible due to prevailing economical constraints (e.g., prohibitive costs and institutional constraints in the energy markets) passes for *economical potential*. Note that, for the present purpose, the first of the three can be considered constant in time, whereas the others prevail only at a certain moment.

Consequently, "practicable potential" can be defined as somewhere in between technical and economical potential. This is because the two do not hold independently, but are interlinked in time; e.g., technical potential is enlarged by investments that stimulate technological progress. The opposite – technical development affecting economic viability – is equally possible; e.g., the impact of improvements of silicon films in photovoltaic systems on the price of solar energy is obvious.

TABLE 6.5 Estimates of Current[1] and Future Costs of Renewable Energy Technologies (U.S. ¢ per KWh)

Source	Solar	Wind	Hydro	Geothermal	Ocean	Biomass Electric	Biomass Fuel ($/GJ)
IEA/OECD (1987)[2]	0–14[a] 7.6–41.9 (15–174)[b] 5.2–26 (22.61)[c] 5(50)[g]	3.5–4.2[h] (4.48–7.62) (20)[i]		(3.6–9.2)	5–20[m3] (11.5–50)[q] 6.7–8[n]		7.58–12.80 (1.85–16.68)[q] 12.70–20.85[t] 15.64–23.70[u]
Johansson et al. (1993)[2]	4.5–11.7 (7.5–32.8)[c] 4.9–9 (8.5–28)[g]	3.13–4.46 (4.29–8.4)		3–12[k] 0.15–2.5[l]	(5–30)[m] 12–25[n] 22–30[o]		1.86–2.73 (2.73–3.86)
Swisher et al. (1993)	5–10[d](12) 4–8[e](25) 4–8[g](30)	3–6(7)	5–10(5)[j]		8–10[m]	5	6[r](8) 7[s](13) 10[t](15) 1[v]
WEC (1993)	no storage[w] 0.4–2.5 (0.5–10)[c] 1–11 (1.2–28)[f] 4–14 (28–45)[g]	3–9 (5–10)		NA	(5–12)[m] (5–7)[n] (12)[o] (10–14)[p]	NA	NA

Note: NA denotes not assessed.

[1] Current costs in parentheses.

[2] 1984 cents.

[3] UK pence per KWh.

[a] passive solar; [b] active solar; [c] solar thermal; [d] solar thermal; [e] solar thermal (line focus); [f] solar thermal (point focus); [g] photovoltaic; [h] small/medium wind energy conversion systems; [i] large wind energy conversion systems; [j] small hydro; [k] electric; [l] direct heat; [m] tidal; [n] wave; [o] salt gradient; [p] ocean thermal; [q] ethanol from corn; [r] ethanol from sugar; [s] ethanol from wood; [t] methanol from wood; [u] methanol from herbage; [v] methanol from biomass; [w] costs exclude storage systems.

Source: Jepma et al. (1996).

and traditional biomass, which, according to the data presented, are exploited at about a quarter to half of probable capacity. Judging by these figures only, the potential contribution for renewables is promising. However, a truly comprehensive assessment must also consider the costs involved.

Table 6.5 gives a selective overview of cost estimates of renewable energy technologies. As usual, the figures diverge widely. This variation is due mainly to (a) the calculation method used or (b) to the inherent peculiarities of the technology. As for (a), the time horizon adopted, the level of discount rate chosen, and the assumed capacity and useful lifetime are important factors. As for (b), costs are strongly influenced not only by the site specificity and temporal variability of supply, as already mentioned, but also by the form of final energy delivered.[23]

Other aspects relevant to cost behavior are learning effects, economies of scale, and the need for immediate storage or transport of the energy generated (the costs of which are very difficult to assess with any precision). Immediate storage or transport needs occur not only when the timing of supply fails to coincide with that of demand, as is commonly the case with solar and wind, but even more when sources and points of end use are far apart. Preferably, generated electricity should be fed into a linked distribution system of sufficient capacity to handle its intermittent supply. Different, but equally difficult to assess, are the problems of location and transportation associated with storable biofuel.

On the basis of the prices in Table 6.5, it has been concluded that hydroelectric, wind, and some solar and biomass technologies are already becoming more competitive with conventional sources. Although many wind and solar power applications are still subsidized or legislatively supported, substantial cost reductions are to be expected within the next few decades.[24] Whether these technologies actually become competitive, however, will also depend on local conditions that shape a renewable's attractiveness and complementarity between renewables and nonrenewables.

Renewable energy at the moment is less portable than fossil fuels: consumption currently seems to be more strongly bound to the production location. Whereas fossil fuels can be stored relatively easily or transported using the existing infrastructure, similar exploitation of new renewables would in

---

[23] E.g., wind energy costs highly depend on wind speed, and solar energy costs on solar irradiance, features that are not equally favorable in different countries (or locations) and/or during different seasons (or hours in the day).

[24] It is not possible to describe cost developments for individual subclasses of technologies on the basis of the figures listed, since they have been aggregated into ranges of similar technologies. The same applies to disparities stemming from differences in sites. This significantly hampers a straightforward comparison. However, greater detail has been avoided for the sake of clarity.

most cases require new investments. Furthermore, the competition among renewables is generally more complex than that among fossil fuels. Solar and geothermal energy, for example, can be produced only on the basis of complementarity by using temporal variation of supply.

Conversely, what often makes up the main part of a renewable's promise is its large potential and modest current price relative to the availability and prices of conventional sources. Moreover, by using local renewables, countries could reduce their dependence on imported fossil fuels and also reduce foreign exchange constraints. In addition, in the case of biomass, local communities could significantly benefit from small-scale applications and their net positive side effects. In this respect, local renewables, like energy efficiency measures, offer a basis for no-regrets policies.

There is reason to believe that a new generation of renewable energy technologies now under development could become commercially viable in the near future. For example, a variety of promising photovoltaic technologies designed to reduce commercial building demand during peak load periods is under active consideration in the United States and elsewhere and might become commercially feasible in the near future (Wenger, Hoff, and Perez 1994; Byrne et al. 1994).

As the preceding discussion implies, the precise future role of renewables is hard to predict. Although some scenarios are more optimistic than others, the share of renewables in the 2020 energy mix will probably not exceed 25%.[25] However, most studies agree that the new renewable mix will tend to be a hybrid that will exploit a variety of renewable energy sources backed up by fossil fuels, which will remain dominant for decades to come.

6.4.5 Nuclear Energy

Nuclear energy now accounts for about 5% of all primary energy production, or 17% of the world's electricity generation.[26] Its production, like that of renewables, involves the emission of relatively little CO_2.[27] Moreover, its technology has passed the demonstration stage, except for the large and unresolved issue of nuclear waste storage. Yet further dissemination could be strongly

---

[25] Estimates are 21.3 to 29.6% in 2020 by WEC (1993), 15% (6% hydroelectric) in 2020 by Grübler et al. (1993), and close to 43% in 2025 by Johansson et al. (1993). According to Grübler et al. (1993a), the latter estimate is probably too high. It would imply an unprecedented rate of change in technology and infrastructure. In comparison, it took no less than 80 years for the market share of oil to grow to 40% of the global primary energy supply (Grübler et al., 1993, p. 580). In the past, the mean duration for replacing most technological systems was about 30 to 40 years.

[26] For a detailed discussion of the nuclear option, see Watson, Zinyowera, and Moss (1996).

[27] Only relatively minor fossil fuel inputs are used to support the overall functioning of the breeder reactor.

prohibited by lack of public acceptance due to major concerns about reactor safety, the risk of theft of nuclear technologies or materials, the proliferation of nuclear weapon capabilities, and the final treatment and disposal of fission products.

Despite these limitations, nuclear energy, if evaluated on the basis of the engineering efficiency approach, can be competitively applied, and in various countries it is, albeit to largely different degrees. (For a comparison with gross costs, see Box 6.3; for an estimate of the UK cost-effective potential, see Jackson [1991], who used data from the mid-1980s.) Because of the long design and construction time (up to 10 to 15 years) and the enormous per plant investment costs, the nuclear option is rather inflexible now. According to Mills et al. (1991, p. 540) the costs of producing electricity with nuclear energy (U.S. dollars per kilowatt-hour) appear to fall within the range of renewable options, though nuclear costs seem to have been rising rather than falling (MacKerron 1992).

Social opportunity costs will remain high until a full and credible investigation of the safety aspects of nuclear power plants is completed. However, if the nuclear option is assessed from the welfare economic point of view, the final assessment becomes much more uncertain because the lack of public acceptance and the risks, advantages, and uncertainties have to be taken into account explicitly. This holds not only in industrialized countries, but also in developing countries and countries in transition. In addition, any future use of nuclear energy, like any switch from fossil to nonfossil fuels, will depend on the underlying cross-price elasticities and energy price assumptions, inflation, public policy, and technological progress. Taking into account these complicating factors – namely, that there is no established technology for decommissioning nuclear plants, that there are hidden external costs regarding nuclear-power-related damage, and that efforts are being made to develop intrinsically safe nuclear reactors – the IEA projects that the share of nuclear energy in total energy use will be 6.1% by 2010; WEC scenario C (see note 10) projects that the share of nuclear energy will be 6.2% in 2020.

6.4.6 CO_2 Capture and Disposal

Carbon dioxide capture and disposal is understood to be any sequence of processes in which carbon is recovered in one form or another from an energy conversion process and disposed of at sites other than the atmosphere. It should be noted, though, that disposal capacity is ultimately limited, both for technical reasons and because not all forms of disposal prevent carbon from eventually reentering the atmosphere. However, assuming sufficient and feasible disposal, the further development of these technologies in combination with coal gasification is thought to have significant potential in the medium term, especially for coal-rich countries such as China, India, the United States, and the Russian Federation (see Nakicenovic and Victor 1993). Since places of recovery do not generally coincide with places of disposal, transport of the recovered carbon is an additional process. In principle, carbon can be recovered from each fossil fuel conversion process. However, recovery is most attractive at energy-intensive stationary point sources, such as steel manufacturing, fertilizer, and power plants.[28] To date, most research effort has been spent on power plants. For these, two types of recovering technologies exist: those that combine separation of CO_2 from flue gases (scrubbing) with modifications to the energy conversion process and those that rely on CO_2 scrubbing only.[29] Modifications of the energy conversion process, which are now in experimental use, include an integrated coal gasifier combined cycle (ICGCC) system, modification of boilers, and modification of gas turbines.[30] The main separation options are chemical or physical absorption, the use of membranes, and cryogenic fractionation. Of these, chemical and physical absorption are the most highly developed, and membrane separation and cold distillation the least.[31]

Depending on the place of disposal, transport takes place onshore or offshore. Onshore, pipelines are most economical. Estimated transport costs vary between U.S.\$1 and \$4/tCO_2 over 100 km, depending on the flow rate (Hendriks 1994). Offshore, tankers can compete with pipelines. For larger distances, tanker transport is likely to be cheaper. Pipeline transport costs are more or less proportional with distance and decrease with increasing flow rate of the gas and decreasing ambient temperatures. Estimates of offshore costs therefore vary between somewhat more than one-half and three times the onshore costs (TNO 1992; Hendriks 1994). After the carbon is recovered, it has to be handled, so that reentry into the atmosphere is prevented or at least delayed as much as possible – that is, so that the mean retention time is large

---

[28] The advantage of recovering carbon at power plants is that carbon can be removed from an energy product before it is distributed to highly dispersed end users.

[29] Because the use of coal and natural gas is predominant in power plants, almost no technologies are based on oil.

[30] In an ICGCC, coal is converted before combustion. After several intermediate steps, CO_2 and H_2 are obtained. CO_2 can be extracted by absorption at a 98% rate, and the latter can be used either directly at the power plant to generate electricity or as a carbon-lean fuel to be distributed to end-user sectors like households, industry, or transport. Modifying a conventional gas or coal-fired boiler amounts to changing the oxidant from air to pure oxygen. The gas turbine of an ICGCC or a steam-injected gas turbine can be modified by changing the combustion medium into an O_2/CO_2 medium.

[31] Cost information suggests that absorption and oxyfuel combustion are most attractive. It appears that absorption is cheaper for conventional coal-derived flue gases than for natural gas flue gases. An ICGCC is promising, although it is not clear yet whether it will replace proven conventional pulverized coal-fired installations.

compared with the residence time of CO_2 in the atmosphere (since not all applications ensure entire or long-term storage of carbon).[32]

Disposal can take place in two ways. The gas can be utilized for the production of long-lived materials,[33] or it can be stored underground, either in aquifers (which, technically, have almost unlimited storing potential) or in the ocean.[34] Environmental risks seem to be involved in the latter cases.

6.4.7 Enhancing Sinks: Forestry Options

Unlike removal options, sink enhancement options involve the removal of carbon after it has been dispersed into the atmosphere. All sources seem to agree that much more carbon is stored in soils than in forests.[35] This suggests that significant attention be given to measures that promote soil conservation, reduce carbon mobilization from soil to air, and increase soil storage of atmospheric carbon through the action of soil microorganisms. Nevertheless, the main options for enhancing carbon sinks – except for iron fertilization and weathering rocks, both of which are still in the experimental stage – are forestry measures. Their importance is due to their expected large storage potential and relatively modest costs. The enhancement of forest sinks is also

[32] In enhanced oil recovery, part of the injected CO_2 reenters the atmosphere, and in food packaging CO_2 is released within days or weeks. Obviously, insofar as CO_2 is released into the atmosphere, these applications, though commercially interesting, are of no significant long-term interest from an abatement point of view.

[33] Applications of carbon (dioxide) storage exist in food industry, chemical manufacturing, metal processing, and enhanced oil recovery, the last having the highest potential. In enhanced oil recovery, carbon dioxide is pumped into the production well to be used as a miscible liquid. This enhances oil recovery rates.

[34] The aggregate potential of the other applications is limited to several hundred megatons of carbon per year. The storage capacity of the ocean is very uncertain, since it already contains nearly 40,000 GtC as (dissolved) CO_2 (compare with some 750 GtC in the atmosphere) and for environmental reasons. What is more, most of the injected carbon will escape after 50 to several hundred years, depending on the depth and manner of injection.

[35] In this context, forestry differs from biomass as a renewable energy resource, especially if forests are grown for conservation purposes. Forests or biomass *can* be classified as renewable energy resources if harnessed for energy purposes and harvested in such a way that supply is practically unlimited and no additional energy is required. This means that the plantation must be rotated after harvest and that either *net* energy inputs for the energy extraction and conversion process are negative or *gross* energy inputs are easily paid out of the extracted energy. Inherently, net carbon emissions (or removal) will be zero for such applications, since carbon emitted by combustion is offset exactly by the carbon removed in the next generation of plantations. However, insofar as forest or biomass energy replaces fossil energy, and/or timber products replace energy-intensively produced substitutes for fossil fuels, net carbon emissions *are* reduced, which explains why the forestry option is distinguished from renewable options.

one of the lowest-risk options and offers substantial positive side effects in the environmental and sometimes (but not always) in the socioeconomic sphere.

The contribution of deforestation to GHG emissions is sizable. After fossil-energy-related activities, deforestation and other land-use changes are the second-largest source of carbon emissions. The net annual flux of carbon into the atmosphere as a result of land-use changes and deforestation ranged between 0.6 and 2.8 GtC during the 1980s, compared with global emissions of slightly less than 6 GtC from the burning of fossil fuels, manufacture of cement, and flaring of natural gas (Grübler et al. 1993, p. 499; see also Houghton 1990). The large amount of uncertainty about the net quantity of carbon released by deforestation and land-use change relates to the extent of the area undergoing land-use change, the carbon content of biota and soils in the deforested land, and the dynamic release profile of biotic and soil carbon after disturbance. The following subclasses of forestry measures are commonly distinguished:

1. Halting or slowing of deforestation
2. Reforestation and afforestation[36]
3. Adoption of agroforestry practices
4. Establishment of short-rotation woody biomass plantations
5. Lengthening of forest rotation cycles
6. Adoption of low-impact harvesting methods and other management methods that maintain and increase the amount of carbon stored in forest lands
7. Sustainable forest exploitation cum sequestration of carbon in long-lived forest products[37]

The first six measures sequester carbon by increasing the standing inventory of biomass or by preventing its decrease. This amounts to a once-and-for-all uptake of carbon. In contrast, the seventh measure aims at continuing to break the carbon cycle, thereby, in principle, enabling permanent application. This option becomes even more efficient and attractive if the timber substitutes on a large scale for such products as brick, concrete, steel, and plastics whose manufacture releases much greater quantities of CO_2.

However, in practice all forestry measures are ultimately limited: the first six by the area available in competition with other potential land uses and the seventh by saturation of demand for timber and other long-lived wood prod-

---

[36] "Afforestation" refers to land that has not been covered by forests for the past 50 years. In contrast, "reforestation" applies to land that was denuded less than 50 years ago.

[37] For a detailed assessment of the subclasses of forestry measures, the potential quantity of carbon to be conserved and sequestered by forestry measures, the effects of climatic and demographic changes on the potential amount of carbon conservation and sequestration, and the new research directions that are required to improve the assessment and development of practical forestry strategies, see Brown et al. (1996).

ucts and the eventual decay of wood. Therefore, forestry measures, like removal options, are to be seen as an intermediate response policy. At least in the long term and also depending on the discount rate employed, trees grown on fairly short rotations (harvested at maximum mean annual increment) are generally more effective carbon sinks than trees that are allowed to mature in old-growth forests. This fact has large implications, especially for developing countries, where the largest demand for wood is for fuelwood and small construction poles that can be grown on short rotations.

In assessing the global potential for the halting or slowing of deforestation and for reforestation, there are four main sources of uncertainty:

1. The potential for slowing deforestation depends on resolving complex problems linked to societal and economic pressures, such as large-scale settlement on forestlands, the sale of timber for export earnings in tropical countries, and policy distortions (e.g., below-cost sales of timber on government lands) in industrialized countries.

2. The potential of the option depends on the amount of land globally available for some kind of forestry measure (see also Brown et al. 1996).

3. The potential for halting or slowing deforestation also depends on the incremental (i.e., annual) and net cumulative carbon uptake per hectare for the main forest species.[38]

4. Large-scale monoculture forestry may not be acceptable to many environmentalists; moreover, local ecosystems may be destabilized.

Societal and Economic Pressures   There is a near consensus in the literature that most deforestation in tropical countries occurs because standing forests are converted to crop- and pastureland. This happens because those encroaching on the forests consider them to have lower economic value than crop- and pastureland. The potential for slowing deforestation is therefore hard to estimate. Furthermore, slowing deforestation requires effective solutions to highly politicized problems such as inequitable land distribution and lack of secure land tenure. It also requires effective means of increasing the per hectare productivity of crops and livestock. The solutions to these problems are partly technical but mostly economic in nature and include im-

---

[38] In addition to carbon stored in the forest wood itself, soil carbon and carbon in other biomass growing in the forest are commonly significant and should be included.

"Cumulative uptake" refers to the total amount of carbon stored after a certain period. Usually after the forest has reached maturity no (net) carbon is absorbed; i.e., incremental uptake is minimal. "Annual uptake" is usually described by one figure only, the average annual uptake rate. However, for plantations to be logged before maturity, the absorption rate depends on the age of the forest. Contrary to common belief, annual uptake of a newly planted forest generally is not highest in the first years, but only after the forest has reached an intermediate age. More precisely, accumulated uptake is an S-shaped growth function of time (Nilsson 1982; Cooper 1983).

proved price structures for farmers (e.g., higher crop and livestock prices vs. lower prices for inputs such as fertilizer) and greater access to markets. Economy-wide (i.e., indirect) policies may crucially affect tropical deforestation (see, e.g., a study of Costa Rica using a CGE model by Persson and Munasinghe 1995).

Availability of Land Estimates of tropical deforestation rates vary widely, partly due to different definitions of both tropical forests and deforestation (for a discussion, see Jepma 1995). According to the FAO (1991), annual tropical deforestation for the late 1980s amounted to some 17 million ha; other estimates vary between 3 and 20 million ha. Estimates of global annual biotic carbon fluxes from closed forests in the late 1980s show an equally large variety, ranging between 600 MtC (IPCC 1992) and 2,800 MtC (WRI 1990). For an overview, see Grübler et al. (1993b).

A similar discussion has arisen on the issue of the global land area that would be suitable and available for carbon-sequestering plantations. One study of the maximum worldwide potential of this approach, by Sedjo and Solomon (1989), suggests that 2.9 Gt of atmospheric carbon could be sequestered annually by approximately 465 million ha of fast-growing plantation forests at a cost of about U.S.$186 to $372 billion. Without employing fast-growing species, the area needed would be several times larger, but many factors will determine whether such high rates of uptake can be achieved.

Keeping these limitations in mind, one can compare the preceding findings with the estimate of Grübler et al. (1993b) that at present 265 million ha globally would be available and suitable for forest plantations and 85 million ha for agroforestry. Other sources (Winjum, Dixon, and Schroeder 1992) suggest significantly larger areas (some 400 to 1,200 million ha). Clearly, a large potential for enhancing forests exists in the tropics. A recent survey, carried out under the auspices of the Asian Development Bank in eight Asian countries (Pakistan, India, Sri Lanka, Bangladesh, Indonesia, Malaysia, Vietnam, and the Philippines) indicates not only that climate change is likely to have large and generally adverse effects on forests and forest ecosystems in the Asia–Pacific region, but also that "forest conservation and afforestation can often be judged to be cost-effective and excellent opportunities for limiting net greenhouse emissions" (Qureshi and Sherer 1994). However, it should be emphasized once again that much of the land availability will depend on the willingness of the local population to cooperate, given its perceptions of the most appropriate use of the land. The figures for the nonindustrialized regions are in sharp contrast with the potential in the OECD countries, amounting to 15 to 50 million ha in the future in the EU, mainly due to redundancy of farmlands, and 30 to 60 million ha in the United States.

Carbon sequestered in global annual wood production is currently estimated to be about 1 GtC (TNO 1992). Such a high rate may not represent the actual "net" addition to the wood product pool and may not be sustainable in the future, but it gives some indication of the potential of annual carbon sequestration through wood products, depending on price, product lifetime, and timber demand (which is commonly projected to rise).

Some authors argue that a considerably higher demand for wood could be achieved if it were used to produce electricity (and liquid fuels such as methanol). Unlike fossil fuel systems, which produce CO_2 emissions, this system would produce zero net emissions, provided that wood were grown on a sustainable basis. (For a feasibility study, see BTG 1994; for a discussion of the institutions needed to monitor and account for the global realization of this concept, see Read 1994a.)

Incremental and Net Cumulative Carbon Uptake Only rough estimates exist for the carbon content of biomass and soils in disturbed areas. Estimates of average annual carbon uptake per hectare vary considerably, depending on, among other factors, plantation age (for a correlation, see Cannell 1982), timber species, soil/climate conditions, and management practices (see also Houghton, Unruh, and LeFèbvre 1991). However, most estimates are in the range of 1 to 8 tC/ha per annum. Cumulative carbon uptake would as a maximum be somewhere between 100 and 150 tC/ha for the main forest types. (Note that the vegetation and soils of undisturbed forests can hold 20 to 100 times more carbon than agricultural systems.)

6.4.8 Costs

Mitigation policies involving forests are generally considered relatively cost-effective (especially if applied in developing countries). Richards, Moulton, and Birdsey (1993a; see also Richards et al. 1993b) even estimated that the overall costs of stabilizing U.S. carbon emissions could be reduced by as much as 80% by forestry options. An early U.S. estimate of $100/tC for carbon sequestration via forestry (Nordhaus 1990) turned out to be rather high, but that was because it apparently ignored changes in soil carbon levels through tree planting and underestimated the carrying capacity and length of productivity of forest plantations. Since then, cost estimates of the forestry option have become much lower.

Most of the studies, which deal with the costs of afforestation or of halting or slowing deforestation, take the engineering efficiency approach rather than the welfare economic approach. With respect to afforestation, the assumption is commonly made that the forests would *not* be harvested but would be left alone to mature (so-called carbon cemetery forests). The emphasis, therefore, is on the assessment of the costs of afforestation (plantations), including maintenance and protection, of administration, and of land availability.[39]

The second option, halting or slowing deforestation, is probably one of the most urgent and most cost-effective options (Grübler et al. 1993b). However, experience with the closest alternative, reforestation, has so far produced mixed results. On the one hand, some success stories can be told about reforestation projects in Sweden, Finland, and parts of Canada. On the other hand, there were large losses in Angola, Nigeria, Morocco, and several other countries, and in China the rate of survival of reforestation efforts is estimated to be no higher than 20% (Nakicenovic and John 1991).

Two crucial factors appear to determine the feasibility of the afforestation option in actual practice. First, it matters a great deal whether the forest can be harvested sustainably and forest products sold at commercial rates, or whether the forest should be left alone. Second, much depends on the acceptance of the newly planted forests by the local population, as might be expected if the population were to derive economic benefit from it.

If, for instance, the forest can be harvested through the exploitation of timber and nontimber products and if, in addition, carbon sequestration credit can be given to trees that are harvested, then the net costs of afforestation could easily be negative. In that case, afforestation could become a no-regrets option, and initiatives could be left to the market (except possibly for the problem of the payback period).[40] However, not all externalities could be incorporated into the prices of the timber and nontimber products (e.g., trees might provide environmental benefits but could also contribute to productive losses by shading adjacent field crops or competing with them for water).

Even if afforestation with sustainable exploitation offers a net positive return, many other factors may still form an obstacle to its implementation. Ac-

---

[39] Note that even in well-developed market economies, the costs of program administration can rise to as much as 15% of total costs of land rental, establishment, and maintenance (Richards et al. 1993a). This percentage can be much higher in developing countries or countries in transition.

Studies of carbon sequestration costs in the United States applied several methods to identify the social costs involved in the conversion of land into forestlands: the use of land rental rates from surveys, the use of market prices adjusted for the elasticity of demand for agricultural land, the use of the estimated lost profits from removing the land from agriculture production, and the use of consumer surplus loss from increases in food prices due to the constriction of agricultural land availability. However, land required for establishing carbon plantations may be considered free, implying that opportunity costs would be zero (e.g., Winjum et al. 1992).

[40] Some have argued that afforestation is a no-regrets option anyway. For example, Xu (in press) suggested that a no-regrets potential might be associated with some Chinese carbon sequestration practices.

TABLE 6.6 Costs of Sequestering Carbon through Forest Projects: Some Selected Cases ($/tC)

Source	Tropical		Temperate Plantation	Boreal	
	Agroforestry	Plantation		Plantation	Protection
Andrasko (1993)	3–5	3–6	0–2	—	—
Dixon et al. (1993)	4–16	6–60	2–50	3–27	1–4
Krankina and Dixon (1993)	—	—	1–7	1–8	1–3
Houghton et al. (1991)	3–12	4–37	—	—	—

Source: Adapted from Dixon et al. (1993).

tual experience has made it abundantly clear that even if both environmental quality and economic productivity in a certain area are low, those who use the land may still be unwilling to convert it to forest. Some even argue that for at least a decade social, political, and infrastructural barriers will keep reforestation rates very modest (Trexler 1991). Indeed, "tropical forestry programs undertaken with global climate change mitigation in mind will need to be integrated into the social, environmental, and economic contexts and needs of the countries [and local communities] in which they are undertaken. Failure to understand this has brought about the failure of many tropical forestry efforts intended to solve fuelwood and other problems. The same could easily occur with forestry efforts intended to mitigate global climate change" (Brown et al. 1993). Estimates on the cost effectiveness of forestry measures involving the engineering efficiency approach are also subject to uncertainties about availability of land, carbon uptake per hectare, and costs of establishment and maintenance per hectare. In addition, figures diverge depending on the method adopted. In this respect, two problems should be discussed: (a) the derivation of point estimates or of cost function and (b) forestry cost function methodology.

With respect to (a), most efforts so far have been devoted to deriving point estimates from average costs. Some selected cases are given in Tables 6.6 and 6.7. Note that the estimates generally assume a "tree cemeteries" approach. These tables show relatively low carbon-sequestering costs through tree planting, in many cases less than U.S.$10/tC and rarely more than $30/tC. Other studies with similar results, stressing various aspects of the problem, include Trexler, Faeth, and Kramer (1989), Swisher (1991), Faeth, Cort, and Livernash (1993), and Winjum and Lewis (1993). For a detailed assessment see Turner et al. (1993) and Brown et al. (1996).

Though these point estimates may give a satisfactory description of cost effectiveness for small areas and single-plantation programs, they are bound to

TABLE 6.7 Establishment Costs of Cost-Efficient Practices

Forest Type/Practices	Median U.S.$/tC[a]	Median U.S.$/ha[a]
Boreal		
Natural regeneration	5	93
	(4–11)	(83–126)
Reforestation	8	324
	(3–27)	(127–455)
Temperate		
Natural regeneration	1	9
	(<1–1)	(9–100)
Afforestation	2	259
	(<1–5)	(41–444)
Reforestation	6	357
	(3–29)	257–911)
Tropical		
Natural regeneration	1	178
	(<1–2)	(106–238)
Agroforestry	5	454
	(2–11)	(254–699)
Reforestation	7	450
	(3–26)	(303–1,183)

[a]The numbers in parentheses are interquartile ranges (middle 50% of observations).
Source: Turner et al. (1993).

lack validity in the case of large areas. From a global perspective, the costs in terms of economic welfare are likely to rise with the scale of the effort. Four forces underlie this cost pattern:

1. diminishing uptakes as less suitable or less well managed land is forested, resulting in a lower carbon uptake per hectare;
2. increasing public resistance, and social and legal objections by the local population to interference with present land use;
3. rising opportunity costs as fallow land is used up and plantations move on to land suitable for alternative uses;[41] and
4. no or negligible economies of scale in operating and maintenance costs.

Together, these factors generally mean that marginal costs will rise as the area being forested increases. Exceptions to this rule may occur only if the amount of land needed for agriculture shows a declining trend. Clearly, this is almost nowhere near the case in developing countries, but it might hold for

[41] Theoretically, opportunity costs for land would be largely reflected in land market rents.

TABLE 6.8 Estimates of Costs (Dollars) of Carbon Sequestered by Tree Planting: Some Comparative Results for the United States

Study	Total Carbon Sequestered (Mt)			
	140	280	420	700
Moulton and Richards (1990)	16.57	20.69	23.24	34.73
Adams et al. (1993)	18.50	25.11	37.21	95.06
Parks and Hardie (1992)	175.00	NA	NA	NA

Note: NA denotes not assessed.
Source: Jepma et al. (1996).

parts of the Western world. Only recently have a number of somewhat more sophisticated studies begun to appear that take increasing marginal costs explicitly into account (see, e.g., Adams et al. 1993; Parks and Hardie 1992; and Richards et al. 1993a,b). Such studies also do more justice to the welfare economic point of view by explicitly recognizing that an expansion of the area forested will most likely increasingly interfere with the expanding domestic demand for agricultural land. Table 6.8 illustrates this feature.

In comparing Tables 6.4 and 6.5 with Table 6.6, it is apparent that the figures correspond roughly only for low levels of sequestering effort. For higher levels, divergence grows rapidly. Therefore, the conclusion seems justified that point estimates, though valid for small areas, seriously fail to describe actual costs for larger areas.

With respect to (b), forestry cost function methodology, a number of more sophisticated forestry studies have recently begun to appear. These include Moulton and Richards (1990), Parks and Hardie (1992), Adams et al. (1993), Richards et al. (1993a), and Read (1994b). These studies refine the approach to estimating the cost of establishing carbon-sequestering tree plantations in three ways. First, they estimate a cost function, not a point. Second, they refine the cost estimates for establishing tree plantations by recognizing differences associated with location and site considerations. Third, they build discounting procedures into the methodology – a common practice in the assessment of other options, but until recently one that was virtually ignored with respect to this option (see Richards 1993). With all this in mind, it is clear that both the methodology and the empirical estimates of the various studies require further revision. Probably a justified generalization is that more recent research is tending to find a somewhat steeper increase in costs than did the earlier studies, with the marginal costs per tonne of carbon roughly doubling, from about U.S.$30 to $60 for large annual uptakes.

Finally, it is increasingly recognized that there are probably limits to the extent to which the global system can maintain forest stocks. Nevertheless, sustainable forest management can make an important long-term contribution to providing a continuous flow of substitutes for net emitting energy sources such as coal.

6.4.9 Methane

Methane emission currently accounts for about 20% of expected warming from climate change. This contribution is a result of the potency of CH_4 as a GHG and dramatically increased anthropogenic emissions. Currently, about 70% of global CH_4 emissions are associated with human-related activities, such as energy production and use (coal mining, oil and natural gas systems, and fossil fuel combustion); waste management (landfill and wastewater treatment); livestock management (ruminants and wastes); biomass burning; and rice cultivation. Technologies and practices for reducing CH_4 emissions from their major anthropogenic sources have been identified and reviewed in a number of meetings of and studies by experts, many under the IPCC. Many of the technological options currently available are cost-effective in many regions of the world and have been implemented to a limited extent. The available options represent different levels of technical complexity and capital needs and therefore should be adaptable to a wide variety of situations. In total, it appears to be technically feasible to reduce CH_4 emissions by about 120 Tg (75 to 170 Tg) per year through reductions in emissions from the following CH_4 sources.

Coal Mining Techniques for removing CH_4 from gassy underground mines have been developed primarily for safety reasons, because CH_4 is highly explosive in air in concentrations between 5 and 15% and is the cause of mining accidents. Some of the same techniques can be adapted to recover CH_4 in concentrations of 30% or more, so the energy value of this fuel can be put to use. Emissions of CH_4 into the atmosphere can be reduced by up to 50 to 70% at gassy mines using available techniques such as gob gas recovery (IPCC 1990a,b,c; van Amstel 1993; U.S. EPA 1993).

Oil and Natural Gas Systems Methane is the primary constituent of natural gas, and significant quantities of it can be emitted to the atmosphere from components and operations throughout a country's natural gas system. The technical nature of emissions from natural gas systems is well understood, and emissions are largely amenable to technological solutions through enhanced inspection and preventive maintenance, replacement of equipment with new designs, improved models, improved rehabilitation and repair, and other changes in routine operations. Reductions in emissions of the order of 10 to 80% are possible at particular sites, depending on site-specific conditions (IPCC 1990b; van Amstel 1993; U.S. EPA 1993).

Landfills The CH_4 generated in landfills as a direct result of the anaerobic decomposition of solid waste can be reduced by recovering this medium-BTU gas for use in electricity generation equipment or for direct use in heating or cooking equipment. At many sites reductions of up to 90% are possible. Additional benefits that result from landfill CH_4 recovery include improved air and water quality and reduced risk of fire and explosion, and the recovery of a clean and convenient fuel (IPCC 1990b; van Amstel 1993; U.S. EPA 1993).

Ruminant Livestock Many opportunities exist for reducing CH_4 emissions from ruminant animals by improving animal productivity and reducing CH_4 emissions per unit of product (e.g., CH_4 emissions per kilogram of milk produced). In general, a greater portion of the energy in animals' feed can be directed to useful products instead of wasted in the form of CH_4. As a result, herd size can be reduced while productivity remains the same. Current technologies and management practices can reduce CH_4 emissions per unit of product by 25% or more in many animal management systems (IPCC 1990a,c; van Amstel 1993; U.S. EPA 1993).

Livestock Manure Methane emissions from anaerobic digestion of animal manures constitute a wasted energy resource that can be recovered by adapting manure management and treatment practices to CH_4 (biogas) collection. This biogas can be used directly for on-farm energy or to generate electricity for on-farm use or for sale. The other products of anaerobic digestion, contained in the slurry effluent, can be used as animal feed and aquaculture supplements or as a crop fertilizer. In addition, managed anaerobic decomposition is an effective method of reducing the environmental and human health problems associated with manure management. Current reduction measures can reduce CH_4 emissions by as much as 25 to 80% at particular sites (IPCC 1990c; van Amstel 1993; U.S. EPA 1993).

6.5 SUMMARY

The overview of technical response options in this chapter has shown that there is a wide variety of strategies for helping to solve the greenhouse problem. At the same time, there is clear evidence that no single mitigation option is capable of limiting GHG emissions enough to stabilize concentrations; in that respect, the climate change issue differs from many other environmental problems. This explains why in practice various types of economic measures will be required to tackle different problems, but these may have – albeit indirectly – greenhouse implications as well. It is important to keep this in mind.

Information about the feasibility and potential of mitigation options is increasing rapidly. A major issue with regard to their cost effectiveness is that the picture can be altered dramatically by technological breakthroughs that bring the rate of technological progress to a level that is higher than the current trend of technological progress: costs are rather time-specific. A related issue is that costs of particular mitigation options often seem to be scale- and location-specific as well. By implication, from an efficiency point of view, the issue is not so much what the best technology would be, but rather how, where, and at what scale the optimal mix of technologies would best be implemented.

The feasibility of a technological response option can be assessed meaningfully only if one takes into account the welfare implications, including the social, political, and environmental acceptability of the option, or the wider implications of implementing such an option given existing levels of technology, lifestyles, or income distribution. Although this factor is extremely important, it severely complicates the assessment of the optimal mix of options.

Finally, insight not only into the cost effectiveness of various options (in terms of dollars per ton of carbon), but also into their no-regrets potential, is increasing rapidly. Indeed, the no-regrets potential of various options seems to be significant.

REFERENCES

Abdel-Khalek, G. 1988. Income and price elasticities of energy consumption in Egypt: A time series analysis. *Energy Economics,* **10**(1), 47–57.

Adams, R. M., D. M. Adams, C. C. Chang, B. A. McCarl, and J. M. Callaway. 1993. Sequestering carbon on agricultural land: A preliminary analysis of social cost and impacts on timber markets. *Contemporary Policy Issues,* **11**(1), 76–87.

Barnes, P., and J. A. Edmonds. 1990. *An evaluation of the relationship between the production and use of energy and atmospheric methane emissions.* Washington, DC: U.S. Department of Energy.

Blok, K., J. Farla, C. Hendricks, and W. Turkenburg. 1991. Carbon dioxide removal: A review. Paper presented at the International Symposium on Environmentally Sound Energy Technologies and Their Transfer to Developing Countries and European Economies in Transition, October, Fandazione ENI and Enrico Mattei.

Blok, K., E. Worrell, R. Culenaere, and W. Turkenburg. 1993. The cost-effectiveness of CO_2 emission reduction achieved by energy conservation. *Energy Policy,* **21** (June), 656–67.

Brown, S., C. A. S. Hall, W. Knabe, J. Raich, M. C. Trexler, and P. Woomer. 1993. Tropical forests: Their past, present, and potential future role in the terrestrial carbon budget. *Water, Air, and Soil Pollution,* **70**, 71–94.

Brown, S., et al. 1996. Management of forests for mitigation of greenhouse gas emissions. In R. T. Watson, M. C. Zinyowera, R. H. Moss, and D. J. Dokken (eds.), *Climate change 1995: Impacts, adaptations and mitigation of climate change – Scientific-technical analyses.* Contribution of Working Group II to the Second Assessment Report of the Intergovernmental Panel on Climate Change. Cambridge University Press, pp. 773–98.

BTG (Biomass Technology Group). 1994. *Potential woodfuel production in developing countries for power generation in the Netherlands.* University of Twente, Enschede, The Netherlands.

Byrne, J., S. Letendre, R. Nigro, and Y.-D. Wang. 1994. PV-DSM as a green investment strategy. In *Proceedings of the Fifth National Conference on Integrated Resource Planning.* Washington, DC: National Association of Regulatory Utility Commissioners, pp. 272–85.

Cannell, M. G. R. 1982. *World forest biomass and primary production data.* London: Academic Press.

Chandler, W. U. 1990. *Carbon emissions control strategies.* Washington, DC: World Wildlife Fund.

Cooper, C. F. 1983. Carbon storage in managed forests. *Canadian Journal of Forestry Research,* **13,** 155–66.

COSEPUP (Committee on Science, Engineering and Public Policy). 1991. *Policy implications of greenhouse warming: Report of the Mitigation Panel.* Washington, DC: U.S. National Academy of Sciences, U.S. National Academy of Engineering and Institute of Medicine, U.S. National Academy Press.

DeCicco, J., and M. Ross. 1993. *An updated assessment of the near-term potential for improving automotive fuel economy.* Washington, DC: American Council for an Energy-Efficiency Economy.

Dixon, R. K., K. J. Andrascok, F. A. Sussman, M. A. Livinson, M. C. Trexler, and T. S. Vinson. 1993. Forest sector carbon offset projects: Near-term opportunities to mitigate greenhouse gas emissions. *Water, Air, and Soil Pollution,* **70,** 561–77.

Elkins, P. 1996. The secondary benefits of CO_2 abatement: How much emission reduction do they justify? *Ecological Economics,* **16,** 13–24.

ESCAP (Economic and Social Commission for Asia and the Pacific). 1991. *Energy policy implications of the climatic effects of fossil fuel use in the Asia–Pacific Region.* Bangkok: ESCAP.

Ettinger, J. van, T. H. Jansen, and C. J. Jepma. 1991. Climate, environment and development. *European Journal of Development Research,* **3**(1), 108–33.

Ewing, A. J. 1985. *Energy efficiency in the pulp and paper industry with emphasis on developing countries.* Technical Paper no. 34. Washington, DC: World Bank.

Faeth, P., C. Cort, and R. Livernash. 1993. *Evaluating the carbon sequestration benefits of sustainable forest projects in developing countries.* Washington, DC: World Resources Institute.

FAO (Food and Agriculture Organization). 1991. FAO's 1990 reassessment of tropical forest cover. *Nature and Resources,* **27**(2), 21–6.

Geller, H. S. 1994. *Review of long term energy efficiency potential throughout the world.* Washington, DC: American Council for an Energy Efficient Economy.

Geller, H., and S. Nadel. 1994. Market transformation strategies to promote end-use efficiency. *Annual Review of Energy and Environment,* **19,** 301–46.

Geller, H. S., and D. Zylbersztajn. 1991. Energy intensity trends in Brazil. *Annual Review of Energy and Environment,* **16,** 179–203.

Goldemberg, J., T. B. Johansson, A. K. N. Reddy, and R. H. Williams. 1988. *Energy for a sustainable world.* New Delhi: Wiley Eastern.

Grübler, A., S. Messner, L. Schrattenholzer, and A. Schäfer. 1993a. Emission reduction at the global level. In *Long-term strategies for mitigating global warming.* Special Issue of *Energy: The International Journal* (N. Nakicenovic, special ed.), **18**(5), 539–81.

Grübler, A., S. Nilsson, and N. Nakicenovic. 1993b. Enhancing carbon sinks. In *Long-term strategies for mitigating global warming.* Special Issue of *Energy: The International Journal* (N. Nakicenovic, special ed.), **18**(5), 499–522.

Gupta, S., and N. Khanna. 1991. India country paper. In P. Ghosh, A. Achanta, P. Bhandari, M. Damodaran, and N. Khanna (eds.), *Collaborative study on strategies to limit CO_2 emissions in Asia and Brazil.* New Delhi: Asian Energy Institute.

Guzman, O., A. Yunez-Naude, and M. S. Wionczek. 1987. *Energy efficiency and conservation in Mexico.* Boulder, CO: Westview Press.

Hendriks, C. A. 1994. *Carbon dioxide removal from coal-fired power plants.* Dordrecht: Kluwer.

Houghton, R. A. 1990. The future role of tropical forests in affecting the carbon dioxide concentration of the atmosphere. *Ambio,* **18**, 204–9.

Houghton, R. A., J. Unruh, and P. A. LeFèbvre. 1991. Current land use in the tropics and its potential for sequestering carbon. In D. Howlett and C. Sargent (eds.), *Proceedings of the Technical Workshop to Explore Options for Global Forest Management.* London: International Institute for Environment and Development, pp. 279–310.

Huang, Jin-Ping. 1993. Industry energy use and structural change: A case study of the People's Republic of China. *Energy Economics,* **15**(2), 131–6.

IEA (International Energy Agency) / OECD (Organization for Economic Cooperation and Development). 1987. *Renewable sources of energy.* Paris: IEA / OECD.

IEA (International Energy Agency) / OECD (Organization for Economic Cooperation and Development). 1991. *Energy efficiency and the environment.* Paris: IEA / OECD.

Imran, M., and P. Barnes. 1990. *Energy demand in the developing countries: Prospects for the future.* Staff Working Paper no. 23. Washington, DC: World Bank.

IPCC (Intergovernmental Panel on Climate Change). 1990a. *Methane emissions and opportunities for control.* Subgroup for Methane Emissions and Opportunities for Control, workshop results of the Response Strategies Working Group, September.

IPCC (Intergovernmental Panel on Climate Change). 1990b. *Energy and Industry Subgroup report.* Energy and Industry Subgroup, workshop results of the Response Strategies Working Group, September.

IPCC (Intergovernmental Panel on Climate Change). 1990c. *Greenhouse gas emissions from agricultural systems.* Agriculture, Forestry and Other Human Activities Subgroup, workshop results of the Response Strategies Working Group, September.

IPCC (Intergovernmental Panel on Climate Change). 1991. *Climate change: The IPCC response strategies.* Washington, DC: Island Press.

IPCC (Intergovernmental Panel on Climate Change). 1992. *1992 IPCC Supplement,* Geneva, February.

IPCC (Intergovernmental Panel on Climate Change). 1995. *Greenhouse gas inventory: IPCC guidelines for national greenhouse gas inventories,* vol. 3. Reference manual. Bracknell: United Kingdom Meteorological Office, England.

Jackson, T. 1991. Least-cost greenhouse planning: Supply curves for global warming abatement. *Energy Policy,* **19**(1), 35–46.

Jepma, C. J. 1995. *Tropical deforestation: A socio-economic approach.* London: Earthscan.

Jepma, C. J., M. Asaduzzaman, I. Mintzer, R. S. Maya, and M. Al-Moneef. 1996. A generic assessment of response options. In J. P. Bruce, H. Lee, and E. F. Haites (eds.), *Climate change 1995: Economic and social dimensions of climate change.* Contribution of Working Group III to the Second Assessment Report of the Intergovernmental Panel on Climate Change. Cambridge University Press, pp. 225–62.

Johansson, T. B., H. Kelly, A. Reddy, R. H. Williams, and L. Burnham. 1993. *Renewable energy: Sources for fuels and electricity.* Washington, DC: Island Press.

Kassler, P. 1994. *Energy for development.* Shell Selected Paper. London: Shell International Petroleum Company Ltd.

Kaya, Y. 1989. *Impact of carbon dioxide emission control on GNP growth: Interpretation of proposed scenarios.* Intergovernmental Panel on Climate Change, Response Strategies Working Group.

Kaya, Y., et al. 1991. Assessment of the technological options for mitigating global warming. Paper presented at the Intergovernmental Panel on Climate Change Energy and Industry Subgroup meeting, Geneva, August 6–7.

Krankina, O. N., and R. K. Dixon. 1993. Forest management options to conserve and sequester terrestrial carbon in the Russian Federation. *World Resources Review,* **6**(1), 88–101.

Levine, M. D., A. Gadgil, S. Meyers, J. Sathaye, and T. Wilbanks. 1991. Energy efficiency, developing nations and Eastern Europe. Paper presented to the U.S. Working Group on Global Efficiency, Washington, DC.

Li, J., R. M. Shreshtha, and W. K. Foell. 1990. Structural change and energy use: The case of the manufacturing sector in Taiwan. *Energy Economics,* **12**(2), 109–15.

MacKerron, G. 1992. Nuclear costs: Why do they keep rising? *Energy Policy,* **20**(July), 641–52.

Makarov, A. A., and I. Bashmakov. 1990. The Soviet Union. In W. U. Chandler (ed.), *Carbon emissions control strategies.* Washington, DC: World Wildlife Fund.

Matsuo, N. 1991. Japan country paper. In P. Ghosh, A. Achanta, P. Bhandari, M. Darnodaran, and N. Khanna (eds.), *Collaborative study on strategies to limit CO_2 emissions in Asia and Brazil.* New Delhi: Asian Energy Institute.

Meier, A. 1991. Supply curves of conserved energy. In *Proceedings of the IEA International Conference on Technology Responses to Global Environmental Challenges: Energy Collaboration for the 21st Century,* vol. 1. Kyoto: Inter Group Corporation.

Mills, E., D. Wilson, and T. B. Johansson. 1991. Getting started: No-regrets strategies for reducing greenhouse gas emissions. *Energy Policy,* **19**(June), 527–42.

Mongia, N., R. Bhatia, J. Sathaye, and P. Mongia. 1991. Costs of reducing CO_2 emissions from India. *Energy Policy,* **19**(10), 978–86.

Moulton, R. J., and K. R. Richards. 1990. *Costs of sequestering carbon through tree planting and forest management in the United States.* USDA Forest Service General Technical Report WO-58. Washington, DC: Department of Agriculture.

Munasinghe, M. 1995. *Sustainable energy development.* Washington, DC: World Bank.

Munasinghe, M. 1996. *Environmental impacts of macroeconomic and sectoral policies.* Washington, DC, and Nairobi: World Bank and UNEP, chap. 9.

Nakicenovic, N., and A. Grübler. 1993. Energy conversion, conservation and efficiency. In *Long-term strategies for mitigating global warming.* Special issue of *Energy: The International Journal* (N. Nakicenovic, special ed.), **18**(5), 421–35.

Nakicenovic, N., and A. John. 1991. CO_2 reduction and removal measures for the next century. *Energy: The International Journal,* **16**(11–12), 1347–77.

Nakicenovic, N., and D. Victor. 1993. Technology transfer to developing countries. In *Long-term strategies for mitigating global warming.* Special issue of *Energy: The International Journal* (N. Nakicenovic, special ed.), **18**(5), 523–38.

Nakicenovic, N., D. Victor, A. Grübler, and L. Schrattenholzer. 1993. Introduction. In *Long-term strategies for mitigating global warming.* Special issue of *Energy: The International Journal* (N. Nakicenovic, special ed.), **18**(5), 401–19.

Nilsson, N.-E. 1982. An alley model for forest resources planning. In B. Ranneby (ed.), *Statistics in theory and practice: Essays in honour of Bertil Matrn.* Umea: Swedish University of Agricultural Sciences.

Nordhaus, W. D. 1990. An intertemporal general-equilibrium model of economic growth and climate change. In D. O. Wood and Y. Kaya (eds.), *Proceedings of the Workshop on Economic/Energy/Environmental Modeling for Climate Analysis.* Cambridge, MA: MIT Center for Energy Policy Research, pp. 415–33.

Ogawa, Y. 1992. Analysis of factors affecting carbon dioxide emissions due to past energy consumption around the world. In *Greenhouse research initiatives in the ESCAP region: Energy.* Bangkok: Economic and Social Commission for Asia and the Pacific.

Parikh, J., and S. Gokarn. 1993. Climate change and India's energy policy options: New perspectives on sectoral CO_2 emissions and incremental costs. *Global Environmental Change* (September), 276–91.

Park, Se-Hark. 1992. Decomposition of the industrial energy consumption: An alternative method. *Energy Economics,* **14**(4), 265–70.

Parks, P. J., and I. W. Hardie. 1992. *Least-cost forest carbon reserves: Cost-effective subsidies to convert marginal agricultural land to forest.* Report sponsored by American Forests, Washington, DC.

Persson, A., and M. Munasinghe. 1995. Natural resource management and economywide policies in Costa Rica. *World Bank Economic Review,* 9(2), 259–85.

Qureshi, A., and S. Sherer. 1994. *Climate change in Asia: Forestry and land use.* Asian Development Bank's regional study on global environmental issues. Washington, DC: Climate Institute.

Read, P. 1994a. *Responding to global warming: The technology, economics and politics of sustainable energy.* London: ZED Books.

Read, P. 1994b. Biofuel as the core technology in an effective response strategy. Economics Discussion Paper, Massey University, Palmerston North, New Zealand.

Reijntjes, C., B. Haverkort, and A. Waters-Bayer. 1992. Farming for the future: *An introduction to low-external-input and sustainable development.* London: Macmillan Press.

Reilly, J., and K. R. Richards. 1993. Climate change damage and the trace gas index issue. *Environmental and Resources Economics,* 3, 41–61.

Richards, K. R. 1993. Valuation of temporary and future greenhouse gas reductions. Paper presented at the Western Economic Association Annual Meeting, Lake Tahoe, NV.

Richards, K. R., R. Moulton, and R. A. Birdsey. 1993a. Costs of creating carbon sinks in the U.S. *Energy Conservation and Management,* 34(9–11), 905–12.

Richards, K. R., D. H. Rosenthal, J. A. Edmonds, and M. Wise. 1993b. The carbon dioxide emissions game: Playing the net. Paper presented at Western Economic Association Annual Meeting, Lake Tahoe, NV.

Robinson, J., M. Fraser, E. Haites, D. Harvey, M. Jaccard, A. Reinsch, and R. Torrie. 1993. *Canadian options for greenhouse gas emission reduction (COGGER).* Final Report of the COGGER Panel to the Canadian Global Change Program and the Canadian Climate Program Board, Royal Society of Canada, Ottawa.

Rogner, H., N. Nakicenovic, and A. Grübler. 1993. Second- and third-generation energy technologies. In *Long-term strategies for mitigating global warming.* Special issue of *Energy: The International Journal* (N. Nakicenovic, special ed.), 18(5), 461–84.

Rubin, E. S., R. N. Cooper, R. A. Frosch, T. H. Lee, G. Marland, A. H. Rosenfeld, and D. D. Stine. 1992. Realistic mitigation options for global warming. *Science,* 257(July), 148–9, 261–6.

Schipper, L., R. B. Howarth, and H. Geller. 1990. United States energy use from 1973 to 1987: The impacts of improved efficiency. *Annual Review of Energy,* 15, 455–504.

Sedjo, R. A., and A. M. Solomon. 1989. Climate and forests. In N. J. Rosenberg, W. E. Easterling III, P. R. Crosson, and J. Darmstadter (eds.), *Greenhouse warming: Abatement and adaptation.* Washington, DC: Resources for the Future, pp. 105–20.

Simpson, V. J., and C. Anastasi. 1993. Communication: Future emissions of CH_4 from the natural gas and coal industries. *Energy Policy,* 21(August), 827–30.

Sitnicki, S., K. Budzinski, J. Juda, J. Michna, and A. Szpilewica. 1990. Poland. In W. U. Chandler (ed.), *Carbon emissions control strategies.* Washington, DC: World Wildlife Fund, pp. 55–80.

Springmann, F. 1991. *Analysis of the ecological impact of demonstration projects in the field of rational use of energy: Development of evaluation criteria.* Study on behalf of the Commission of the European Communities, Directorate General for Energy (DG XVII), Regio-Tec GmbH, Starnberg, Germany.

Swisher, J. 1991. Cost and performance of CO_2 storage in forestry projects. *Biomass and Energy,* 1(6), 317–28.

Swisher, J., D. Wilson, and L. Schrattenholzer. 1993. Renewable energy potentials. In *Long-term strategies for mitigating global warming*. Special issue of *Energy: The International Journal* (N. Nakicenovic, special ed.), **18**(5), 437–59.

TNO (Netherlands' Organisation for Applied Scientific Research). 1992. *Confining and abating CO_2 from fossil fuel burning: A feasible option?* Report to the European Community, TNO, Apeldoorn.

Train, K. 1985. Discount rates in consumers' energy-related decisions: A review of the literature. *Energy,* **16**(12), 1243–53.

Trexler, M. C. 1991. Estimating tropical biomass futures: A tentative scenario. In *Proceedings of the Technical Workshop to Explore Options for Global Forestry Management*. London: International Institute for Environment and Development.

Trexler, M. C., P. E. Faeth, and J. M. Kramer. 1989. *Forestry as a response to global warming: An analysis of the Guatemala agroforestry and carbon sequestration project*. Washington, DC: World Resource Institute.

Turner, D. P., J. J. Lee, G. J. Koperper, and J. R. Barker (eds.). 1993. *The forest sector carbon budget of the United States: Carbon pools and flux under alternative policy options*. Corvallis, OR: U.S. Environmental Protection Agency.

UNEP (United Nations Environment Program). 1993. *UNEP greenhouse gas abatement costing studies: Analysis of abatement costing issues and preparation of a methodology to undertake national greenhouse gas abatement costing studies*. Phase Two Report, UNEP Collaborating Centre on Energy and Environment/Risö National Laboratory, Denmark.

U.S. Congress, OTA (Office of Technology Assessment). 1991. *Changing by degrees: Steps to reduce greenhouse gases*. OTA-O-842. Washington, DC: U.S. Government Printing Office.

U.S. Congress, OTA (Office of Technology Assessment). 1992. *Fuelling development: Energy technologies in developing countries*. OTA-E-516. April. Washington, DC: U.S. Government Printing Office.

U.S. EPA (U.S. Environmental Protection Agency). 1993. *Options for reducing methane emissions internationally*. Volume 1 of K. B. Hogan (ed.), *Technological options for reducing methane emissions,* report to Congress. Washington, DC: Office of Air and Radiation.

van Amstel, A. R. (ed.). 1993. *Proceedings of International IPCC Workshop on Methane and Nitrous Oxide: Methods in national emissions inventories and options for control,* Amersfoort, the Netherlands.

Watson, R. T., M. C. Zinyowera, and R. H. Moss (eds.). 1996. *Climate change 1995: Impacts, adaptations and mitigation of climate change – Scientific-technical analyses*. Contribution of Working Group II to the Second Assessment Report of the Intergovernmental Panel on Climate Change. Cambridge University Press.

WEC (World Energy Council). 1993. *Renewable energy resources: Opportunities and constraints, 1990–2020*. London: WEC.

WEC (World Energy Council). 1994. *New renewable energy resources: A guide to the future*. London: Kogan Page.

WEC (World Energy Council) Commission. 1993. *Energy for tomorrow's world: The realities, the real options and the agenda for achievement*. London and New York: Kogan Page and St. Martin's Press.

Wenger, H., T. Hoff, and R. Perez. 1994. Photovoltaics as a demand-side management option: Benefits of a utility–customer partnership. In *Proceedings of the 15th World Energy Engineering Congress,* Atlanta, October.

Winjum, J. K., R. K. Dixon, and P. E. Schroeder. 1992. Estimation of the global potential of forest and agroforest management practices to sequester carbon. *Water, Air, and Soil Pollution,* **62**, 2131–227.

Winjum, J. K., and D. K. Lewis. 1993. *Forest management and the economics of carbon storage: The nonfinancial component.* Corvallis, OR: U.S. Environmental Protection Agency, Environmental Research Laboratory.

World Bank. 1993. *Energy efficiency and conservation in the developing world.* Washington, DC: World Bank.

WRI (World Resources Institute). 1990. *World Resources, 1990–91: A guide to the global environment.* New York: Oxford University Press.

Xu, P. In press. The potential for reducing atmospheric carbon by large-scale afforestation in China and released cost/benefit analysis. *Biomass and Bioenergy.*

7

COSTS OF IMPLEMENTATION

In the preceding chapter, a number of response options were discussed. In this chapter we investigate how the information regarding these options and their costs, potential, and feasibility can be employed in order to form a basis for decision making.

The fundamental understanding derived from Chapter 6 is that the costs of a particular option depend on three main factors: the scale of application, the place of application, and the timing of application. Therefore, decision making on greenhouse policies should always take these into account. In this chapter, we provide information on the costs of mitigation options from the perspective of these three factors.

In Section 7.1, we focus on the factor of scale by assuming that greenhouse policies are considered now, based on the information available on present cost functions for a particular set of regions. In Section 7.2, we deal with the factor of place by investigating how costs can be reduced by carrying out policies in which the cost effectiveness of options is largest. In Section 7.3, we focus on the factor of time by considering the impact on the current discounted costs of delayed implementation of some policies. Finally, in Section 7.4, we integrate the three elements in order to see what might constitute an ideal policy package.

7.1 THE OPTIMAL BLEND OF RESPONSE OPTIONS AND THE SCALE OF APPLICATION

On the basis of the discussion in Chapter 6, it seems fair to assume that cost functions of response options are generally characterized by internal diseconomies of scale leading to increasing marginal costs (at least for large-scale applications; see Box 7.1). Empirical evidence supports this assumption.[1] This explains why a blend of options will be the most likely optimal outcome of a response strategy. According to economic theory, the optimal mixture of options – assuming that the cost functions are known and that place and time of implementation are irrelevant to present policy decision making – should be characterized by equal marginal costs for the individual options.[2] This is logical because if this condition were not satisfied it would make sense from an economic point of view to shift the focus from the option with the higher marginal costs to an option with lower marginal costs. Thus, the overall costs of the response strategy would be reduced.

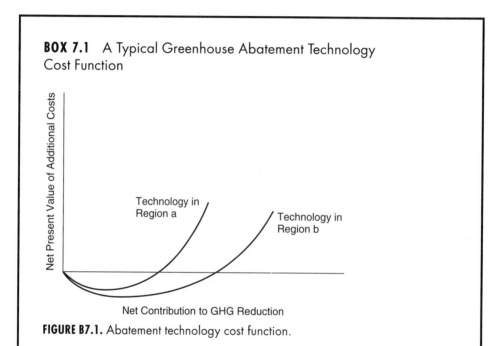

BOX 7.1   A Typical Greenhouse Abatement Technology Cost Function

FIGURE B7.1. Abatement technology cost function.

An example of a typical abatement technology cost function – assuming a given technology – is represented in Figure B7.1. Note that the horizontal axis reflects the net contribution (beyond what would have happened in the base scenario) to greenhouse gas reduction, and the vertical axis reflects the net present value of the additional costs (compared

---

[1] For some evidence with respect to forestry options, see, e.g., Moulton and Richards (1990), Parks and Hardie (1992), Adams et al. (1993), and Qureshi and Sherer (1994); with respect to energy technologies, see, e.g., Southern Centre/Risö (1993) and Kram (1994b); for broader analyses, see, e.g., TNO (1992).

[2] Other approaches to decision making are also possible. One can rely, e.g., on the concept of safe minimum standards (which may be particularly important in evaluating investments in nuclear power plants).

with the costs involved in a traditional technology) or life cycle costs involved in starting a new technology.

The graph has a number of characteristics:

1. At a relatively small-scale application of the new technology, the average costs in terms of dollars of investment needed to enhance an extra emission reduction of 1 ton of coal (beyond the emission reduction that would have been realized anyway in the base scenario) tend to decline. This is due to the economies of scale.
2. If the technology is applied at a modest scale, the average costs of the first applications are negative. This is the no-regrets potential of the technology.
3. If the technology is applied on a larger scale, the average costs tend to increase due to diseconomies of scale – for example, because increasing opportunity costs or political resistance surpass any remaining technical economies of scale.
4. Marginal costs surpass the average costs once the cost function moves upward.
5. At a certain stage, that is, if the technology is applied on a very large scale, public resistance or other societal obstacles can become so large that the marginal costs will tend to rise very steeply; any further expansion of the technology will be virtually impossible.
6. The cost functions differ depending on the region where the technology is applied.

The literature on response options focuses mainly on individual technologies and their cost effectiveness. Options are assessed in isolation rather than on the basis of mutual comparison. The main explanation for the predominance of the "partial" approach is the emphasis on engineering aspects and the limited availability of reliable and accepted data on the costs and benefits of the options.[3]

However, a generic assessment would require a framework that allows for a simultaneous evaluation of technologies or options. If emission reduction targets are to be achieved in an economically optimal way, and in terms of flexibility and spreading of risks, a full picture of all the alternatives must be available so that an integrated portfolio of options can be determined that minimizes the costs of a given level of reduction of the carbon concentration.[4] This is especially relevant at the country level, where national energy plans

---

[3] Moreover, it is increasingly recognized that the costs of options depend on the assumptions made about the efficiency of the baseline scenario used in the analysis (see also Section 7.3).

[4] Note that an assessment may take into account noneconomic considerations. However, here we assume economic optimization.

must be developed on an integrated basis (see Chapter 3). Ultimately, any climate change response strategy has to be implemented by nation-states, so that the process demonstrated in Figure 3.3 plays a key role (for a practical application to Sri Lanka, see Meier, Munasinghe, and Suyambatapiye 1993).

A fairly small number of studies taking an integrating approach have been carried out. Some of these studies have been performed in the top-down tradition, aiming at completeness and based on generalized estimates of the options' cost functions (e.g., McKinsey & Company 1989; Nordhaus 1991; Jepma and Lee 1995; Richels et al. 1996). Other studies are more in the bottom-up tradition (e.g., Jackson 1991; Mills, Wilson, and Johansson 1991; Rubin et al. 1992; Kram 1994b; UNEP 1994). For some remarks on the controversy over top-down versus bottom-up modeling approaches, see Box 7.2.

BOX 7.2   The Top-Down versus Bottom-Up Modeling Controversy

Researchers have tried to analyze, predict, explore, and assess the feasibility of alternative future developments in various ways. As far as the greenhouse gas emission issue is concerned, there are two broad analytical traditions, even though this distinction has become increasingly blurred by recent developments in modeling: top-down and bottom-up modeling. Top-down models are economically driven, have a predictive orientation, are based on historically derived parameter values, and consider the energy sector as part of the overall economic system. They are typical tools of economists. Bottom-up models, however, are more commonly applied by researchers in the tradition of technology and engineering and tend to focus explicitly on the technology factor in particular sectors; in this respect these models are generally more detailed.

Although there is no a priori reason why the two approaches would give different results in terms of future energy scenarios, in practice they do. This is one of the main reasons for the controversy over the two approaches. The greater emphasis in top-down models on economic feedback loops has caused greater pessimism about the costs of emission reductions among representatives of this approach than among bottom-up modelers. Usually, the latter group has more faith in the ability of society to adopt new technologies more quickly and points out the success of, for instance, earlier government initiatives to foster technological innovation and improvements in energy efficiency or the potential of demand-side management programs to speed up new initiatives. The result is that where top-down modelers feel uneasy about adopting parameter values that have not emerged from analyzing historical data sets, bottom-up modelers more daringly point out certain success stories. Top-down modelers rely on historical rules; bottom-up modelers trust the successful exceptions.

An example involving two crucial variables in energy modeling may clarify this distinction: the autonomous energy efficiency trend and the elasticity of substitution (which describes the degree to which capital and labor can be substituted for energy if the relative price of energy, i.e., vis-à-vis the price of the other inputs, changes). Obviously, both variables play a crucial role in future energy (and overall economic) development. Top-down modelers tend to use the parameter values for these two variables as they have been derived from historical data: why would things be different in the future? If, however, bottom-up modelers were to convincingly demonstrate that future parameters can easily deviate from past values, the parameter values could be adjusted in top-down models. This would obviously affect predictions. If not, the controversy remains. The main reason why it has become so fierce has to do with the models' different assessments of no-regrets potential. Top-down modelers tend to be skeptical about such potential and argue that environmentally benign technologies that are commercially feasible will be implemented automatically. That is, if technologies are not implemented, they are obviously not commercially feasible (yet). Bottom-up modelers, in contrast, argue that there is considerable scope for no-regrets strategies in energy efficiency – about 15 to 30%. What is needed to put this negative cost potential into effect is simply the application of certain policies that remove market distortions, lower transaction costs and risks generally assumed by firms and households, set mandatory standards, manage the energy demand side, and so on.

In sum, what the bottom-up vs. top-down controversy boils down to is whether one is prepared to accept major deviations from past trends (bottom-up) or instead tends to argue along the lines of "seeing is believing" (top-down): optimism (or naïveté?) versus pessimism (or unwarranted skepticism?).

Before we discuss some of the research outcomes concerning the ideal blend of options for a response strategy, several sensitivities of the options' costs should be reemphasized:

- Often, studies are not entirely clear as to the degree to which opportunity costs, social and institutional barriers, and other environmental side effects have been included in the cost functions employed. Obviously, this highly affects the form of certain cost functions. Welfare economic considerations may cause the cost curves to become more or less steep or to shift upward or downward depending on what would otherwise have been the case.
- The global costs of achieving the generally acknowledged long-term emission reduction targets – commonly estimated at several tens of billions of dollars per annum – turn out to be sensitive to the degree of no-regrets

policies, especially in energy conservation, efficiency improvement, and fossil fuel switching.

• Many other factors make the costs of response options very dependent on the context, which may explain why the costs of response strategies in the literature differ widely. Box 7.3 illustrates this point.

Thus, caution is required in drawing conclusions on the basis of studies integrating the cost functions of different options.

BOX 7.3   The Wide Variety of CO_2 Abatement Costs

What would the costs be in terms of (future) GNP if CO_2 emissions were reduced by a specific percentage from the baseline projection? This is one of the key questions policy makers would like to answer in order to prepare policy initiatives. Obviously, the answer is: it all depends. It depends on the baseline projection, technological progress, elasticities, timing, the types of policies, the progress of international policy coordination, and so on. Many of these factors will be discussed in greater detail in this chapter. To start with, however, Figure B7.2 illustrates the wide variety of outcomes of different CO_2 abatement cost studies, all related to the United States.

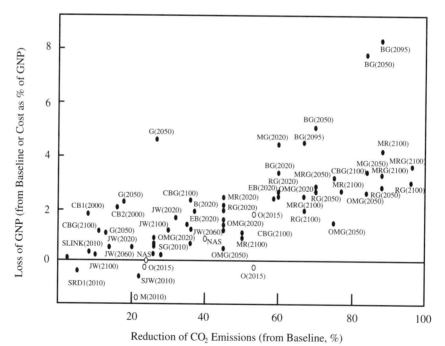

FIGURE B7.2. U.S. studies: cost of CO_2 abatement relative to baseline projection. Solid ovals, top-down, economic model estimate; open ovals, bottom-up, technology-based estimate. For a key to the studies used, see Hourcade et al. (1996), p. 304, Table 9.1. From Grubb et al. (1993).

1. Cost estimates vary widely but suggest that a considerable reduction of CO_2 emissions from the baseline can be achieved in the United States only at formidable cost, amounting to a few percentage points of (future) GNP (which itself may have doubled or tripled by the time the reductions occur). To illustrate, if CO_2 emissions were to be reduced by less than 40% from the baseline level, most studies indicate that the U.S. GNP loss would vary from a rather small amount (or might even be negative) to about 2%; an additional 60% reduction would cost about 2 to 4% of U.S. GNP.

2. Larger reductions incur greater costs. There are no clear signs that costs will rise exponentially if higher targets are set, but this may be due to the fact that higher levels of abatement generally assume a longer time interval to carry out the policy measures. This increases the probability that society will adapt.

3. There is a difference of opinion concerning the scope for no-regrets strategies. Some studies – especially bottom-up studies (see Box 7.2) – suggest that there is a no-regrets potential of 15 to 30% of projected CO_2 emissions; however, other studies – the majority being in the top-down tradition – argue that any emission reduction from the baseline would lead to a GNP loss.

The key insight to be derived from the figure is that many factors play a role in determining the costs of abatement strategies, given the existing response options.

To illustrate a straightforward integrated approach, we present the results of an optimization procedure in Table 7.1. By means of linear programming, starting from a predetermined emission reduction target, the procedure could be applied to the cost functions of the various options per region featuring stepwise-increasing marginal costs and based on a combination of several sources (McKinsey & Company 1989; Jackson 1991; Mills et al. 1991; Rubin et al. 1992).

The table shows the optimal mixture of options in terms of type of option (and region of application) if a medium-term emission reduction target of -2.4 GtC is to be achieved. (In parentheses are the outcomes if marginal costs of the renewable option are assumed to be 50% of those in the database and if marginal costs of the forestry option are doubled compared with the database, respectively.)[5]

The outcomes suggest that the largest potential in overall emission reduction at current cost estimates is in forestry (especially in developing countries), energy conservation and efficiency improvement (especially in the

[5] This sensitivity test suggests that the outcomes are rather robust. Obviously, other sensitivity tests, e.g., on the impact of changing lifestyles, could be carried out.

TABLE 7.1 Optimal Mixture of Options and Regions of Application for −2.4 GtC Emission Reduction Target

| Option | Level of Emission Reduction (MtC) | | | |
	OECD	Eastern Europe	Rest of the World	Total
Improvements in energy conservation and efficiency	250 (250;250)	250 (250;250)	100 (100;100)	600 (600;600)
Fuel switching	50 (50;50)	50 (50;50)	50 (50;50)	150 (150;150)
Removal and disposal	100 (100;150)	50 (50;100)	0 (0;50)	150 (150;300)
Nuclear energy	50 (50;50)	50 (50;50)	0 (0;50)	100 (100;150)
Renewable energy	50 (100;50)	50 (100;100)	100 (150;150)	200 (350;300)
Forestry	250 (250;250)	250 (250;250)	700 (550;400)	1,200 (1,050;900)
Total	750 (800;800)	700 (750;800)	950 (850;800)	2,400 (2,400;2,400)

Note: Marginal costs in optimum, $50/tC. Numbers in parentheses are figures for 50% reduction in marginal costs of renewables over all intervals and for a doubling of marginal costs of forestry over all intervals, respectively.

Source: Jepma and Lee (1995).

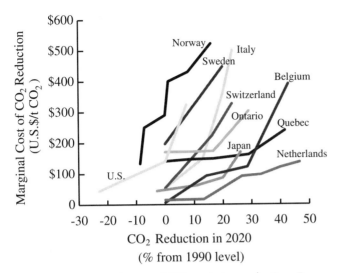

FIGURE 7.1. Marginal costs of CO_2 emission reduction. From Kram (1994a).

OECD and Eastern Europe), and to a lesser extent in renewable energy (especially in developing countries) and fuel switching (especially in Eastern Europe, if CH_4 leakages can be restricted). Needless to say, the ideal mix can change easily if cost functions turn out to have a different shape than assumed here and/or due to ongoing technological progress. The results in the table simply serve as an illustration.

An example of a detailed integrated response study in the bottom-up tradition is that of Kram (1994b).[6] Here an overall assessment is made on the basis of long-term bottom-up country models (MARKAL) for nine Western countries with more than 70 types of technologies (more than 30 supply technologies and more than 40 end-use technologies), which are then integrated. A range of targets for CO_2 emission reductions by 2020 were tested to determine the mixture of energy technologies that would result in reductions at the least total energy system cost. The results revealed considerable diversity among these countries with respect to the ideal mixture of energy technologies applied to CO_2 emission reduction and to the cost and volume of reductions that could be achieved. This diversity depended on the future energy needs of the countries, their existing energy system, natural resources, technology options, and energy policies, especially with regard to hydroelectric and nuclear power.

In addition, the marginal costs of CO_2 reduction during 1990–2020 have been calculated for various countries (Figure 7.1). The results clearly show

[6] There seem to be no similar detailed integrating response studies for developing countries. Given that the level of economic development and other circumstances differ widely among developing countries, the marginal costs of emission reduction are likely to be rather context-specific.

that these costs differ widely among the countries. If one accepts the fact that the most efficient allocation of emission reductions is at the point of equal marginal costs, this provides an argument for implementing options on a joint or cooperative basis. This is the subject of the next section.

7.2   THE OPTIMAL BLEND OF RESPONSE OPTIONS AND THE PLACE OF APPLICATION

The need for an integrated approach is reinforced by the evidence that the cost functions of options also differ depending on where the options are applied. The underlying rationale is the same as that for the differences in production costs of the same goods and services among regions of the world: there are differences in supply conditions, levels of technology, infrastructure, and so on. Evidence suggests that even within a relatively homogeneous area like the European Union, marginal emission reduction cost curves differ significantly (COHERENCE 1991); a fortiori, it can be hypothesized that some options can also be significantly more cost-effective if applied elsewhere – meaning in countries in transition or developing countries.[7]

This raises the question of whether there are further a priori arguments for the fact that the cost functions of response options in economies in transition and in developing countries may differ from those in the industrialized world. The next subsections deal with this issue.

7.2.1   Emission Reduction Costs in Economies in Transition: A Special Case?

It is not easy to develop models that provide insight into the costs of emission reduction strategies in economies in transition, precisely because these economies are undergoing thorough transition processes and usually face economic crises along the way. A complicating factor is the long-standing tradition of highly subsidized energy prices in these economies, which has contributed to the fact that they are among the most energy-intensive economies in the world. Since the transition process entails not only structural adjustments (away from a strong emphasis on heavy industries) but also price reforms and the removal of energy subsidies, different factors may work together to increase energy efficiency. This has caused several researchers to hypothesize that the costs of medium-term stabilization of emissions may be relatively small for economies in transition.

---

[7] For more evidence, see McKinsey & Company (1989); Jackson (1991); Nordhaus (1991); Mills et al. (1991); Rubin et al. (1992); and Richels et al. (1996).

The latter has been corroborated by a number of studies, in both the top-down and bottom-up approaches. In a top-down study by Manne and Oliveira-Martins (1994), using the GREEN and 12RT model, it was projected that, in the former Soviet Union and Eastern Europe, no carbon taxes would be required to stabilize emissions to 1990 levels before 2040 (12RT) or even 2010 (GREEN), because price reforms and the removal of energy subsidies in these areas would suffice.[8] A similar impression arises from a number of indigenous country-specific models in the bottom-up tradition that have provided projections for the medium term (until, e.g., 2010); see also Table 7.2. They tend to indicate that very low or even negative costs (as a percentage of GNP) will be incurred in achieving nontrivial (i.e., roughly between 20 and 50%) CO_2 emission reductions from the baseline in the medium run. In the longer term, however, the costs of abatement strategies may rise if the positive effects of the transition have been fully realized. Abatement costs will probably develop more in line with the rest of the world, as Table 7.2 also suggests.

7.2.2 Emission Reduction Costs in Developing Countries: Is Stabilization Feasible?

In discussing any emission reduction target, there must be no misunderstanding about its meaning: either emission reductions are achieved vis-à-vis a particular benchmark year (e.g., 1990) or emission reductions are targeted as a percent change from a baseline scenario projection (for, e.g., 2025). It will be clear that it matters a great deal which of the two approaches is applied, particularly if emissions are projected to increase rapidly in the base scenario. Figure 7.2 illustrates this point. An emission stabilization at the 1990 level in 2025 corresponds to a in the figure; a 20% reduction from the base-case scenario projection in a world of low emission growth, instead, corresponds to b; if, however, emission growth is projected to be high in the base-case scenario, the same target corresponds to c. Thus, the formulation of targets does indeed matter.

This may explain why, when developing countries are concerned, targets of the latter type are usually employed: it is because of the rapidly increasing CO_2 emissions projected for such countries. Whereas nowadays less than a third of annual emissions are caused by developing countries (see Box 7.4), most scenarios suggest that in the next decades the growth in carbon emissions will increasingly take place in developing countries. According to

---

[8] Due to measurement problems, exact information needed for international comparison often cannot be provided. However, the fact that planned economies use(d) energy rather inefficiently is considered an established fact.

TABLE 7.2 Cost of CO_2 Emission Reductions

Country	Study	Type[a]	Forecast Year	CO_2 Reduction from Baseline (%)	Cost of Reduction (% of GNP)
Former Soviet Union	Burniaux et al. (1992)	TD	2020	45	0.9
	Burniaux et al. (1992)	TD	2050	70	2.3
	Burniaux et al. (1992)	TD	2100	88	3.7
	Kononov (1993)	BU	2005	50	0.3
	Makarov and Bashmakov (1991)	BU	2005	23	0.5
	Makarov and Bashmakov (1991)	BU	2020	44	1
	Manne (1992)	TD	2020	45	3.1
	Manne (1992)	TD	2050	70	6.4
	Manne (1992)	TD	2100	88	5.6
	Oliveira-Martins et al. (1992)	TD	2020	45	1.7
	Oliveira-Martins et al. (1992)	TD	2020	70	3.7
	Rutherford (1992)	TD	2020	45	1.5
	Rutherford (1992)	TD	2050	70	5.8
	Rutherford (1992)	TD	2100	88	4.1
Hungary	Jaszay (1990)	BU	2005	17	−0.1
Poland	Leach and Nowak (1991)	BU	2005	37	−0.1
	Leach and Nowak (1991)	BU	2005	53	−0.1
	Sitnicki et al. (1991)	BU	2005	44	0
	Sitnicki et al. (1991)	BU	2030	62	0.3
	Radwanski et al. (1993)	TD	2010	27	0
	Radwanski et al. (1993)	TD	2030	39	0
Czechoslovakia	Kostalova et al. (1991)	BU	2005	20	0
	Kostalova et al. (1991)	BU	2030	29	0

[a] TD indicates a top-down study; BU, a bottom-up study.
Source: Hourcade et al. (1996).

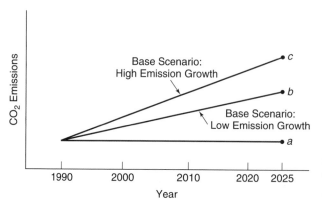

FIGURE 7.2. Implications of the time factor for emissions reduction targets.

Houghton et al. (1995), the mean world annual growth rate of CO_2 emissions until 2020 is 1.56% for more than 20 scenarios; the corresponding mean rate for China is 2.83%, for Eastern Europe and the former USSR 0.76%, and for Africa 3.85%. Consequently, the share of developing countries in global CO_2 emissions in 2020 is expected to be 46% according to Environmentally Compatible Energy Scenario 1992 (cf. OECD 34%; countries in transition 20%). However, according to the various World Energy Council scenarios (see note 10, Chapter 7), the share of developing countries in 2020 would be much larger, more than 60%. In any case, it is most likely that developing countries as a group will become the main CO_2 emitters within a few decades as per capita consumption levels – currently much lower than in Western countries – begin to rise with increasing economic growth. This picture is reinforced if CH_4 emissions from wetland rice cultivation and from enteric fermentation are also taken into account.

BOX 7.4 Energy Use in Developing Countries from an International Perspective

Although industrialized countries account for just 25% of the world's population, they account for 72% of the current global energy-related carbon emissions, as well as for most of the historical cumulative carbon emissions, that is, 80 to 85% (Fujii 1990). Clearly, this is an argument in favor of industrialized countries assuming their historic responsibility, which has been translated into the concept of *common but differentiated responsibilities* mentioned in Article 3.1 of the UNFCCC and elaborated in Principle 7 of the Rio Declaration: "The developed countries acknowledge the responsibility that they bear in the international pursuit of sustainable development in view of the pressures their societies place on the global environment and of the technologies and financial resources they command" (see also Boxes 2.2 and 2.3 on the role of equity and burden sharing).

In Table B7.1 the situation of the main regions of the world in 1990 is compared with respect to their economic development in general and their energy use in particular. This comparison highlights the following striking differences among and within regions and may clarify why the involvement of developing countries in greenhouse policy formulation and implementation is imperative:

1. GNP per capita varies from $440 for Asia to $17,473 for the Western world; the ratio is 1:40. Differences can also be large within regions: sub-Saharan Africa has a per capita GDP of $322 as against $1,507 for North and South Africa together (1:5).

(continued)

TABLE B7.1 Comparison of the 1990 Development and Energy Situations in the Main (Sub)regions

| Region | TR (%) | Fossil Fuels (%) | | | | Al (%) | ΣEner (Mtoe) | Popul (million) | GNP (billion per U.S.$) | GNP (U.S.$ per capita) | toe per capita | toe per thousand U.S.$ |
		Co	Oi	NG	ΣF							
West	1	22	41	19	82	16	4,296	849	14,840	17,473	5.06	0.29
North America	2	22	38	24	83	15	2,289	276	5,738	20,809	8.30	0.40
Western Europe	1	22	43	16	80	19	1,456	430	5,901	13,726	3.39	0.25
JANZ	1	21	51	12	84	16	551	144	3,202	22,279	3.83	0.17
East	1	25	29	36	91	8	1,745	414	3,047	7,367	4.22	0.57
Eastern Europe	1	46	25	19	90	9	352	125	388	3,101	2.81	0.91
Former USSR	1	20	31	40	91	8	1,393	289	2,660	9,215	4.83	0.52
North	1	23	38	24	85	14	6,040	1,263	17,888	14,163	4.78	0.34
Africa	37	21	27	9	57	6	360	642	385	600	0.56	0.93
North and South Africa	5	na	na	na	92	3	154	150	227	1,507	1.02	0.68
Sub-Sahara	61	na	na	na	30	9	206	492	158	322	0.42	1.30
Asia	15	49	25	4	78	7	1,475	2,810	1,237	440	0.53	1.19
China	6	72	15	2	89	5	718	1,139	393	345	0.63	1.83
India	25	41	23	4	68	7	253	853	287	337	0.30	0.88
Other	23	19	39	8	66	11	504	817	557	681	0.62	0.90

Latin America	15	4	45	14	63	22	556	448	842	1,880	1.24	0.66
Brazil	30	5	34	2	41	29	185	150	375	2,495	1.23	0.49
Mexico	5	5	49	22	75	20	142	89	170	1,920	1.60	0.83
Other	9	4	51	20	74	17	228	209	297	1,421	1.09	0.77
Middle East	1	1	64	32	98	1	233	129	370	2,860	1.81	0.63
South	17	31	33	10	74	10	2,624	4,029	2,835	704	0.65	0.93
World	6	25	36	20	81	13	8,664	5,292	20,723	3,916	1.64	0.42

Abbreviations and definitions: Tr, traditional (woodfuel, crop residues, and animal dung); Co, coal; Oi, Oil; NG, natural gas; ΣF, total fossil fuel; Al, alternatives (nuclear, hydro, wind, geothermal, etc.) (all % of total energy); ΣEner, total energy; (M)toe, (million) tons of oil equivalent; popul, population; GNP, gross national product in 1989; na, not available. Western Europe = OECD Europe (includes Turkey); Eastern Europe = non-OECD Europe (includes Cyprus, Gibraltar, and Malta); JANZ = Japan, Australia, and New Zealand.

Source: Ettinger (1994), based on *BP Statistical Review, 1992* and *World Resources, 1992–3.*

BOX 7.4   Energy Use in Developing Countries from an International
Perspective *(continued)*

2. The relative use of traditional energy (woodfuel, crop residues, and
   animal dung) varies from 1% for the East to 37% for Africa; within
   Africa it amounts to 5% for North and South and 61% for the sub-
   Saharan region (1:12).
3. The relative use of fossil fuels varies from 57% of total energy use for
   Africa to 98% for the Middle East; within Africa it amounts to 30% for
   sub-Saharan Africa and 92% for North plus South Africa (1:3).
4. The relative use of nuclear plus renewables varies from 1% for the
   Middle East to 22% for Latin America. Within Africa it amounts to 3%
   for North plus South and 9% for sub-Saharan Africa.
5. The share of the main energy source varies from 36% for the East
   (natural gas) to 64% for the Middle East (oil). For Africa this source is
   traditional, for Asia coal, and for the West and Latin America oil.
6. Energy use per capita varies from 0.53 for Asia to 5.06 toe for the
   West, or 1:10. Between Western subregions it still varies from 3.39 for
   Western Europe to 8.30 toe for North America.
7. The energy intensity (in toe/$1,000 GNP) varies from 0.29 for the
   West to 1.19 for Asia (1:4). Between Western subregions it varies from
   0.17 for Japan/Australia/New Zealand to 0.4 for North America.
   However, energy intensity in toe/$1,000 GNP is an unreliable yard-
   stick for comparisons among regions, especially between North and
   South, because of differences between nominal GNP and real GNP in
   purchasing power parities (PPP). If the energy intensity data in the
   table had been expressed in toe/$1,000 PPP (correction based on
   data from UNDP 1993), the energy intensity of the South would be
   0.35 but that of the North would remain at 0.34. However, this would
   still ignore the relatively higher energy contents of Southern imports
   and the lower energy contents of Southern exports. If these two factors are taken into account, Southern energy intensity is higher than
   that in toe/$1,000 PPP but probably lower than that in nominal terms
   (Ettinger 1994, p. 118, note 9).

As in economies in transition, in developing countries energy subsidies often distort energy prices. This implies a significant potential for improvement
in energy efficiency after the removal of subsidies. To illustrate, in an OECD
study (1994) – using the GREEN model – on the impact of energy subsidy removal in China and India, energy consumption and emissions in 2050 were
projected to decline nearly 40 and 60%, respectively, relative to the 14-fold in-

crease in energy consumption and carbon emissions in the 1985–2050 base-case scenario.

A weakness of energy policies in developing countries is the often significant share of the informal/subsistence sector. This sector is hard to model and reliable data about it are lacking; moreover, it is often beyond real government control. This explains why the informal sector sometimes shows unexpected behavioral patterns, which is highly relevant to our topic. In deciding between the use of commercial energy (gasoline or kerosene) or traditional, noncommercial energy sources (fuelwood or dung) for daily household purposes, income and substitution effects play an important role. As income increases, the share of commercial energy tends to increase, which is generally good from an emission reduction perspective because of the lower carbon content per unit of energy in commercial energy.[9] However, if energy prices were to rise – for example, due to the abolition of energy subsidies – people at the subsistence level would be likely to shift back to noncommercial energy sources, such as fuelwood, which could aggravate deforestation.

In sum, it is clear that although the scope for effectively applying policy options in developing countries seems to be significant (for an evaluation of technical options at the country level, see UNEP 1994), so are the obstacles likely to be encountered. Indeed, the availability of technical options for higher energy efficiency, to give just an example, does not guarantee their adoption on a large scale. A significant stimulus may be needed to achieve widespread efficiency improvements, particularly in markets characterized by high implicit discount rates. But a combination of education, financial incentives, and minimum efficiency standards together with better functioning energy markets (i.e., free of distortionary policies) can effectively transform energy use, so that large energy savings and emission reductions are achieved along with net economic savings (Geller and Nadel 1994).

The literature on the adoption and diffusion of technology clearly indicates that although profitability is probably the most straightforward reason for adopting a new idea, technology, or equipment, other factors may also be important. A review of recent research on the diffusion of energy technologies in developing countries shows that many financial, institutional, and other factors affect the successful adoption of these technologies (Barnett 1990; Ghai 1994). Often the initial awareness of such benefits and new opportunities may be contingent on factors like persuading women to introduce more energy-efficient cooking stoves. To give another example, the diffusion of drought-tolerant crops in Africa was limited by social resistance.

---

[9] Note that the positive impact can be negated by adverse effects if the energy consumption per capita tends to rise due to the shift toward commercial energy.

Moreover, a necessary ingredient for adopting technology, namely a pool of local skills to draw on, may be lacking or inadequate in many cases, so that even proven technologies may spread rather slowly in these countries. For all these reasons, an adequate and timely process of building institutions whose function is to enhance energy efficiency seems imperative, especially in developing countries (see Chapter 3). There is evidence that such institutions have helped, for example, Indonesia, South Korea, and Thailand make greater headway in the scope and coverage of energy efficiency policies and programs (Byrne et al. 1991).

7.3   THE OPTIMAL BLEND OF RESPONSE OPTIONS AND THE TIMING OF APPLICATION

In dealing with the optimal overall timing of greenhouse strategies, one faces a dilemma (see Section 2.3). On the one hand, information on climate change and its potential damage may increase in the course of time. Thus, it may be beneficial to reduce scientific uncertainty before implementing greenhouse policies. On the other hand, the costs of waiting can be considerable if a wait-and-see attitude causes future damage (or adjustment) costs to become excessive.

A similar dilemma with regard to the timing of action is caused by expectations regarding future economic and technological progress (see Section 2.3). In analyzing the optimal timing of abatement strategies, one must realize that in the long term greenhouse gas (GHG) emissions depend not only on factors like the rate of economic growth but also on the structure and physical characteristics of this growth – countries with similar development levels may have very different energy consumption per capita ratios. This is partly due to differences in such factors as technological patterns, consumption and trade patterns, the geographical distribution of activities, and structural changes in the productivity system.

The point, therefore, is that in analyses with a long time horizon – for example, assessing the feasibility and timing of GHG mitigation options – historical trends underlying development patterns can no longer be assumed to correctly describe the major future characteristics of development patterns or the pace and direction of the transformation of these characteristics. In other words, economic parameters derived from the past cannot be easily applied to the prediction of future production and consumption systems. The dynamics of long-term technological development may also be of great importance; for an illustration of the impact of different technology expectations on emission reduction projections, see Box 7.5.

BOX 7.5 Technology Expectations and Emission Reduction Scenarios

In a study conducted by the Energy Modeling Forum of Stanford University (EMF 1993), 14 top-down models were used to analyze U.S. emission reduction scenarios. For the sake of comparison common assumptions were employed for selected numerical inputs, such as the GDP growth rate, population growth, the fossil fuel resource base, and the cost and availability of various long-term backstop technologies. However, models differed in their methodologies, and particularly with respect to technology representation. So the models differed with respect to optimism (or pessimism) toward the potential decline in energy use per unit of output and carbon content per unit of energy (among others, through technology).

Figure B7.3 illustrates the U.S. carbon emissions for the period up to 2030 projected by the various models, assuming no control measures – a variety ranging from some 35 to 80%!

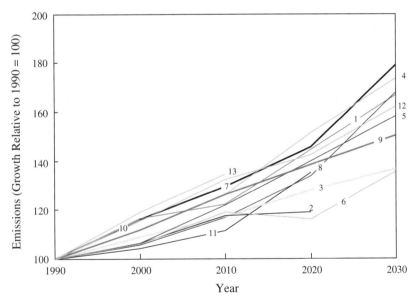

FIGURE B7.3. U.S. carbon emissions. From EMF (1993). Key: 1, CRTM; 2, DGEM; 3, ERM; 4, Fossil 2; 5, Gemini; 6, Global-Macro; 7, Global 2100; 8, Goulder; 9, GREEN; 10, ICA; 11, MARKAL; 12, MWC; 13, TGAS.

Whether a certain technical option is commercially viable in a particular situation often depends to a large extent on the time at which the option is applied. If the option involves a particular technology, this technology may be under development and may not yet be commercially appealing. Through research and development, as well as technological progress, however, this option may become viable a few years later. Thus, options that are now (far) too expensive for commercial application may become standard procedures some time in the future.

The problem, however, is that technological progress is difficult to predict. At a macroeconomic level, the overall productivity increase or energy saving potential due to technological improvements can be characterized more or less by a few trends. At the more detailed level of individual technologies, however, it is very hard to make sound statements about the pattern of technological progress and breakthroughs and their impact on productivity and energy use. Any cost assessment of abatement options as a function of the time at which they are applied, therefore, is rather speculative (for an illustration, see Box 7.6). Still, some general remarks can be made here.

BOX 7.6 Reduced Methane Emissions Through Improved Management in Dairy Production

Table B7.2 gives an example of the CH_4 emission reduction achieved in the United States between 1960 and 1990 as a result of improvements in feed supplementation technologies. It can be inferred from the table that U.S. dairy cows at present produce nearly 10 million tonnes more milk and 170,000 tonnes less CH_4 than dairy cows in 1960.

TABLE B7.2 Emission Reduction Resulting from Improved Livestock Management Techniques

	1960	1990
CH_4/cow/yr (kg)	76.1	114.6
Milk/cow/yr (kg)	3,195	7,000
CH_4/kg milk (g)	24	17
Lactating cows (no.)	17.5 million	10.1 million
Total milk production (tonnes)	55.9 million	65.6 million
Total CH_4 from dairy cows in the United States (Tg)	1.33	1.16

Source: Locklin (1995).

First, technology is not developed in isolation; it should be considered the result of an interactive process among different developments in society. Second, once the decision has been made to continue investments in a certain type of technology, irreversible processes of technological change are easily created for considerable periods of time, thus almost blocking the further development of alternative technologies. Third, the choice of a particular technology may retard the development of new energy-efficient technologies, so that the overall long-term costs of greenhouse abatement will eventually rise. These considerations are elaborated on in the following subsection.

Development Patterns and Irreversibilities

The close connections between technical choices, consumer demand, geographical distribution of activities and human settlements, and institutional setting explain the fact that particular sets of technological and behavioral options can be clustered into consistent packages, which, at least for a long period of time, create foreclosures of options in technology and innovations. These clusters are systematic in industries relying on network structures such as energy, transportation, and telecommunications because of the characteristics of their production function, the need for technical harmonization across the network, and their dynamic interactions with markets.

In this context, a crucial factor seems to be the flexibility (or inertia) of production and consumption patterns. As far as consumption patterns are concerned, the roots of inertia are technical and depend on specific cultures, habits, and social determinants.[10] As far as production patterns are concerned, inertia may occur in the decision-making process with respect to the renewal of technical equipment or other systems due to their long lead times, such as in building, infrastructure, or transportation.

The key point is that this inertia also has a crucial effect on the welfare consequences of specific mitigation strategies, at least during the transition period toward consumption and production systems that emit fewer GHGs. In other words, the welfare implications of a mitigation option depend on its timing. Precisely because of the inertia just mentioned, an abatement strategy that is introduced too quickly can be easily frustrated because transition costs are considered unsurmountable, whereas if the strategy is applied more gradually, it might be readily accepted, and successful as well.

The main impact of inertia on development patterns is that it can create an irreversible process of technological change. This can occur when such factors as learning curves, economies of scale, increasing informational returns, positive network externalities, and barriers to entry are combined so that a particular trajectory of technological change and development is created, which effectively makes impossible alternative choices that were available earlier. This time dependence of technical choices gives rise to technology

---

[10] Recall the comments in Sections 6.5.1, 6.5.2, and 7.2 with regard to the implicit discount rates in consumer decisions.

bifurcation points (for some examples, see Box 7.7). These occur, for instance, when during a particular period the government or private parties in a country have to choose between the introduction of alternative technologies – for example, the expansion of the railway system or the road system for private cars, or the expansion of nuclear energy production or of fossil fuels cum renewables. Once this choice has been made and the bifurcation point has been passed, market forces will reinforce the first choice in a self-fulfilling process (rail tickets will be cheaper due to economies of scale, or there will be less traffic congestion, etc.).

BOX 7.7   Two Examples of a Bifurcation Point

China: The Implications of Different Technologies – A Simulation

China is the world's most populous country and the largest coal producer and consumer. Not surprisingly, China was responsible for 11% of global CO_2 emissions in 1990; under a business-as-usual scenario, this contribution is projected to rise to 17% in 2050 and 28% in 2100 (Manne and Richels 1991). As a fuel, coal dominates CO_2 emissions, accounting for more than 83% of total emissions in 1993.

An obvious way of reducing CO_2 emissions would be to discourage coal production and consumption. However, given the fact that at the end of 1987 China's proven coal reserves were estimated at 859.4 billion tons, or about 50% of the world total, incentives to do so are weak. Moreover, technical capacity constraints limit the possibility of changing the infrastructure of existing power plants in the short and medium term.

In this spirit Zhang (1996) carried out a policy simulation for the Chinese electricity sector. Assuming that China reduced its CO_2 emission level in 2010 by 20% below the baseline CO_2 emissions for that year (which is still 2.46 times the 1992 emission), he estimated that a carbon tax of 205 yuan (current prices) per ton of carbon would be needed, corresponding to ad valorem tax rates on the use of coal, oil, natural gas, and electricity of 64, 14, 46, and 20, respectively. Similarly, a 30% emission reduction in 2010 would require a carbon tax of 400 yuan per ton of carbon, corresponding to ad valorem tax rates of 122, 28, 96, and 38, respectively.

Such a tax regime would stimulate the electricity sector in China to economize on the use of coal and to increase as sources of energy the share of nuclear energy (projected capacity to be installed in the first half of next century, hundreds of gigawatts), hydroelectric power (economically exploitable capacity estimated at 378 GW, the largest in the world; by the end of 1990, 9.5% of the exploitable potential developed), gas-based power (proven recoverable gas reserves in 1992, 1.4 trillion m³; exploration activity aimed at boosting production from 15.3 billion m³ in

TABLE B7.3 Simulation of a 65% Increase in the Real Coal Price in China: Main Results for 2010

	Large Coal Power		Hydroelectric Power		Nuclear Power	
	2000	**2010**	**2000**	**2010**	**2000**	**2010**
Reduction in CO_2 emissions relative to the baseline (%)	1.3	1.6	3.1	5.7	2.5	4.7
Capital investment in the electricity sector as a share of GNP (%)	2.34	2.25	2.38	2.38	2.34	2.34
Gross coal consumption (gce/kWh)[a]	348.37	333.46	351.58	336.96	352.19	337.64
Total generating capacity installed (GW)	275.69	539.63	277.21	545.01	275.32	538.27
Share of coal power in total capacity (%)	72.86	69.93	70.51	65.70	71.31	67.06
Share of hydropower in total capacity (%)	26.04	26.22	28.40	30.49	26.08	26.29
Share of nuclear power in total capacity (%)	0.76	3.45	0.76	3.41	2.28	6.24
GNP relative to baseline (%)		−1.535		−1.546		−1.542
Price of electricity		+23.19		+23.49		+23.40

[a]gce denotes grams coal equivalent.

1990 to 30 billion m³ by 2000), and renewables (abundantly available, present supply about 0.04% of total national commercial energy supply).

Zhang translated these shifts into three simulated policies entailing several relatively modest changes from the baseline scenario, assuming a linear increase in real coal prices of only 65% above the 1990 level by 2010:

1. moving to large coal-fired units (>300 MWe);
2. switching to hydroelectric power (>25 MWe); and
3. switching to nuclear power (600 to 1000 MWe).

A computable general equilibrium model was used to assess the GNP effects of these shifts. The overall results are presented in Table B7.3.

(continued)

BOX 7.7 Two Examples of a Bifurcation Point *(continued)*

They show that relatively modest shifts in policies in the medium term may already have a significant impact on CO_2 emissions; however, they also show that China's GNP is projected to decline in the policy scenarios relative to the baseline even if carbon tax revenues are expected to be recycled. This is because more capital is diverted to the relatively capital intensive electricity sector.

It can be concluded from the simulation that the hydroelectric power plant option is optimal from the perspective of CO_2 emission reductions. However, the price to be paid is that the same option has the highest cost in terms of capital investments in electricity required, GNP loss, and increased electricity prices.

Sri Lanka at the Crossroads of the Expansion of the Power System

By 1996 Sri Lanka was essentially a hydro-based power-generating economy with thermal power plants serving as a backup during dry periods (capacity of hydropower stations, 1,135 MW; thermal power stations, 250 MW; total, 1,385 MW).

Sri Lanka has no proven resources of fossil or nuclear fuels but has a certain amount of hydropower, 1,135 MW of which is already developed and 200 MW of which is under development; an economic potential of just 300 MW remains. Therefore, most of the future power production will have to come from thermal power plants. The overdependence on hydropower has already proved costly, with lower than average rainfall resulting in heavy power cuts and resulting economic losses in 1996.

In 1996, the demand for electricity in Sri Lanka had grown at an average annual rate of 7.8% since the mid-seventies; since 1992, the growth rate even increased to about 10%. Demand will continue to grow at this rate in the medium term. Therefore, a significant expansion of electricity generation is needed: according to projections of the Ceylon Electricity Board, some 1,150 MW during the next decade, involving investments of the order of U.S. $1.9 billion.

According to calculations of the Ceylon Electricity Board, the cheapest generation expansion plan would involve a significant amount (some 600 MW) of coal power plants, and in addition 190 MW diesel-based and 136 MW combined power plants. However, this option meets severe environmental opposition because of the delicate agricultural systems and the high population density of the island. For this reason, natural gas instead of coal is preferred from an environmental perspective, because it does not contain sulfur, it has a high percentage of hydrogen, and it generally produces the lowest amount of polluting gases – including CO_2 – for a given output.

Therefore, the Sri Lankan government, acknowledging there are no technological impediments to gas generation in Sri Lanka, after having taken the policy decision to open up thermal generation to the private sector, indicated that it will offer every encouragement to the private participants to make gas generation a reality in Sri Lanka.

Source: Based on a speech of Sri Lanka Energy Minister Anuruddha Ratwatte at an energy conference in Singapore, from the *Sri Lanka Journal Daily News,* March 21, 1996, p. 18.

Moreover, bifurcation points can also be reached by a broader set of decisions. Once certain choices have been made regarding key characteristics of a region's socioeconomic development (see Box 7.8) – usually for reasons that have nothing to do with energy policy or climate change issues, and often even without explicit public policy – to a large extent the baseline emission levels for the medium to long term are determined, and thus so are the cost functions of the emission reduction or adaptation options. Two examples will illustrate this. If (in the 1970s) Brazil had not selected a large-scale biomass ethanol program, current mitigation costs would be quite different. Similarly, the mitigation cost functions for France would be different if in the 1970s France had not chosen to embark upon a large-scale nuclear energy program.

BOX 7.8 Long-Term Cost Effectiveness of Response Options

The following are key characteristics that pinpoint the long-term cost effectiveness of various response options:

1. The raw material intensity of the economy. This has dropped significantly in industrialized countries – for example, by the increasing share of services and the increase in information intensity and use of telecommunications. On the other hand, raw-material- and energy-intensive activities have been shifted partly to other countries.

2. The links among transport, infrastructure, and urban planning. Since towns and cities and their infrastructures are formed and change over periods measured in decades and centuries, the full effect of alternative transportation developments will manifest only in the relatively long term, and a high uncertainty stems from the long-run impact of short-term decisions (or nondecisions).

3. Land-use and human settlement. This factor is particularly relevant to developing countries where land-use patterns can be largely explained by trends in population growth, urbanization trends, and rural–rural migration. These factors play a dominant role in deforestation processes and more generally in agricultural land use. Policies aimed at changing these patterns, for instance with an eye to

(continued)

BOX 7.8   Long-Term Cost Effectiveness of Response Options *(continued)*

GHG abatement, are often very difficult to apply, given the long-term trends already mentioned.

4. The overall development patterns and technological choices in production and consumption in developing countries. Developing countries may be in a position to adopt strategies directed at the long-term avoidance of problems currently faced by industrial societies. Examples are leapfrogging, energy-wasteful technologies, and speeding up the use of modern technology with respect to renewables. However, access to superior technologies may be limited, for example, for economic reasons or due to the increase in the import of obsolete technology and products.

Future scenarios with regard to GHG emissions are often determined by (short-term) decisions, behavior, and expectations that together shape the above-mentioned structural characteristics of socioeconomic development – each with its own bifurcations and irreversibilities. Therefore, making long-term economic projections involves great uncertainty, very much complicating the assessment of the costs and impact of long-term abatement strategies. This makes it hard to carry out (energy) modeling activities with respect to policies in the distant future. This also applies to modeling in the top-down or bottom-up tradition (for a description of the two types of modeling approaches, see Box 7.2).

The best one can do in analyzing long-term abatement strategies is to construct various scenarios without attaching probabilities to them. This notion has inspired many researchers to develop multiple scenarios and alternative energy modeling analyses. All scenarios have their own patterns of mitigation costs, which can differ widely and are very difficult to compare. Because of the different technology bifurcation points, modelers have to decide without much guidance which technology assumptions to use. Therefore, their mitigation cost estimates are relevant only in the framework of the baseline scenario of which they are a part. This can be illustrated by means of an example. If, for some reason, experts expected the commercial feasibility of renewables to improve more rapidly than anticipated earlier, this could be built into a scenario and would crucially affect the (present) assessment of the cost functions of the different mitigation options. Box 7.9 presents an example of such a (rather unorthodox) scenario that attributes a larger role to renewables as an energy source than do many of the other projections and scenarios in the literature. The example illustrates that once the possibility of an unorthodox technological evolution pattern of a particular energy technology is accepted, the scenario outcome can become quite different from previous expectations. If this new perception were to become generally accepted, it could easily affect the decisions concerning the (optimal) timing of abatement strategies.

BOX 7.9 A Scenario Based on a Fossil-Fuel-Free Energy Future

TABLE B7.4 Results of the IPCC 91 and U.S. EPA Reference Scenarios and the FFES and U.S. EPA Policy Scenarios

	1988	2000	2010	2030	2100
IPCC 91					
CO_2 (PgC)	5.9	7.3	9.4	12.8	22.6
Primary energy (EJ)	349	460	471	797	1,641
EPA Reference					
CO_2 (PgC)	5.1	6.6	7.7	9.9	17.7
Primary energy (EJ)	302	384	451	585	1,067
Renewable (%)	7	8	10	13	19
Solar/wind (%)	0	0	1	2	7
Biomass (%)	0	0	1	3	5
FFES					
CO_2 (PgC)	5.3	5.7	5.6	2.6	0.0
Primary energy (EJ)	338	396	400	384	987
Renewables (%)	13	21	29	62	100
Solar/wind (%)	0	5	9	31	79
Biomass (%)	7	10	13	24	18
EPA "rapid reductions"					
CO_2 (PgC)	5.1	5.5	4.5	2.5	1.5
Primary energy (EJ)	302	334	408	545	799
Renewables (%)	7	10	38	68	77
Solar/wind (%)	0	1	2	4	9
Biomass (%)	0	0	26	54	58

Note: For the U.S. EPA and IPCC studies, the 1988, 2010, and 2030 values were interpolated from the 1985, 2000, and 2050 results. Results shown for IPCC for 1988 are actually 1990 values.
Source: SEI/Greenpeace (1993).

An alternative and daring scenario for 1985–2100 of a fossil-fuel-free future was designed by the Stockholm Environment Institute in Boston and carried out for Greenpeace International (SEI/Greenpeace 1993). The scenario is called FFES and tries to

(continued)

BOX 7.9   A Scenario Based on a Fossil-Fuel-Free Energy Future
(continued)

- meet ambitious CO_2 emission reduction targets;
- phase out nuclear power by 2100; and
- achieve a greater degree of economic equity worldwide.

Two reference scenarios were projected: a high-projection case corresponding to the IPCC 1991 scenario (22.6 PgC in 2100) and a low-projection case based on the average of two projections by the U.S. Environmental Protection Agency (17.7 PgC in 2100). Table B7.4 shows that, according to FFES, CO_2 emissions could be zero by 2100 if the share of renewables grows to 100% by this time.

The underlying technological assumptions of the FFES scenario are

- a breakthrough in solar photovoltaic electricity production costs during 2010–2030 and a similar breakthrough in wind power; and
- the development of advanced storage facilities enabling intermittent solar and wind resources to follow a greater part of electric system load.

Such optimism with regard to the future role of renewables is not shared by many others, however. In its projections, the World Energy Council considers a share of 50% for renewables by 2100 as the most optimistic figure; even if this share is technically larger by then, supplies of fossil fuels will still be abundant enough to make it beneficial to mix fossil fuels in the energy supply system.

7.4   THE OPTIMAL BLEND OF RESPONSE OPTIONS AND THE SCALE, PLACE, AND TIMING OF APPLICATION

The preceding sections have clearly shown that the overall costs of response strategies can be reduced if intelligent use is made of economies and diseconomies of scale and of the fact that cost functions are place- and time-specific. Thus, it seems obvious that the overall costs of abatement strategies can be reduced if the most appropriate blend of options is determined at the most appropriate places and within the most appropriate time frame. It is clear that society is not wholly free to determine the optimal response strategy, however. Still, economic models can provide some guidance by exploring the direct and indirect effects of alternative options on the (global) economy and their overall costs. Figure 7.3 presents the results of such an exploratory modeling exercise. The figure illustrates the possible impact of a strategy to reduce CO_2 emis-

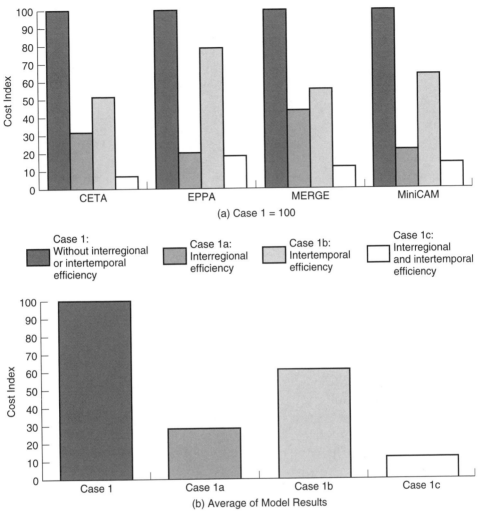

FIGURE 7.3. Global costs in four alternative cases. From Richels et al. (1996).

sions in OECD countries 20% under 1990 levels by 2010 and to keep them at the same level; no constraints are placed on non-OECD emissions.

Four different models incorporating features from both bottom-up and top-down energy modeling were employed – EPPA, MERGE, CETA, and Mini-CAM – to analyze how overall costs will be affected by the place and time of application. International cooperation in the form of joint implementation (JI or AIJ)[11] could reduce costs by increasing the prospects of carrying out strategies in those regions where marginal costs are lowest; moreover, addi-

---

[11] At the first Conference of the Parties in Berlin (1995) the decision was made to use "activities implemented jointly" (AIJ) instead of "joint implementation" (JI) as the official term during the pilot phase (lasting no longer than until 2000) for cooperation among parties to the UNFCCC to implement their commitments jointly under the Convention.

tional cost reductions could be achieved if parties were allowed to choose the optimal moment during a certain time interval to carry out emission reduction measures (timing). The models were used to simulate the impact of optimization with regard to place and time in comparison with a base case in order to get an impression of the cost savings that could be achieved through these strategies.

To allow for such an optimization in terms of place and time, trading in emission rights was assumed to facilitate international cooperation (see Chapter 8), and at the same time the OECD region's carbon budget was defined as the cumulative permissible emissions during 2000–2050, leaving the decision on timing emission reductions during this interval to the region.

This approach has two obvious weaknesses. A global system based on the exchange of emission permits using a JI/AIJ type of international cooperation does not exist (yet) and may be difficult to realize in the near future. The benefit of shifting emission reductions to the future depends on the assumptions made vis-à-vis (future) technological progress, which is precisely one of the variables that is so hard to predict, as argued in the preceding section.

Nevertheless, this modeling exercise is interesting as a thought experiment, because it at least provides some insight into the possible contribution of an optimization process to a reduction in overall abatement costs – in terms of *where* and *when* emission reduction is cheapest. The results of the simulations (average of model results) are presented in Figure 7.3, in which a cost index is provided for the base case, a scenario with the optimal location of emission reductions, the optimal timing, and the combination of these aspects. Even if the results are only indicative, it is apparent that the total costs of abatement strategies very much depend on the location and timing of application; in the optimum, the overall discounted costs turn out to be about 10% of the overall costs in the baseline!

7.5 SUMMARY

Costs of abatement technology options are scale-, place-, and time-specific. With regard to the scale factor, a clear distinction must be made between the internal economies of scale, which is an engineering concept; the external economies of scale, which is a welfare economic concept; and the implications of scale related to public acceptance, institutional capacity, and so on, which is a broad societal or, if you like, political concept. Optimism with regard to the feasibility of applying a particular option on a large scale may well depend on the scale elements one tends to focus on. This may also explain, at least in part, why such a strong controversy has emerged between bottom-up and top-down modelers.

With regard to the factor of place, it has been convincingly demonstrated empirically that the application of the same technology in different parts of the world may have extremely different cost implications. Because the greenhouse implications of actions do not depend on *where* the actions are carried out, this opens the possibility for efficiency gains through international cooperation. This is especially so when costs differ to a great extent, as is often the case between industrialized and nonindustrialized countries. The UN Framework Convention on Climate Change contains an opening for such cooperation, usually referred to as joint implementation, which is now in the process of being explored.

With regard to the time factor, the key issue is that despite some technological long-term trends that can be traced statistically, it seems very hard to predict any clear patterns in the timing and impact of technological breakthroughs that could alter these long-term trends. What we do know, however, is that technologies will never develop in isolation, and that once society has chosen a particular line of technology, it will not be easy to switch to a different one. For some, all this is reason for optimism about the potential of technological breakthroughs in dealing with the greenhouse problem in the long run; for others, however, it is cause for concern, because decisions taken now (think of the huge energy investment to be made in the near future in developing countries and countries in transition) may propel technological progress for the next generations in the wrong direction from the greenhouse perspective. This is another argument for exploring how countries can cooperate in order to collectively reap the efficiency gains due to international differences in costs of options.

REFERENCES

Adams, R. M., D. M. Adams, C. C. Chang, B. A. McCarl, and J. M. Callaway. 1993. Sequestering carbon on agricultural land: A preliminary analysis of social cost and impacts on timber markets. *Contemporary Policy Issues,* **11**(1), 76–87.

Barnett, A. 1990. The diffusion of energy technology in the rural areas of developing countries: A synthesis of recent experience. *World Development,* **18**(4), 539–53.

Burniaux, J.-M., G. Nicoletti, and J. O. Martins. 1992. GREEN: A global model for quantifying the costs of policies to curb CO_2 emissions. *OECD Economic Studies,* **19**(Winter), 49–92.

Byrne, J., Y. Wang, K. Ham, I. Han, J. Kim, and R. Wykoff. 1991. Energy and environmental sustainability in East and Southeast Asia. *IEEE Technology and Society Magazine* (Winter).

COHERENCE, 1991. *Cost-effectiveness analysis of CO_2 reduction options.* Synthesis report for the CEC CO_2 Crash Programme. Brussels: Commission of the European Community.

EMF (Energy Modeling Forum). 1993. *Reducing global carbon emissions: Costs and policy options.* EMF-12, Stanford University, Stanford, CA.

Ettinger, van J. 1994. Sustainable use of energy, a normative energy scenario: 1990–2050. *Energy Policy,* **22**, 111–18.

Fujii, Y. 1990. *Assessment of CO₂ emission reduction technologies and building of a global energy balance model.* Department of Electrical Engineering, Kaya Laboratory, University of Tokyo.

Geller, H., and S. Nadel. 1994. Market transformation strategies to promote end-use efficiency. *Annual Review of Energy and Environment,* **19,** 301–46.

Ghai, D. 1994. Environment, livelihood and empowerment. *Development and Change,* **25**(1), 1–11.

Grubb, M., J. Edmonds, P. ten Brink, and M. Morrison. 1993. The costs of limiting fossil-fuel CO₂ emissions: A survey and analysis. *Annual Review of Energy and Environment,* **18,** 397–478.

Houghton, J. T., et al. (eds.). 1995. *Climate change 1994: Radiative forcing of climate change and an evaluation of the IPCC IS92 emission scenarios.* Cambridge University Press.

Hourcade, J. C., et al. 1996. A review of mitigation cost studies. In J. P. Bruce, H. Lee, and E. F. Haites (eds.), *Climate change 1995: Economic and social dimensions of climate change.* Contribution of Working Group III to the Second Assessment Report of the Intergovernmental Panel on Climate Change. Cambridge University Press, pp. 297–366.

Jackson, T. 1991. Least-cost greenhouse planning: Supply curves for global warming abatement. *Energy Policy,* **19**(1), 35–46.

Jaszay, T. 1990. Hungary. In W. Chandler (ed.), *Carbon emissions control strategies: Case studies in international cooperation.* Washington, DC, and Oak Ridge, TN: Conservation Foundation and Oak Ridge National Laboratory, Carbon Dioxide Information Center.

Jepma, C. J., and C. W. Lee. 1995. Carbon dioxide emissions: A cost-effective approach. In C. J. Jepma (ed.), *The Feasibility of Joint Implementation.* Groningen: Kluwer, pp. 57–68.

Kononov, Y. 1993. Impact of the economic reforms in Russia on greenhouse gas emissions, mitigation and adaptation. Paper presented at the International Workshop on Integrated Assessment of Mitigation, Impacts and Adaptations to Climate Change, International Institute for Applied Systems Analysis, Laxenburg, Austria, October.

Kostalova, M., J. Suk, and S. Kolar. 1991. *Reducing greenhouse gas emissions in Czechoslovakia.* Richland, WA: Pacific Northwest Laboratories, Advanced International Studies Unit, Global Studies Program.

Kram, T. 1994a. *Boundaries of future carbon dioxide emission reduction in nine industrial countries* (executive summary). Energy Technology Systems Analysis Programme, Annex IV: *Greenhouse gases and national energy options.* Netherlands Energy Research Foundation ECN/International Energy Agency, Petten.

Kram, T. 1994b. National energy options for reducing CO₂ emissions. In *The international connection,* vol. 1. Report of the Energy Technology Systems Analysis Programme, Annex IV (1990–1993): *Greenhouse gases and national energy options.* Netherlands Energy Research Foundation ECN/International Energy Agency, Petten.

Leach, G., and Z. Nowak. 1991. Cutting carbon dioxide emissions from Poland and the UK. *Energy Policy,* **19**(10), 918–25.

Locklin, K. R. 1995. Ruminant methane emissions reductions and JI. *Joint Implementation Quarterly,* **1**(3), 9–10.

Makarov, A., and I. Bashmakov. 1991. An energy development strategy for the USSR: Minimizing greenhouse gas emissions. *Energy Policy,* **19**(10), 987–94.

Manne, A. S. 1992. *Global 2100: Alternative scenarios for reducing emissions.* OECD Working Paper no. 111. Paris: OECD.

Manne, A. S., and R. B. Richels. 1992. Global CO₂ emission reductions: The impacts of rising energy costs. *Energy Journal,* **12**(1), 87–107.

Manne, A. S., and J. Oliveira-Martins. 1994. Comparisons of model structure and policy scenarios: GREEN and 12RT. Draft. Annex to the WP1 Paper on Policy Response to the Threat of Global Warming, OECD Model Comparison Project (II). Paris: OECD.

Meier, P., M. Munasinghe, and T. Siyambalapitiya. 1993. *Energy sector policy and the environment: A case study of Sri Lanka.* Washington, DC: World Bank.

McKinsey & Company. 1989. Protecting the global environment: Findings and conclusions, appendices. Paper presented at the Ministerial Conference on Atmospheric Pollution and Climate Change, Noordwijk.

Mills, E., D. Wilson, and T. B. Johansson. 1991. Getting started: No-regrets strategies for reducing greenhouse gas emissions. *Energy Policy,* **19**(July–Aug.), 526–42.

Moulton, R. J., and K. R. Richards. 1990. *Costs of sequestering carbon through tree planting and forest management in the United States.* General Technical Report WO-58. Washington, DC: USDA Forest Service.

Nordhaus, W. D. 1991. The cost of slowing climate change: A survey. *Energy Journal,* **12**(1), 37–65.

OECD (Organization for Economic Cooperation and Development). 1994. *GREEN: The reference manual.* Working Paper no. 143. Paris: OECD.

Oliveira-Martins, J., J.-M. Burniaux, J. P. Martin, and G. Nicoletti. 1992. *The cost of reducing CO_2 emissions: A comparison of carbon tax curves with GREEN.* OECD Economics Working Paper no. 118. Paris: OECD.

Parks, P. J., and I. W. Hardie. 1992. Least-cost forest carbon reserves: Cost-effective subsidies to convert marginal agricultural land to forest. Report sponsored by American Forests, Washington, DC.

Qureshi, A., and S. Sherer. 1994. *Climate change in Asia. Forestry and land use.* Asian Development Bank's regional study on global environmental issues. Washington, DC: Climate Institute.

Radwanski, E., A. Gromadzinski, E. Hille, P. Skowronski, and S. Szukalski. 1993. *Case study of greenhouse gas emission: Final report.* Washington, DC: Polish Foundation for Energy Efficiency for the Pacific Northwest Laboratories.

Richels, R., J. Edmonds, H. Gruenspecht, and T. Wigley. 1996. *The Berlin Mandate: The design of cost-effective mitigation strategies.* Subgroup on the Regional Distribution of the Costs and Benefits of Climate Change Policy Proposals, Energy Modeling Forum-14, Stanford University, Stanford, CA.

Rubin, E. S., R. N. Cooper, R. A. Frosch, T. H. Lee, G. Marland, A. H. Rosenfeld, and D. D. Stine. 1992. Realistic mitigation options for global warming. *Science,* **257**(July), 148–9, 261–6.

Rutherford, T. 1992. *The welfare effects of fossil carbon reductions: Results from a recursively dynamic trade model.* Working Papers no. 112, OECD/GD(92)89. Paris: OECD.

Sitnicki, S., K. Budzinski, J. Juda, J. Michna, and A. Spilewicz. 1991. Poland: Opportunities for carbon emissions control. *Energy and Policy,* **19**, 995–1002.

SEI (Stockholm Environment Institute)/Greenpeace. 1993. *Towards a fossil free energy future: The next transition.* A technical analysis for Greenpeace International. Boston: SEI.

Southern Centre/Risö. 1993. *UNEP greenhouse gas abatement costing studies.* Zimbabwe Country Study, Phase Two. Southern Centre for Energy and the Environment, Risö National Laboratory Systems Analysis Department, Denmark.

TNO (Netherlands' Organization for Applied Scientific Research). 1992. *Confining and abating CO_2 from fossil fuel burning: A feasible option?* Report to the European Community. TNO, Apeldoorn.

UNDP (United Nations Development Project). 1993. *Human development report.* New York: Oxford University Press.

UNEP (UN Environmental Program). 1994. *UNEP greenhouse gas abatement costing studies.* Part 1: Main report; Part 2: Country summaries. UNEP Collaborating Centre on Energy and Environment, Risö National Laboratory, Denmark.

Zhang, Zhong Xiang. 1996. *Integrated economy–energy–environment policy analysis: A case study for the People's Republic of China.* Wageningen: Agricultural University of Wageningen.

8

SUMMARY AND CONCLUSIONS

In this final chapter, we bring together the main threads of the material presented in the book. As a backdrop, the chapter begins with a brief summary of the most likely range of climate change scenarios over the next century and beyond, including a midrange scenario as well as the best- and worst-case outcomes, with and without human intervention. Next, the main conclusions of the most recent studies are summarized and several recommendations are offered. Finally, the priority issues and questions that remain are identified, and ways of addressing them are discussed.

8.1 RANGE OF CLIMATE CHANGE SCENARIOS

8.1.1 Scenarios without Intervention (IS92)

A range of variations in climate change are summarized in the top half of Table 8.1, based on the IS92 scenarios, which assume no systematic human response strategy or intervention to address the climate change problem.

Carbon Emissions and Concentrations Experts would agree that the IPCC IS92 emission scenarios provide the most thoroughly analyzed and understood basis for predicting outcomes. As indicated in Table 8.1, the midrange IS92a case assumes that net carbon emissions from anthropogenic sources will increase from about 7 GtC per year in 1990 to more than 20 GtC per year by 2100. The best and worst cases are IS92c and IS92e, respectively, where corresponding emissions are about 5 and 36 GtC per year in 2100. The basic assumptions that underlie these scenarios – regarding population growth, economic activity, and energy supplies – are also summarized in the table.

Atmospheric concentrations of carbon are about 700 ppmv and still rising in 2100 for the IS92a emission scenario. In the best case, carbon concentrations have already begun to stabilize below 500 ppmv by 2100, while in the

TABLE 8.1 Range of Likely Climate Change Scenarios (with and without Intervention)

Scenario and Year	Net Anthropogenic Carbon Emissions (Gt/yr)	Carbon Conc. (ppmv)	Cumulative Carbon Emissions from 1990 to 2100 (GtC)	Temperature Rise (°C)	Sea Level Rise (cm)
Actual (in 1900)	<0.5	315	—	0.3–0.6 below 1990	10–25 below 1990
Actual (in 1990)	7	350 (rising)	—	0	0
Midrange (IS92a), nonintervention (in 2100)	20	680 (rising)	1,500	2 (rising)	50 (rising)
	IS92a			IS92a profile with increasing aerosol concentrations and moderate climate and ice-melt sensitivities	
	Economic growth (per year) 2.9%: 1990–2025 2.3%: 1990–2100 *Energy supplies* 12,000 EJ oil 13,000 EJ natural gas *Population* 11.3 billion in 2100				
Best (IS92c), nonintervention (in 2100)	5	480 (stabilizing)	770	0.9 (rising)	13 (rising)
	IS92c			IS92c profile with increasing aerosol concentrations and low climate and ice-melt sensitivities	
	Economic growth (per year) 2.0%: 1990–2025 1.2%: 1990–2100 *Energy supplies* 8,000 EJ oil 7,300 EJ natural gas *Population* 6.4 billion in 2100				

Scenario					
Worst (IS92e), nonintervention (in 2100)	36	1,000 (rising)	2,190	4.5 (rising)	94 (rising)
		IS92e *Economic growth (per year)* 3.5%: 1990–2025 3.0%: 1990–2100 *Energy supplies* 18,400 EJ oil 13,000 EJ natural gas Phase-out nuclear investments by 2075 *Population* 11.3 billion in 2100			IS92e profile with constant 1990 aerosol concentrations and high climate and ice-melt sensitivities
Midrange (S450/S650) with intervention (in 2100 and 2500)	2.5–9 in 2100; 1–2 in 2500	450–600 in 2100; 450–650 in 2500	630–1,190	0.8–1.3 in 2100; 1.1–2.2 in 2500	25–30 in 2100; 85–145 in 2500
		S450 to S650 range of profiles			S450 to S650 range of profiles with constant 1990 aerosol concentrations and moderate climate and ice-melt sensitivities
Best (S450), with intervention (in 2100 and 2500)	2.5 in 2100; 1 in 2500	450 in 2100; 450 in 2500	630–650	0.5 in 2100; 1.7 in 2500	10 in 2100; 10 in 2500
		S450 profile			S450 profile with constant 1990 aerosol concentrations and low climate and ice-melt sensitivities
Worst (S650), with intervention (in 2100 and 2500)	9 in 2100; 2 in 2500	550–600 in 2100; 650 in 2500	1,030–1,190	2 in 2100; 3.5 in 2500	70 in 2100; 325 in 2500
		S650 profile			S650 profile with constant 1990 aerosol concentrations and high climate and ice-melt sensitivities

worst case the concentrations in 2100 have soared above 1,000 ppmv. Cumulative CO_2 emissions from 1990 to 2100 are 770 GtC in the best case, and two to three times larger in the midrange and worst-case outcomes, respectively. We note that between the years 1990 and 2100, the increases in atmospheric concentrations for the best and worst IS92 cases, respectively, differ by a factor of 5 – that is, 130 ppmv (or 480 to 350) versus 650 ppmv (1,000 to 350). The corresponding ratio of emissions in the year 2100 is 7 (5 vs. 36 Gt/yr).

Global Mean Temperature Increase Based on the foregoing results and other supplementary calculations, the magnitude of global warming is likely to be about 2°C in the next century, corresponding to the midrange IS92a emission projection, a moderate value of global climate sensitivity, and increasing future aerosol concentrations. If aerosol concentrations are held constant at 1990 levels, the corresponding temperature increase is higher (about 2.4°C). Combining the low IS92c emission scenario with a low climate sensitivity value yields the best-case temperature increase of 1°C by 2100, while the high IS92e case and a high climate sensitivity result in the worst case, a 3.5°C temperature rise. The temperature increases even further for this worst case – by 4.5°C – when aerosol concentrations are maintained at 1990 levels. Thus, there is an approximately fourfold variation between the nonintervention worst- and best-case outcomes for global warming in 2100.

In every case, the mean rate of warming (ranging from about 0.1 to 0.4°C per decade) would exceed the natural rate of temperature increase observed during the past 10,000 years. The superposition of natural and local variations on this broad trend would give rise to both yearly and regional temperatures that could differ significantly from the global average. Only about 50 to 90% of the ultimate equilibrium temperature increase will have been realized by 2100, because of the thermal inertia of oceans. Thus, even if GHG concentrations were stabilized by 2100, global temperatures would continue to rise.

Global Mean Sea Level Rise The IPCC has reviewed the results of all currently available studies and concluded that the best estimate of the rise in global mean sea level will be about 50 cm between 1990 and 2100, based on the midrange IS92a emission projection, moderate values of both climate and ice-melt sensitivities, and increasing future aerosol concentrations. The IS92c case, combined with low climate and ice-melt sensitivities, yields a best-case estimate of a 14-cm sea level rise by 2100. Correspondingly, the high emission IS92e scenario and the high climate and ice-melt sensitivities result in a large 94-cm increase in sea level. Thus, the ratio of sea level rise between the worst and best cases is over 6.

The expected rate of sea level rise for the next century (about 5 cm per decade) is higher than the corresponding rate during the twentieth century, and significantly exceeds the natural rate of increase observed during the past 10,000 years. Because of built-in time lags in the chain of causality from radiative forcing, through temperature increase, to sea level rise, the increase in mean sea level is relatively independent of the choice of emission scenario for the next 50 years or so. By the same token, the global mean sea level will continue to rise long after GHG concentrations and global mean temperatures have stabilized. Local and regional sea level changes will vary about the global mean, due to vertical movements in land masses and shifts in ocean currents. Uncertainties in the estimates of sea level rise are greater than for temperature increases. In particular, the behavior of polar ice sheets is highly uncertain. Present models are not adequate for making reliable predictions about more geographically detailed patterns of climate change.

8.1.2 Scenarios with Intervention (S450 and S650)

Two cases (S450 and S650), involving stabilization of CO_2 concentrations at 450 and 650 ppmv, respectively, are examined in this subsection. They will require significant measures to restrict future net emissions of GHGs. These cases facilitate the analysis of climate change as long-term equilibria are reached (or approached) after several centuries.

Carbon Emissions and Concentrations The information summarized in the lower half of Table 8.1 shows that the projections for all the intervention cases result in a range of CO_2 emission reductions that could be matched only by the rather conservative IS92c scenario, up to 2100. The cutbacks required by the S450 profile are quite severe, leading to a mere 2.5 GtC of annual emissions in 2100 and even more drastic reductions by 2500 (i.e., down to 1 or 2 GtC per year). Stable CO_2 concentrations are achieved in the period from 2100 to 2200. For S450 and S650, accumulated carbon emissions from 1990 to 2100 are in the range of 630 to 1190 GtC, which is comparable only with the best-case IS92c nonintervention scenario.

Global Mean Temperature Increase Estimates of the global mean temperature rise for the two intervention cases examined are in the range 0.5 to 2°C in 2100, with the midrange values being 0.8 to 1.3°C. These values are comparable with the outcome of the best IS92c nonintervention scenario (0.9°C in 2100). With the S450 and S650 cases, even in 2500 the midrange estimate of temperature rise is only 1.1 to 2.2°C, while the best- and worst-case outcomes are 0.7 and 3.5°C, respectively. In all these cases, global mean temperatures have generally stabilized by 2500, albeit several centuries after atmospheric CO_2 concentrations have become constant.

Global Mean Sea Level Rise In 2100, the midrange mean sea level rise for the S450 and S650 cases is a modest 25 to 30 cm, and the best- and worst-case estimates are 10 and 70 cm, respectively. However, the lagged effects become important in later years. Only in the best case (S450, L) has the sea level rise stabilized at around 10 cm by 2500. In all other cases, the global mean sea level rise is quite considerable in 2500 (i.e., 85 to 145 cm and 325 cm in the midrange and worst-case outcomes) and is expected to increase for several centuries after global mean temperatures have stabilized.

Practical Prospects for Mitigation The selected profiles that lead to stabilization of CO_2 concentrations at 450 and 650 ppmv straddle the benchmark value of 560 ppmv for which much of the climate change impact assessments have been carried out (see Chapters 1 and 5). Although the S450 case is a useful reference, its requirements (e.g., CO_2 emissions of 2.5 GtC per year in 2100) appear to be practically infeasible in view of current emission trends (i.e., CO_2 emissions of about 7 GtC in the early 1990s). Even meeting the constraints of the S650 profile will require very significant mitigation measures during the next century.

8.2 MAIN FINDINGS

8.2.1 Importance of Climate Change and Problems of Analysis

The main reason for the extraordinary importance of global climate change to human society emerges from the key conclusion of the recent Second Assessment Report of the IPCC that "the balance of [scientific] evidence suggests a discernible human influence on the global climate" (Houghton et al. 1996, p. 4).

The extent of global warming, sea level rise, and related effects like extreme weather events are of particular concern. Despite gaps in our knowledge the possible socioeconomic and environmental consequences of global climate change have been studied on a preliminary basis, corresponding to the benchmark atmospheric CO_2 concentration equivalent to a doubling of the preindustrial level (i.e., 560 ppmv). The focus has been on the vulnerability of ecosystems and natural habitats, hydrology and water resources, food and agriculture, human health, and human infrastructure and habitats. The overall worldwide impact is likely to be significantly negative, although some regions might actually benefit.

At the same time, there are many formidable scientific, economic, social, and technological uncertainties that make it difficult to make detailed predictions about the future impact of global climate change. To begin with, climate change issues and trade-offs have to be analyzed in the context of sustainable development – an important goal for society that has economic, social, and environmental dimensions. Decision makers have to address many other pressing problems, such as poverty, malnutrition, and localized environmental degradation. In the special case of climate change, generic complications that analysts face include the global and centuries-long scale of events, the complexity of interactions among large geophysical, ecological, and socioeconomic systems, and the potential for irreversible, nonlinear, and catastrophic effects. Special problems are caused by uncertainty – in particular, scientific uncertainty arising from lack of knowledge about natural systems, and socioeconomic and technological uncertainty associated with human systems.

There are also a host of sociopolitical issues, especially the difficulties of ensuring equity and fairness to all (living as well as future generations), in both the climate change decision-making process and its outcome. Procedural equity is important in decision making to ensure that all stakeholders and affected parties (especially developing countries) are fairly represented in the process. At the same time, in-country institutional capacities must be strengthened to facilitate the participation of all groups, particularly the poor. Consequential equity requires that the decision-making process result in fair choices. In this context, spatial or intragenerational equity requires that the burdens of both GHG mitigation and effects be shared fairly among those living today. Two fundamentally different approaches to allocating future emission rights involve (a) equal per capita emissions for everyone (which would favor developing countries but require sharp cutbacks among the industrialized nations) and (b) equal percent reductions from a baseline year, or "grandfathering" (which would severely restrict the growth prospects of developing countries whose current per capita consumption levels are low). A practical compromise solution might involve a gradual transition to an equal per capita rights regime over a period long enough for industrialized countries to adjust without major shocks, coupled with an emission rights trading scheme among emission-surplus and -deficit countries.

At the same time, temporal or intergenerational equity suggests that burdens should be shared fairly among present and future generations. Here the social rate of discount will play a critical role in determining policy over the centuries-long time horizons involved. The lower the discount rate, the greater will be the impact of future benefits and the lesser will be the effect of current costs in determining climate change policies. As a practical guideline, a range of 1 to 6% per annum might be used for long-term real discount rates, with a midrange estimate of about 3 to 4%. Needless to say, some analysts may dispute these ranges.

8.2.2 Making Decisions

Given the serious risks posed by climate change, a precautionary approach suggests that it would be prudent and necessary (although problematic) to determine the desirable target conditions required to stabilize the global climate, as well as the most effective human actions necessary to achieve such results. In the words of the UN Framework Convention on Climate Change (UNFCCC), the fundamental challenges are to determine what might constitute "dangerous anthropogenic interference with the climate system" and how to achieve "stabilization of greenhouse gas concentrations" (UNFCCC 1992). A framework for making such decisions is beginning to emerge. It consists of three key elements and seeks especially to disentangle issues concerning efficiency (i.e., overall global well-being) and equity (i.e., fairness and distribution).

First, *global optimization* attempts to provide efficient benchmark responses (i.e., without considering equity) to issues such as desirable levels of total future GHG emissions, as well as a cost-effective portfolio of remedial measures. Ideally, GHG mitigation efforts should be pursued up to the point where the marginal costs of such measures are equal to the marginal benefits from avoided climate change damage. In practice, neither the costs nor benefits of GHG mitigation are known with sufficient accuracy. Under such conditions, it is more practical to base the target level of GHG emission reduction on a safe minimum standard, to be determined on the basis of what mitigation measures are broadly deemed affordable and what effects might be considered unacceptable in scientific terms.

Second, *collective decision-making principles* (including both procedural and consequential equity) would help to ensure fairness in allocating both the rights to emit and the responsibilities for undertaking future GHG abatement measures based on a variety of considerations, including the widely differing responsibilities for past emissions, as well as current and future emissions, the uneven incidence of future climate change effects, and differences in vulnerability, affordability, and distribution of resources. It is recognized that while climate change policies should avoid aggravating existing disparities in well-being at all levels, they cannot be expected to resolve all equity issues.

Third, practical *procedures and mechanisms* are necessary for reaching and implementing a global consensus; these include incentives to encourage international cooperation. The preliminary framework that has emerged in recent years includes the UNFCCC and its Conference of Parities (CoP) for international negotiations, the IPCC for scientific work, and the Global Environment Facility (GEF) and Montreal Protocol for financing. Other new and innovative financing mechanisms, as well as insurance schemes, are under consideration. A sustainable energy development framework is an essential mechanism for facilitating implementation within individual countries, when internationally agreed-upon GHG abatement measures ultimately have to be consistent with other national energy programs.

Among the decision criteria available, utility-based approaches that favor economic efficiency, and the Rawlsian view, which leans more toward greater equity, provide useful insights. In particular, economic efficiency suggests that cost–benefit analysis (CBA) could play a useful role. However, the requirements of traditional CBA, including limited scale, geographic extent, and time span, as well as discreteness, make it difficult to apply in the case of climate change decisions. The centuries-long time lags, irreversibilities, nonlinearities (that threaten potentially catastrophic outcomes), and great uncertainty associated with climate change further complicate the analysis. As a response, modern CBA has evolved into a more generic approach that includes a family of decision techniques for overcoming many of the shortcomings just mentioned.

Modern decision techniques relevant to the climate change issue include (a) economic cost–benefit analysis (ECBA), which compares all costs and benefits expressed in terms of a common numeraire (usually monetary units) and relies heavily on environmental economic analysis to value environmental effects; (b) cost-effectiveness analysis (CEA), which is designed to identify the lowest-cost method of achieving an objective already justified using other criteria; (c) multicriteria analysis (MCA), which attempts to find desirable options when some or all of the costs and benefits are not measurable in monetary terms; and (d) decision analysis (DA), which is useful when choices have to be made under conditions of uncertainty. This group of decision techniques can help to provide better responses to several of the key public policy questions posed earlier, including: (1) How much should GHG emissions be reduced? (2) When should emissions be reduced? (3) How should emissions be reduced? Another important question, as to who should reduce emissions, involves equity. Therefore, it is not amenable to resolution by ECBA alone, although the broader range of decision techniques (especially MCA) can elucidate the trade-offs between economic efficiency and equity.

The complications posed by uncertainty merit special attention. In general, strictly quantitative application of decision techniques to climate change analysis and undue reliance on numerical projections into the distant future (e.g., 100 years) are not practical due to uncertainty. However, these methods could help to structure thinking and provide a more systematic basis for developing qualitative insights, especially concerning the rough magnitudes and directions of change. Quantitative analysis may be more successfully applied to the evaluation of shorter-term alternatives like energy efficiency and fuel switching in the existing portfolio of options. Further (unresolvable) uncertainties must be handled through strategies that include sequential approaches and hedging, insurance, portfolios of options, robust solutions, and better information gathering and research. Such an approach will require a decision-making framework that is both flexible and heuristic, given that a conventional approach based on deterministic optimization in the long term is not possible in the face of high levels of uncertainty. At the same time, the precautionary approach indicates that uncertainty should not be used to justify inaction, since a business-as-usual scenario results in a high level of climate change risk. A stepwise decision framework would facilitate the making of short-term decisions (such as no-regrets measures) that are sensible and robust but adjustable in the longer term, as more accurate information becomes available.

8.2.3   Assessing Mitigation Costs and Benefits

Preliminary studies indicate that the ultimate stabilization of atmospheric GHG concentrations will require reductions in future GHG emissions below present levels. Such reductions could entail significant adjustments and costs, since the primary causes of GHG emissions – energy and land use – lie at the heart of modern economic activities. Nevertheless, there are sound economic and scientific grounds for undertaking at least the no-regrets measures – that is, activities, such as energy conservation, that would be undertaken anyway because they are relatively costless. More costly measures also appear to be justified, involving even further climate change mitigation and adaptation measures. However, the scope of such measures cannot be defined in greater detail, until their implementation costs can be compared with the value of averted global warming damage. Under these circumstances, it would be prudent to rely on a flexible overall climate change response strategy that could be adjusted systematically in the light of new information, especially to deal with the high levels of uncertainty. Such a strategy would include a portfolio of mitigation, adaptation, and other measures based on the coordinated application of a variety of market-based, regulatory, and other instruments.

The benefits of climate change mitigation policies are basically the avoided costs of future damage plus the net secondary benefits that may be associated with these policies. Recent estimates of such damage have been made for a scenario involving the doubling of preindustrial CO_2 concentrations (i.e., up to 560 ppmv) and a global mean temperature rise of about 2°C by 2100. While such estimates vary widely, globally they are in the range of a few percentage

points of world GDP, although developing countries would suffer disproportionately and might lose a much higher share of their GDP.

Chapter 5 provided information on possible greenhouse damage and on
the benefits of abatement strategies – in particular adaptation – which are
measured by the net damage prevented. Concern over this damage, as well as
the many uncertainties and associated risks, has resulted in a general awareness of the need for both adaptation and mitigation abatement strategies that
go beyond the possible no-regrets potential of response options. In other
words, society will probably face costs if it is serious about addressing the
greenhouse problem. Chapter 6, therefore, provided a survey of mitigation
response options that might be considered and delineated the complications
that arise in assessing the costs of these options.

So far, the basis for the present discussion has been the notion that the
ideal set of greenhouse policies can be determined on the basis of a rather
straightforward efficiency criterion. According to this criterion – which is not
uncontroversial – the optimum level of response to the greenhouse problem
is determined by the point at which marginal abatement benefits (i.e., damage avoided plus net secondary benefits) and marginal abatement costs (including the costs of climate change uncertainty and associated risks) are
equal. In other words, if less abatement were to be carried out, further action
would provide an additional net gain, while abatement action beyond this optimum would cause a net loss. A figure representing this equilibrium and
the marginal cost and benefit curves of abatement strategies was presented in
Box 3.1.

The scenario containing marginal abatement cost and benefit curves would
be acceptable as a guiding framework for presentation (assuming the acceptability of the efficiency criterion) if perfect information on the costs and
benefits of abatement were available and if the world were a single entity with
one supranational government that could impose policies and achieve full
compliance. Unfortunately, the world is characterized by great uncertainty
with respect to both the costs and benefits of abatement strategies. Moreover,
it consists of a large number of sovereign nation-states, which (without
sufficient incentives) seldom comply with weak international agreements, implicitly or explicitly.

As explained in Chapter 3, since the climate change problem has become a
global issue, nations will have to cooperate in the design of policies if real
progress is to be made. The UNFCCC was organized from this point of
view – that is, in order to involve many countries in the decision-making process on greenhouse policies as much as possible and as soon as possible. In the
end, however, individual countries are responsible for implementing internationally agreed-upon policies by means of policy instruments applied at home.
This raises several questions: What is the optimal set of instruments that can

be implemented in participating countries, how can the use of policy instruments be coordinated internationally, and how is the choice of instruments affected by the existing uncertainty concerning costs and benefits of response strategies?

8.2.4   Comparing Costs and Benefits of Abatement

After having read Chapters 5 through 7, the reader may well be uncertain as to whether there are stronger arguments for taking abatement action now or for maintaining a wait-and-see attitude. Much information has been provided on the benefits of abatement action: future damage will be avoided and early secondary benefits can be reaped (Chapter 5). Increasing information has also been provided on the costs associated with abatement technologies, and these are often determined by their scale, location, and time of application (Chapters 6 and 7). But how can all of this information be lumped together to provide an overall picture?

As was extensively argued, particularly in Chapters 2 through 4, assessment of the feasibility of abatement action must be based on both equity and efficiency criteria. What has also become clear, though, is that these criteria cannot easily be separated. In order to apply the efficiency criterion, costs and benefits have to be made comparable – that is, expressed in monetary terms. However, as soon as one starts to do so, equity issues arise. For instance, it is necessary to discount in order to "translate" future costs or benefits into present valued costs or benefits. By doing so, one implicitly uses some criteria regarding the concern of present generations for the welfare of future generations, which is an equity matter, at least in part. Furthermore, if one tries to value certain types of climate change damage (such as loss of biodiversity or of human lives), the question arises as to whether such matters can be translated into monetary values at all and, if so, whether such values made in a specific context (e.g., countries) retain their meaning when one tries to aggregate them over different countries. In the latter case it is clear that one faces a severe intragenerational international equity issue.[1]

Another example of how difficult it is to disentangle efficiency from equity aspects involves the concept of opportunity costs (i.e., the benefits forgone that determine the costs of action). Various options will have opportunity costs having to do with income distribution. For instance, a forestry project may be effective as a carbon store, but may force poor landless peasants to migrate to less productive areas. How might one value this intragenerational national equity issue?

---

[1] The issue of equity also exists when valuation takes place within the context of, say, a country but is less conspicuous from the perspective of international negotiations.

One can nevertheless try to put the pieces of information together in order to see if some broad patterns emerge. From the damage-avoided data (for the $2 \times CO_2$ scenario), the general impression arises – acknowledging the valuation issue and other problems – that damage may well amount to a few percentage points of (future) GDP in industrialized countries, and even considerably more (up to 9%) in some developing countries and economies in transition. For the world as a whole, damage might amount to a few percentage points of GDP.

Broadly speaking, from the data on costs of control, one can infer that global abatement costs of, for instance, stabilizing emissions at 1990 levels may rise to a level of the order of a few percentage points of (future) world GDP in the next decades – in the wording of the IPCC Summary for Policymakers: ". . . costs of substantial reductions below 1990 levels [of emissions] could be as high as several per cent of GDP" (Bruce et al. 1996, p. 14). The combination of these broad, aggregate, and rough data on damage and costs of control have led some to conclude that since (future) global costs and benefits of abatement do not seem to differ widely (and costs may even be higher than actual damage), there is no strong argument for taking considerable abatement action now. However, such a conclusion would be terribly wrong, for a number of reasons.

First, the data on damage and on costs of abatement are actually not comparable. The data on damage, as presented in Chapter 5, relate to a hypothetical situation wherein the atmospheric concentrations of CO_2 will have reached a level that is twice the industrial level, but as if this were occurring *now* and would therefore relate to the present world economic system. The data on costs, in contrast, generally relate to emission stabilization at 1990 levels (which is quite distinct from a doubling of atmospheric concentrations), as they would occur at some point in the next several decades. The two sets of data are clearly not directly comparable in the way they have been defined. Note also that some studies suggest that the costs of stabilizing long-term atmospheric concentrations *with* geographic and temporal flexibility are of the same order of magnitude (as a percentage of world GDP) as the costs of substantial reductions of emissions below 1990 levels for the relatively short term, assuming *no* geographic or temporal flexibility.

Second, and even more fundamental, the efficiency rule tells us that it is the marginal costs and benefits rather than total costs and benefits that matter in assessing the optimal policy. The reason is simply that total costs or benefits are the result of adding up all the costs incurred or benefits received from a set of actions. But costs of actions differ: some actions can already be implemented cost effectively, or have clear positive welfare implications (no-regrets measures), whereas other actions show positive net costs. Together, they form the basis of a cost function: if one lists the actions with the highest return first, followed by those yielding progressively lower returns (e.g., in terms of increasing dollars per ton of carbon), then one gets the familiar cost function with increasing marginal costs. Similarly, one can construct a declining marginal benefit (or avoided damage cost) function. As indicated in Box 3.1, the optimal policy is now determined by the point where marginal (discounted) costs equal marginal (discounted) benefits for all of the abatement actions taken separately (rather than lumped together). There is no a priori reason why total costs and benefits would be equal at this point. In other words, even if the total costs of a set of actions surpassed their benefits, there is no reason why one should not start carrying out those actions that yield net benefits, up to the point where the incremental costs of an extra effort equal the incremental benefits that result from that effort.

Third, the benefits of a climate change mitigation measure are not only the avoidance of future damage, but also secondary benefits (some of which may be felt immediately). This makes quite a difference. Let us assume – in line with some figures from the IPCC Second Assessment Report, Working Group III, and adopted in the Summary for Policymakers (Bruce et al. 1996) – that marginal damage avoided ranges anywhere between $5 and $125/tC and that $20/tC is a notional figure for a rough calculation. (The latter figure is also the central value in the range of notional shadow values assigned to the benefits derived per ton of avoided carbon in a recent GEF study; see Box 3.5.) If one took only damage avoided into account, according to the decision rule just mentioned to reach the optimal investment level, one would continue investing in abatement actions up to the point where the associated costs reached the $20/tC level. If, however, secondary benefits were included in the benefits as well, and if these benefits amounted to another $20/tC avoided, marginal benefits would, instead, be $40/tC. So the optimal level of investment in abatement actions would then increase up to the level where marginal costs were $40/tC. It will be clear from the abatement cost data presented in Chapters 6 and 7 that if $40/tC is an acceptable rough estimate of marginal abatement benefits (including secondary benefits), a significant amount of abatement action can be undertaken before such a level of marginal costs are reached.

Finally, marginal abatement costs and benefits have a strong time dimension, which severely complicates the issue of assessing optimal policies. Costs incurred now will have benefits that will not be felt until sometime in the future, in some cases in the distant future. This boils down to the fact that both costs and benefits have to be discounted, which is difficult not only because of the choice of the proper discount factor (as discussed in Chapter 2), but also because the fact that future economic structures and technologies will be different from the present ones has to be taken into account – which gives some people cause for optimism, and others for pessimism. However, precisely because of the many uncertainties involved with climate change and the associated damage, one has to decide now about how cautious to be about potential future developments. As discussed in Chapter 3, the precautionary approach may call for a strategy that aims at reducing future risks below the level at which (discounted) marginal benefits equal marginal costs.

To conclude, the present knowledge about climate change damage and abatement costs, combined with the methodological issues raised in this section, does not allow for the making of clear and unambiguous decisions based on quantitative assessment. However, there is strong evidence that if the marginal abatement costs equal marginal benefits rule is applied, a significant amount of abatement action can be justified even today, on the basis of present technological knowledge, starting with no-regrets options.

8.2.5 Making Trade-offs

The foregoing discussion has focused on the monetary valuation of costs and benefits of abatement. While such estimates must be improved and will be refined in the years to come, it might be useful to examine the usefulness of other techniques such as multicriteria analysis (described in Chapter 4) in the more near-term decision-making process. More specifically, we examine here the possibility of making a trade-off between the monetary costs of abatement measures and some (nonmonetary) physical index of climate change – as a proxy for avoided future damages.[2]

Figure 8.1 is a hypothetical trade-off curve of the costs of alternative abatement responses (e.g., measured as the average percentage of GDP lost per year) plotted against a climate change severity index (e.g., the global mean temperature rise at equilibrium in the distant future).[3] To simplify the presentation, we will specify alternative abatement paths in terms of the target concentration and the earliest date of stabilization. Thus, a point such as 450/2100 indicates that the atmospheric GHG concentration will be stabilized at a CO_2 equivalent concentration of 450 ppmv from 2100 onward. For a given stabilization date, lower target concentrations would imply higher abatement costs (e.g., 450/2100 vs. 500/2100), and vice versa. Furthermore, for a given target concentration, an earlier stabilization date would also be associated with greater abatement costs (e.g., 500/2100 vs. 500/2150), and vice versa. The data in the figure are illustrative only and are plotted as ellipses to indicate the considerable margin of uncertainty.

[2] The authors are grateful to Eric Haites for suggesting this example.
[3] Preliminary evidence tends to indicate that the actual economic damage will rise more than linearly as the target level of stabilized GHG concentration is increased.

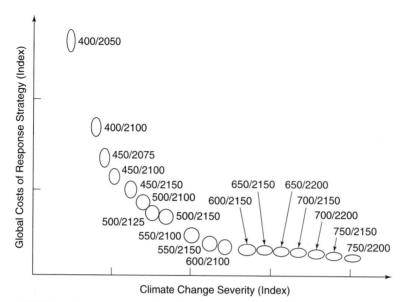

FIGURE 8.1. Hypothetical trade-off curve of an index of response costs versus an index of climate change severity. The pairs of numbers (*m/n*) represent the stable atmospheric GHG concentration (*m*) and the year of stabilization (*n*).

Preliminary information about such a trade-off curve suggests that abatement costs might rise rather slowly with decreasing concentrations up to about 600 or 650 ppmv, and then begin to increase more steeply – especially for concentrations below about 450 ppmv. Even with rather uncertain estimates about damage and imprecise knowledge of what might constitute "dangerous anthropogenic interference with the climate system," the shape of the curve may help us to develop a better climate change response strategy, especially if this curve exhibits a sharp bend or "knee."

First, the hypothetical numbers suggest that aiming for a stable concentration of less than about 500 ppmv could become very costly. At the same time, adopting a target level above 600 ppmv is unlikely to reduce abatement costs significantly relative to the sharply rising risk of damage. This aspect resembles the discussion in Chapter 3, in the context of Figure 3.1b and the affordable safe minimum standard. Second, delaying the stabilization date may not decrease abatement costs a great deal, except for the lower concentration targets (below about 500 ppmv). The latter conclusion has important implications for the debate regarding "putting off" decisions. Another relevant factor is that monetary damage is likely to be an exponentially increasing function of the climate change index (e.g., global mean temperature rise). The foregoing information, combined with affordability, risk aversion, and the precautionary principle, suggests a reasonable concentration target in the range 450 to 650 ppmv, and little justification for unduly delaying the stabilization date.

8.2.6 National Response Options and Costs

A variety of mitigatory, adaptive, and other response measures are available. It is clear that international cooperation is essential to achieve a globally coherent response strategy, and could both significantly reduce the costs and increase the effectiveness of GHG mitigation measures. Implementation costs might be further reduced by introducing efficient policies that provide correct economic signals through such policy instruments as carbon taxes, specific subsidies, quota schemes, and tradable emission permits, as well as international incentive programs like joint implementation schemes.

In this and the following subsections we will focus on these policy instruments – that is, economic instruments used by governments to affect the behavior of people and firms in such a manner that fewer GHGs are emitted and/or more are stored. Among these are command and control measures, taxes and subsidies, tradable emission rights, voluntary agreements, international cooperation arrangements, and so on.

Before domestic policy instruments for emission abatement are discussed in more detail, it should be mentioned that the most appropriate mixture of policies depends on the coverage of GHGs and whether or not carbon sinks are included. Carbon dioxide is the main source of past and present greenhouse concerns, though not the only one: methane and nitrous oxide, for example, are also significant in radiative forcing (see Chapter 1). However, including non-CO_2 GHGs in a policy strategy may cause complications, because the weights of these gases in relation to their radiative forcing potential are unknown. Including carbon sinks in domestic abatement strategies (e.g., through afforestation or reduced deforestation) seems quite logical, especially in view of the relatively low estimates of the costs of carbon sequestration through tree planting in certain areas. However, this option may seriously complicate the assessment of greenhouse policy instruments, mainly because a generally accepted system of measuring carbon stored in trees is not yet available due to numerous technical problems (see Chapter 6). In order to avoid these various complications in determining policy instruments, it will be assumed here that the policies aim exclusively at CO_2 emission reduction.

We can distinguish among three sets of domestic policy instruments to combat climate change. The first set consists of regulatory instruments; specific rules and standards are established in order to reduce GHG emissions either directly or indirectly. The second set consists of market-based instruments, which means that governments try to alter price signals so that GHG emissions are lowered. These instruments are mainly taxes and subsidies. The final set of measures aims at influencing consumer or producer behavior in a different way – for example, by providing information (through education, eco-labeling, home energy ratings, etc.) and supporting research on and development of new energy technologies. To the latter, several other policies may be added whose main target is not energy use or GHG abatement, but which may nevertheless have a significant impact on GHG emissions. Examples are policies on family planning, migration, and international trade and finance (see Section 6.1.3).

What can be said about the ideal blend of instruments (not to be confused with the ideal blend of options discussed earlier)? For the moment, we will assume that a country facing this issue will not be concerned with how its new policy will affect the rest of the world, nor will it be concerned with how policies in the rest of the world might affect its own climate change policy. In other words, we shall assume an autarkic world in which one country tries to achieve a predetermined emission reduction target – for example, the stabilization of emissions by 2000 at the 1990 level.

Individual national governments will usually consider more than one criterion in order to distinguish among competing policy instruments. Moreover, each government can attach its own weights to different criteria. Some key criteria include

- the effectiveness of the policy instrument, or the degree to which the instrument helps achieve environmental targets;
- the efficiency or cost effectiveness of the instrument, including the costs involved in administration, capacity building, control, and monitoring (including a long-term impact on innovation, technical progress, learning, etc.); and
- the instrument's political acceptability to the constituency (including equity considerations).

If the economy is not autarkic, these criteria will have to be amended, as discussed in Section 8.2.7.

Effectiveness If one government knew exactly how emission levels were affected by emission standards and controls or by the costs of implementing a given policy instrument (e.g., an emission tax), it would make no difference with respect to the end result whether emissions were controlled directly (quantity control) or indirectly (through emission taxing). This is because in a world of certainty there is always one combination of price and volume that satisfies the market equilibrium.

Clearly, this certainty is lacking, as explained earlier. There is uncertainty about potential climate change implications, as discussed in Chapter 2.[4]

---

[4] However, this uncertainty does not affect the conclusion drawn in the preceding section once an emission reduction target has been accepted. This is because the interchangeability of a volume- or price-based instrument depends only on certainty with respect to market behavior.

Moreover, the government is not fully informed on the behavior of the public or market facing taxes/subsidies or standards in terms of supply and demand. For example, what would the emission reduction be if an emission tax (or emission reduction subsidy) of $1 per tC were introduced? Would the impact per $1 tax/subsidy change at higher tax rates? (See Box 8.1.) Furthermore, public behavior may change in the course of time; for example, people might try to evade the tax or the effects of standards. Thus, something that is effective now may be relatively ineffective after some time; for example, a fuel efficiency standard for cars may have a much stronger impact on promoting energy-saving technologies than a gasoline tax or catalyzer subsidy, whereas a gasoline tax may have a stronger impact on the overall use of cars.

BOX 8.1 Carbon Taxes and Their Effectiveness

An EMF (1993) study, the reference case projections of which were presented in Box 7.5, was used to estimate what tax rates (dollars per ton of carbon for a carbon tax) would be required to stabilize U.S. emission levels in 2010 at 1990 levels and to reduce these emissions by 20% below 1990 levels, respectively. The results are represented in Table B8.1.

Estimates of carbon taxes required to reduce emissions by 20% range from $50 to $330 per ton! Much of this difference can be attributed to the fact that it was left to the discretion of the modelers to introduce a price elasticity of demand and a factor determining capital stock adjust-

TABLE B8.1 Comparison of Carbon Taxes and GDP Losses in 2010

Model	Carbon Tax ($ per Tonne of C)		GDP Loss (% of GDP)	
	Stabilization	20% Reduction	Stabilization	20% Reduction
CRTM	150	260	0.2	1.0
DGEM	20	50	0.6	1.7
ERM	70	160	0.4	1.1
Fossil 2	80	250	0.2	1.4
Gemini	120	330		
Global 2100	110	240	0.7	1.5
Global macro	20	130		
Goulder	20	50	0.3	1.2
GREEN	80	170	0.2	0.9
MWC	70	180	0.5	1.1

Source: EMF (1993).

(continued)

> **BOX 8.1** Carbon Taxes and Their Effectiveness *(continued)*
>
> ment to higher energy prices. However, if the modelers already have such a wide variety of views, how would a government determine whether specific carbon taxes would be effective?
>
> Note, finally, that taxes tend almost to double if one switches from the stabilization target toward the −20% target. This suggests that the extra tax that must be raised to achieve an extra unit of emission reduction will increase with the level of emission reduction already (expected to be) achieved. Moreover, as was said before, the public's behavior may change over time, for example, by trying to evade taxes, so that what is effective now may be relatively ineffective in the longer term. The same effect may occur when environmental standards are set.

In general, one can argue that if a government wants to realize a particular emission (reduction) target in the short term, it could apply a quantity-based policy instrument, even if it is generally assumed that the same target could be achieved more efficiently by altering prices rather than setting standards (the theoretical considerations will be discussed in the next subsection). The rationale here is that relying on only the right price signals may not guarantee that all market parties will consume or produce the optimum amount. When the government is more relaxed about achieving emission targets, it may decide to concentrate on taxes, or on mixed systems involving taxes/subsidies and standards, or it may take indirect measures or measures that induce private parties to solve the problem by means of voluntary agreements.

If the government – with an eye to achieving the maximum effect – decided to apply uniform regulatory standards as part of its climate change strategy, it would still have to choose between technology-based and performance-based standards. In the first case, standards would require specified equipment, processes, or procedures, while in the latter case, acceptable levels of pollutant emissions or polluting activities would be specified, leaving the decision of how to achieve such targets to the regulated entities. Thus, performance-based standards offer private participants more freedom to determine the best way to meet standards and are generally considered more cost-effective and efficient than technology-based standards. Extreme forms of standards are bans, like the one in the Montreal Protocol regarding ozone-depleting substances. Obviously, bans can be highly effective in reducing levels of polluting substances, and if a ban also stimulates large-scale, low-cost production of substitutes, it may be rather efficient.

Efficiency or Cost Effectiveness   There is a considerable literature arguing for the use of market-based policy instruments to meet GHG emission targets.

The argument is based on the view that such instruments – particularly taxes and tradable emission quotas or tradable permit schemes – best satisfy the criterion of efficiency or cost effectiveness. In addition, it is argued that such instruments promote dynamic efficiency by providing a continuous incentive for research and development in emission abatement technologies in order to avoid tax or quota purchases. Tradable domestic permit schemes are considered superior, at least theoretically, from an efficiency point of view. If applied under perfect competition, such schemes – a bit of experience with which has been built up through their application to other environmental problems like local air pollution – could theoretically guarantee equilibrium permit prices in the market, which are the same everywhere. Consequently, the ex post allocation of permits will be at minimum costs of reducing emissions, because firms will continue to buy (or sell) permits as long as permit prices are lower (or higher) than marginal abatement costs.

Because of this characteristic of tradable permit systems and because of their proclaimed effectiveness in terms of greenhouse abatement, many have suggested their use in GHG abatement strategies. This is not to say that such schemes would be entirely without complications. Issues that arise concern how permits should be distributed to individual firms; whether permits should be valid permanently or be time-limited; whether the banking of permits should be allowed (see Box 8.2); whether the permit system should be left to private participants or controlled by the government; and whether permit revenues should be used for redistributive purposes, and if so, how. Most important, one has to determine the transaction costs of such systems, particularly if they are to be applied internationally. This may be a serious matter, because there is anecdotal evidence with respect to some U.S. tradable permit schemes (mainly dealing with local pollution) that transaction costs can be considerable. Even if a hypothetical tradable permit scheme to control global GHG emissions were not entirely comparable with existing domestic U.S. permit schemes, the issue of transaction costs should still be given serious attention.

BOX 8.2 Forward Sales in Emission Quotas

The United States has an auction system for forward sales in emission quotas, as part of its provisions for controlling sulfur dioxide (SO_2) emissions from fossil-fuel-fired power plants. By the year 2000, total SO_2 emissions from the U.S. electricity power sector are required to be about 50% under the 1980 level. As of January 1, 1995, each of the 111 power plants directly affected during the first phase of implementation was to meet tradable quotas covering its total annual emission target, capped at about

(continued)

BOX 8.2 Forward Sales in Emission Quotas *(continued)*

50% of the 1980 level. Currently, allowances may be traded to any party, or they may be credited for future use. During the second phase, starting January 1, 2000, most electric power utilities will be brought into the system. Quotas for excess emissions – that is, beyond the annual target allowance, may be bought directly from other plants or through auctions organized by the Chicago Board of Trade for the U.S. Environmental Protection Agency.

Two of the key issues of an international tradable permit scheme, as with national tradable permit schemes, are the transaction costs involved and the potential abuse of market power. These are already a matter of concern with respect to the existing national schemes in the United States, although the evidence is rather mixed. Theoretically, transaction costs will decline relatively and the market power of specific agents on the permit market will disappear as the number of potential trading sources and the number of transactions per source increase. However, this applies only to the final stage rather than to the process leading up to it, which may take quite a while. Therefore, the competition on the permit market and the transaction costs involved will be issues to consider if a greenhouse tradable permit system is considered for introduction on an international scale.

Next, we will offer some more general comments on the costs and benefits associated with different greenhouse policy instruments in order to arrive at a better understanding of the efficiency and cost effectiveness of these instruments. Irrespective of whether efficiency (or the maximization of net benefits) or cost effectiveness (or the minimization of aggregate costs) is applied as a criterion, all possible costs of policy programs should be taken into account and compared with the benefits in order to minimize the costs of achieving a certain GHG emission target. The cost categories include not only the implementation costs usually paid by governments, but also transaction costs paid for mainly by the private sector. Examples of costs incurred by the government are those associated with monitoring and enforcing environmental laws and regulations. Corresponding examples for the private sector include the capital and operating expenditures associated with compliance, legal, and other transaction costs, the effects of altered management priorities or the disruption of production, lower incentive to invest, and delayed innovation. Finally, examples of costs paid by society include transition costs, such as the loss of jobs and economic security.

These cost components have to be weighed against the benefits that may also result from the use of a policy instrument – for example, the impact of a cleaner environment on productivity and the positive effect of the policy on innovation and research and development. Irrespective of whether a policy instrument is a domestic energy or carbon tax, its efficiency or cost effectiveness will depend mainly on whether and how the tax income will be recycled.

It is generally accepted that the societal costs of a carbon tax can be significantly reduced if the tax revenue is recycled. It can even be argued that, under certain circumstances, all abatement costs associated with a carbon tax can be eliminated through revenue recycling by means of income tax cuts. One of the main preconditions for this, however, is that the existing income tax exceeds the optimum level, and therefore is already reducing national income below what it might have been. If this is the case, there clearly is another benefit to be gained from introducing a carbon tax. For an assessment, see Box 8.3.

BOX 8.3 Welfare Implications of a European Union Carbon–Energy Tax Plus Recycling: Mixed Evidence

In the early 1990s the European Commission put forward a proposal to levy new taxes on most major energy sources, or at least to study this idea. Taxes would be based (as an intra-EU compromise) partly on carbon and partly on energy content and would increase stepwise to $10 per barrel of oil in 2000. Various recycling schemes were projected. Several models were then used to analyze the economic implications of such a new tax: the results of two models, HERMES and DRI, are summarized in Table B8.2 (in both cases, tax recycling was foreseen).

The table clearly illustrates that economic modeling can contribute to increasing rather than solving policy controversies: the outcomes lead to opposing conclusions. Whereas HERMES predicts overall welfare improvements due to the introduction of energy taxes, DRI draws the opposite conclusion. In the meantime, the proposal has been dropped from the political agenda (for the time being).

In the discussion on domestic carbon taxes, particularly within the European Union, much attention has been paid to the issue of whether a greenhouse abatement tax will have to be levied on the energy content of fuels, on the value of energy products, or on the carbon content of primary fossil fuels consumed. Although the latter tax is not a perfect proxy for a tax on CO_2 emissions, from an efficiency point of view it is superior to taxes levied on the basis of energy use. This impression is corroborated by research concerning the U.S. economy, indicating that an energy tax could be 20 to 40% more costly and a tax based on the value of energy products two to three times more

TABLE B8.2 Differences in the Macro Effects of the Energy Tax between EC Countries in 2001 and 2005 (Percent Difference from the Baseline)

	Model	FRG	F	UK	IT	NL	B	EUR6
GDP	HERMES[a]	0.22	0.06	−0.72	0.72	−0.16	0.57	0.15
	DRI[b]	−0.26	−0.13	−0.52	−0.39	−0.39	−0.65	—
Private consumption	HERMES[a]	0.27	0.03	−0.57	0.75	0.34	0.23	0.15
	DRI[b]	−0.78	−0.39	−0.78	−1.04	−1.18	−0.78	—
Employment	HERMES[a]	0.79	0.44	0.56	0.79	0.30	0.88	0.64
	DRI[b]	−0.39	−0.13	−0.39	−0.39	−0.26	−0.39	—
Inflation	HERMES[a]	0.45	0.80	2.12	0.86	0.80	0.20	0.95
	DRI[b]	1.96	1.30	2.22	2.10	2.10	2.00	—

Abbreviations: FRG, Germany; F, France; UK, United Kingdom; IT, Italy; NL, Netherlands; B, Belgium; EUR6, the six European countries listed.
[a]Recycling by reduction of social charges in HERMES model (1993). The reference year is 2001.
[b]Mixed policy test in DRI model (1994). The tax revenue is recycled as follows: 30% for reduction of empolyers' social charges, 30% for reduction of personal direct taxes, 10% for reduction of corporate taxes, 30% for incentives to energy conservation. The reference year is 2005.

costly than a carbon tax on primary fuels for equivalent reductions in emissions (Jorgenson and Wilcoxen 1992; Scheraga and Learly 1992). The reason is that an energy tax raises the price of all forms of energy, whether or not their use contributes to CO_2 emissions.

Efficiency problems are often connected with regulatory standards. If at the same time uniform technology standards were imposed on polluting firms (e.g., electricity companies), there is a fair chance that the marginal costs of controlling the pollutant would vary greatly among the firms due to the differences in their age and location. Some sources indicate that these marginal costs might vary by a factor of 100 or more. The point is that a minimum level of aggregate pollution control costs for all firms together could be realized theoretically only if the marginal costs of pollution control were the same among firms. In actual practice, when applying technology standards this condition is seldom fulfilled. So, although uniform technology standards can be quite effective, they are seldom efficient. This is not to say, however, that technology-based standards never have straightforward advantages, especially if they are set and adjusted over time in such a way as to continuously force firms to readily adopt the latest technologies in a manner that is easy to control.[5] The government does not always succeed in achieving this, as illustrated in Box 8.4.

BOX 8.4   Technology Standards Inhibiting Innovation

When in the past vehicle emission standards requiring catalytic converters were adopted by the European Union (nowadays the EU has adopted performance-based standards, just like the United States and Japan), incentives to develop superior lean-burn engines were in fact reduced. This was because lean-burn engines could not be fitted with a three-way catalytic converter. Today, lean-burn technology is available that can meet Japanese and European standards but not U.S. emission standards or the stricter standards to be introduced in the near future. Thus, the introduction of catalytic converters may have delayed the introduction of environmentally superior lean-burn technology.

The Political Acceptability of a Policy Instrument to the Constituency   The aim of most governments is to stay in power. Together with equity and other considerations, this may explain why many countries traditionally subsidize energy, especially that derived from primary fossil fuels and electricity. For ex-

---

[5] In this respect, technology-based standards can be superior to performance-based standards.

ample, in 80% of developing countries, average electricity prices are 30% below long-run marginal costs (Munasinghe 1995). However, in various industrial countries, too, electricity is sold against prices that are considerably lower than marginal costs. The implications of these policies can be significant. Larsen and Shah (1992, 1994) – using border prices as a benchmark – calculated that primary fossil fuel subsidies worldwide are equivalent to a negative carbon tax of U.S.$40 per ton. They estimated that the abolition of all energy subsidies would reduce global CO_2 emissions by 4 to 5%. A similar figure emerged from an OECD study using the GREEN model – removal of energy subsidies would reduce global emissions by 18% by 2050 (OECD 1994). Moreover, the removal of these subsidies could lead to an increase in real income because of the removal of the market distortion, as well as to additional benefits in terms of reducing SO_2 and NO_x. A study of the Norwegian situation indicated that a 20% CO_2 emission reduction would cause SO_2 and NO_x emissions to fall by 21 and 14%, respectively. Consequently, the IPCC Working Group III Summary for Policymakers states, "A number of studies . . . indicate that global emission reductions of 4–18%, together with increases in real incomes, are possible from phasing out fuel subsidies . . . however, subsidies are often introduced and price distortions maintained for social and distributional reasons, and they may be difficult to remove" (Bruce et al. 1996, p. 15).

It is against the background of traditional energy subsidies that emission abatement policy instruments have to be evaluated. This means that a country's overall system of GHG abatement and energy subsidy policies (rather than isolated abatement policies) should to be taken into account.

Governments may subsidize not only energy but also, directly or implicitly, the use of private transport, agricultural production, the exploitation of forests, the use of fertilizer, and so on. It is obvious that such policies will often have an adverse effect on GHG abatement. In other words, in many cases the best thing governments can do in trying to meet their emission reduction targets is simply to remove implicit subsidies.

One way for governments to try to maintain political support for their greenhouse policies is to invite private parties to introduce abatement strategies themselves. By threatening to use mandatory government intervention, the government can "encourage" firms to enter voluntary agreements on controlling GHG emissions. Although there is some debate about the effectiveness of this approach, several experts consider it promising. A major advantage is that the initiative is left with the private participants, and measures are undertaken in a mood of mutual confidence and understanding, which reduces overall transaction costs. In fact, the vast majority of GHG reductions from the actions planned or undertaken by, for example, the U.S. Climate Change Action Plan are brought about by voluntary initiatives aimed at increasing the energy efficiency of the industrial, commercial, residential, and transport sectors.

A final aspect of the political acceptability of market-based measures concerns whether the instruments (e.g., carbon or energy taxes) are regressive or progressive – that is, whether the domestic income distribution becomes less or more equitable, respectively. The literature on industrialized countries shows that carbon taxes are likely to be regressive, since expenditures for fossil fuel consumption as a proportion of current annual income tend to fall as income rises. However, the regressiveness of carbon taxes is less relative to lifetime income or annual consumption expenditures than to annual income. There is evidence that in developing countries the impact of a carbon tax would either be progressive or less regressive than is often suggested: because of certain institutional factors common in such countries, these taxes would be only partially shifted to consumers, leaving the remainder of the tax burden with the producers, who are usually better off.

8.2.7 International Response Options

As argued in Section 3.2, solving the greenhouse problem requires a collective process of decision making with fair representation and treatment of all parties (mainly sovereign nations) and transparent procedures. In the same section, it was pointed out that international negotiations and agreements may not be adopted by all parties, and even if they are, their implementation usually takes a long time. Keeping this in mind, we will now turn to greenhouse abatement policy instruments in an international context. More or less the same categories of policy instruments are distinguished as in the preceding subsection – regulatory measures, market-based instruments, and other policies. To this list can be added the typical international instruments that require international coordination, such as joint implementation, international transfers, international technological cooperation, and international tradable permit schemes. Again, these instruments will be evaluated against the background of the criteria applied earlier: effectiveness, efficiency, and political acceptability.

Effectiveness If abatement strategies are carried out by countries willing to adopt these policies even if other countries will not follow immediately, this will not automatically detract from the effectiveness of these unilateral strategies. The reason is that those countries that ratified the UNFCCC have accepted equity and fairness as important elements of the Convention, as well as the fact that the "Parties should protect the climate system for the benefit of present and future generations of humankind, on the basis of equity and in accordance with their common but differentiated responsibilities and respective capabilities. Accordingly, the developed country Parties should take the lead in combating climate change and the adverse effects thereof" (UNFCCC 1992, Article 3.1). In other words, the parties have agreed that if certain countries do not take action, their inaction should not be thought of as undermining the effectiveness of the actions of the leading countries.

However, this is not the case when countries do benefit from global abatement (or at least can be held responsible for contributing to global climate change) but have *no justified reason* for not or insufficiently contributing to abatement efforts.[6] These countries can be considered free riders, and free riding can undermine existing or potential abatement initiatives, because it may frustrate countries that do take full action. Moreover, free riding always results in abatement that is less than the global optimum. This is because the climate system can be considered an international public good in that changes in the global climate are the result of the combined actions of all countries. (In a similar spirit, the UNFCCC refers to the climate system as a "common concern of humankind.") The preservation of the climate system will always be less than optimal if not all responsible parties cooperate and fully comply with international agreements. There is a serious risk that this will happen, because certain states resisted compulsory use of judicial processes in the UNFCCC. Although the UNFCCC provides several means of settling disputes, including judicial recourse and arbitrage, and although the UNFCCC requires parties to consider the introduction of a "multilateral consultative process" to assist in the implementation of the Convention and to anticipate and prevent confrontations concerning compliance and enforcement, these mechanisms are too weak to prevent free riding completely. Obviously, free riding might pose a problem irrespective of the policy instruments the parties consider applying.

The question remains as to how free riding can be prevented. Clearly, a system has to be designed that will enable cooperating countries to either punish or reward unjustifiably noncooperating countries, or countries that withdraw or threaten to withdraw from the group, or whose performance is less than was agreed on originally. However, this raises another question: Who will decide what the contribution of unjustifiably noncooperating countries ought to be, which countries belong to the category of unjustified free riders, and what punishment (for uncooperative behavior) or reward (for cooperation) is appropriate? This question lies at the very heart of the complexity of international decision making. In Chapter 3, it was argued that a collective decision-making process would be required to reach a useful consensus on the assign-

---

[6] Needless to say, whether the absence of an abatement strategy should be considered unjustified is a highly political and controversial issue.

ment of differentiated responsibilities for implementing abatement measures and thus achieving agreed-upon common emission targets. The key issue will be the formation of a stable group of cooperating countries – that is, a group that will continue to exist despite certain incentives to free-ride. So far the issue of how and under what circumstances a stable group of countries can be formed that also comply fully with the rules agreed upon (a group that former free riders would gradually join) has been paid relatively little attention in research and will definitely not be easy to solve.

The fact that greenhouse policies in cooperating countries will affect energy and other prices and thus, for example, energy use, in noncooperating countries further complicates the international decision-making process. For example, if cooperating countries were to tax carbon emissions and therefore the use of fossil fuels, this would seriously affect noncooperating countries, which might frustrate the good intentions of the cooperating countries. First, declining world demand for fossil fuels might lead to lower energy prices worldwide, and thus to more fossil fuel use in the countries outside the group, as well as slower adoption of fossil-fuel-saving technologies in these countries, and so on. Second, the relatively energy-intensive industries in member countries would become less competitive than the same industries in nonmember countries. These industries might therefore try to relocate in a nonmember country. Third, the net effect of abatement policies in terms of worldwide GHG emission reduction would therefore be smaller than would be the case if all countries joined the group. These effects are commonly referred to as *leakage.*

Leakage is likely to be larger if the demand elasticity of fossil fuels is large (given the fact that the position of member countries in the world market is strong enough to affect world energy prices), if fossil-fuel-intensive products from countries belonging to the group are more competitive than goods from nonmember countries, if energy-intensive firms become more mobile internationally, and if there are more free riders.

Several models have been used to estimate emission leakages that would result if an important group like the EU or OECD were to carry out a carbon emission reduction policy unilaterally. Since these models employed various assumptions, particularly with regard to the substitutability of fossil-fuel-intensive products between member and nonmember countries, it will not come as a surprise that the outcomes of the leakage rates (the increase in emissions by nonmember countries and the reduction in emissions by member countries) differed a great deal. The Whalley–Wigle model estimated that a 20% reduction in carbon emissions within the EU would be associated with a 80% leakage rate (20% reduction in OECD: 70% leakage rate) (Whalley and Wigle 1992). In the GREEN model, stabilization of EU carbon emissions unilaterally at the 1990 level would lead to a leakage rate of about 12% in 1995 only (2.2%

in 2050). A similar policy throughout the OECD would lead to corresponding leakage rates of 3.5 and 1.4%, respectively (OECD 1994). Similarly, Hanslow et al. (1994), using the MEGABARE model, found a 3.8% leakage rate for a 20% reduction in CO_2 emissions in Annex I countries.[7] The outcomes revealed that leakage might be either a very serious or a relatively minor complication of a system in which only a group of countries is prepared to carry out abatement strategies. This is another way of saying that currently there is no consensus among economists about the magnitude of leakage, and that it is a potentially serious problem that might threaten the formation of a stable group willing to comply with greenhouse abatement measures.

Suppose that there is a significant leakage problem once a group of countries has adopted abatement strategies and fully complies with them. In a global free-trade system, the member countries might start to import products produced by carbon-intensive processes from nonmember countries that are similar to products heavily taxed at home. Can anything be done? In other words, is it possible to discourage such leakage imports? The answer depends very much on the trade policy regime that is considered acceptable. Basically, one is faced with two conflicting targets: on the one hand, there is the wish to keep the trade system open and, on the other, there is the desire to fight the leakage in order to contribute to greenhouse abatement. It is difficult to know beforehand what is an acceptable solution to this dilemma. One of the main reasons is that so far the World Trade Organization (WTO), which deals with the international trade system from a multilateral perspective, has tended to argue that trade policy measures against imports should not be allowed on the basis of production techniques. The reason for this is clear; it is to prevent the abuse of such arguments for protectionist purposes (e.g., steel will not be imported in the EU from Poland because the local production process is polluting). Yet the Montreal Protocol includes a provision for restricting trade in products made by means of ozone-depleting substances, such as electronics components that use chlorofluorocarbons as a solvent. However, it is not altogether clear whether such a provision is WTO compatible and could be included in an international GHG regime. In any case, the provision in the Montreal Protocol has not been implemented, and in 1993 the secretariat to this agreement was advised that implementation would not be (technically) feasible.

Efficiency and Cost Effectiveness   What would be the most efficient greenhouse abatement strategy if a group of countries – assuming the group were stable – carried out the agreed-upon abatement strategies, while the remain-

---

[7] For a survey of several global simulation studies providing leakage estimates, see, e.g., Barrett (1994).

ing countries did not, for the sake of equity or other reasons? If the strategy were based on a carbon tax, the most efficient system would be the levying of a uniform tax across all countries belonging to the group. Only then could the efficiency condition – that the marginal costs of investment in carbon saving everywhere equal the marginal benefits in terms of carbon tax avoided – be satisfied. Under these circumstances, a change in the adjustment process would increase the overall costs of the greenhouse strategy.

This condition is assumed to be independent of actions involving international tax revenues. In other words, member countries could use these revenues to deal with international equity considerations by transferring them to poor countries – for example, to induce these countries to adopt greenhouse abatement measures without reducing the efficiency of the tax system. It can be questioned, however, whether this assumption holds in all cases. The reader will recognize the issue, discussed earlier, of the overall tax burden in a country, which will undermine the optimal efficiency of the tax system if it is too high. In other words, if carbon tax revenues collected in a member country are used mainly for transfer to other countries for reasons of international equity and justice, the overall tax burden could rise to such a level that it would conflict with the efficiency criterion. It is important to keep this consideration in mind, although in the following paragraph we will assume that efficiency and equity aspects can be dealt with separately.

As in an autarkic economy, in a system based on international cooperation the instrument of an (international) tradable emission quota scheme is often considered superior from an efficiency (as well as an effectiveness) point of view. Under such a scheme, all countries – member and nonmember – would initially be allocated a quota for emissions based on a particular distributive criterion. Assuming noneternal quotas and the possibility of banking, ideally in each period countries would be free to buy and sell quotas on an international exchange, thus guaranteeing uniform quota prices globally and maximum efficiency. Putting a time limit on the quotas would probably be necessary to account for uncertainty concerning the extent of the greenhouse problem, to give credibility to the system, and, specifically, to avoid a situation in which a current government sold quotas (i.e., part of the nation's wealth) to an extent that would not be appreciated by future governments. In addition, all conditions that together allow for a reliable and effective market organization for quota trade would have to be fulfilled. If all conditions were satisfied, an international tradable quota scheme could prove to be among the most efficient greenhouse abatement instruments, at least theoretically.

An additional factor in favor of a tradable quota scheme is that a forward and/or futures market based on (net) emission quota contracts could be established to provide a way of efficiently reducing greenhouse costs of uncertainty and of risks: by buying futures, a country that invested in a risky

TABLE 8.2 JI/AIJ Pilot Projects

Project	Type	Host Country	Investing Country
Decin	Fuel switching	Czech Republic	U.S., Denmark
Rio Bravo	Forestry	Costa Rica	U.S.
CARFIX	Forestry	Costa Rica	U.S.
ECOLAND	Forestry	Costa Rica	U.S.
Plantas Eolicas	Wind energy	Costa Rica	U.S.
Enersol Electrification	Solar energy	Honduras	U.S.
RUSAFOR-SAP	Forestry	Russian Federation	U.S.
Krokonose/FACE	Forestry	Czech Republic	The Netherlands
Profafor/FACE	Forestry	Ecuador	The Netherlands
Energy saving	Energy efficiency	Hungary	The Netherlands
Compressed natural gas fuel engines	Fuel switching	Hungary	The Netherlands
Landfill	Methane recovery	Russian Federation	The Netherlands
Tyumen Horticulture	Energy efficiency	Russian Federation	The Netherlands
ILUMEX	Energy efficiency	Mexico	Norway, GEF
Coal-to-Gas	Fuel switching	Poland	Norway, GEF
Vologda	Forestry	Russian Federation	U.S.
Klinki	Forestry	Costa Rica	U.S.
El Hoyo-Monte Galan	Geothermal	Nicaragua	U.S.
Bio-Gen Phase I	Biomass power generation	Honduras	U.S.
Dona Julia	Hydroelectricity	Costa Rica	U.S.
Tierras Morenas Windfarm	Wind energy	Costa Rica	U.S.
Aeroenergia	Wind energy	Costa Rica	U.S.
Biodiversifix	Forestry	Costa Rica	U.S.
RUSAGAS-FGC	Fugitive gas capture	Russian Federation	U.S.

Uganda National Parks	Forestry	Uganda	The Netherlands
Bhumtang	Microhydropower	Bhutan	The Netherlands
Renewable Energy Systems	Renewables	Indonesia	Germany, Japan
Ainazi windpower	Wind energy	Latvia	Germany
Virilla river basin	Hydropower/forestry	Costa Rica	Norway
Skoda-Mlada Boleslav	Energy efficiency	Czech Republic	Germany
Bel/Maya Biomass	Biomass power generation	Belize	U.S.
Zelenograd	District heating	Russian Federation	U.S.
Sonora	Halophyte cultivation	Mexico	U.S.
Climate Action forestry	Forestry	Bolivia	U.S.
Chiriqui Province	Forestry	Panama	U.S.
Bio-Gen Phase II	Biomass power generation	Honduras	U.S.
Sustainable energy management	Sustainable energy	Burkina Faso	World Bank
Renel-SEP	Energy efficiency	Romania	The Netherlands

Source: JIN Foundation (1997).

technology transfer project could guard itself against the risk of project failure.

Political Acceptability As argued in Section 2.4, equity is an important element of the collective decision-making framework, for various reasons:

- Principles of justice and fair play are important in their own right.
- Equitable decisions carry greater legitimacy and induce parties to cooperate.
- Equity and fairness are important elements of the sustainable development concept.
- The UNFCCC has several specific references to equity in its substantive provisions.

This is not to say, however, that there is consensus among economists or philosophers about the appropriate ethical responses to the threat of climate change. In the end, this remains a political issue and the subject of international negotiations. For this reason, equity principles should apply to both procedural and consequential issues (on the latter, see Section 2.4).

An instrument that deserves special attention from both an efficiency and an equity point of view is joint implementation (JI; see Box 3.3). JI allows an investing country (most likely a UNFCCC Annex I country with a GHG emission reduction target) to fulfill (part of) its commitments under the Convention in another country (the recipient country) for reasons of cost effectiveness. Instead of undertaking relatively expensive measures at home, an Annex I country could invest in a GHG emission reduction project in, for example, a non–Annex I country where such a measure can be applied at lower marginal costs. Globally, this implies that commitments under the UNFCCC can be met at lower costs, or that more abatement action can be undertaken with the same amount of money, or both. Because the recipient country can be a developing country or a country in transition, a North–South or West–East transfer is usually part of the JI project. This transfer can be financial or technological, or a combination of the two.

At the first Conference of the Parties to the UNFCCC, in Berlin in March and April 1994, it was decided that a JI pilot phase would follow, not to last until after 2000, in order to carry out experiments with JI projects and to solve several technical problems with regard to such issues as baseline establishment, monitoring, and inspection procedures. Since then, various donor countries have taken the initiative to establish and carry out JI projects. Some countries, especially the United States, developed domestic procedures for assessing project proposals as potential JI projects. A survey of official JI projects (as of April 1997) is presented in Table 8.2 and two examples of a typical JI project, including its equity aspects, are presented in Box 8.5.

BOX 8.5 Two Examples of Joint Implementation Pilot Projects

Fuel Switch: Decin, Czech Republic

On September 18, 1995, the JI project in Decin, situated in northern Bohemia in the Czech Republic, was officially started. Earlier in 1995, the project had received approval during the first round of the U.S. Initiative on Joint Implementation (USIJI). Decin is a heavily industrialized center with 55,000 inhabitants. The extensive use of brown coal (since 1992, 75% of the community's square footage was heated by coal) is the main source of the high level of air pollution suffered by the region.

Although large power plants within the Czech Republic are receiving assistance for the abatement of pollutants, district heating plant projects are often too small to attract multilateral funding, despite their devastating effect on the environment. Through the assistance of the U.S. Center for Clean Air Policy (CCAP) additional funds were raised to finance the replacement of a brown coal heating plant with a natural gas plant in one of Decin's five district heating plants. For this purpose three U.S. utility companies were lined up, each contributing $200,000. Because of the project, local air pollution will be reduced, and at the same time the system's efficiency will be improved significantly. In short, the activities undertaken as part of the project are:

- to increase supply-side efficiency by means of a fuel switch in a district heating plant;
- to improve energy efficiency in the district heating network;
- to install energy control equipment in energy supplying units; and
- to use the process of cogeneration instead of production of heat only.

The project's results in terms of pollution reduction are summarized in Table B8.3.

Rural Solar Electrification: Honduras

On February 3, 1995, a JI project involving a solar system project in Honduras received official approval from the U.S. government as part of the USIJI. The project aims at replacing kerosene with solar systems in the Honduran countryside, where around 400,000 homes, or more than 80% of the rural population, have no access to electricity. Early in 1995 the project was among the 15 projects that had received approval by the governments of the host country (Honduras) and the investing country (the United States). The project has been developed by the Honduran nongovernmental organizations COMARCA and ADHEJUMUR and the U.S. nonprofit organization Enersol and Associates.

(continued)

BOX 8.5 Two Examples of Joint Implementation Pilot Projects (*continued*)

TABLE B8.3 Reduction of Pollution through the Decin Project

CO_2	133,827 t (due to fuel switching) (at a cost of \$4.48/t CO_2)
	475,125 t (due to cogeneration)[a]
SO_2	96 t/yr (practically eliminating SO_2 emissions from the district heating plant)
Ash	3,190 t/yr (total elimination of this waste stream)

Note: The figures for CO_2 are totals over the project's lifetime; those for SO_2 and ash are expressed per year.

[a]Cogeneration is the process in which the heat created by the production of energy is reused for additional energy production. So cogeneration refers to the efficiency gains resulting from the new plant equipment.

Source: JIN Foundation (1995) and CCAP/SEVEn (1996).

Global Environmental Benefits   The solar-electric systems directly replace kerosene and batteries that are currently used in rural homes. The project participants estimate that CO_2 displacement occurs at an average total of about 4.7 metric tons of CO_2 per system over a 20-year period, which corresponds to the system's typical design life. Replacing kerosene lamps also reduces emissions of carbon monoxide, sulfur dioxide, nitrous oxides, and particulates within homes. The 750 systems currently installed thus save more than 175 metric tons of CO_2 per year. This concept could be applied in numerous other parts of the developing world; thus, this project could serve as a model.

Benefits for the Host Country   In areas of low energy demand and low population density, solar-electric systems are often the most cost-effective way of supplying electricity. Utility grid infrastructure – transmission and distribution lines – is expensive in these cases, and cost recovery is usually impossible. Consequently, people in the countryside have to use kerosene for lighting and dry-cell batteries for small appliances, and often car batteries have to be charged at the nearest electrified town. Solar-electric systems satisfy the basic energy needs of families, farms, schools, and health clinics in rural areas that are beyond the reach of the electricity distribution network. The quality of lighting and indoor air has improved, and the fire hazard has been reduced dramatically. Furthermore, the efficiency of electric lighting is many times higher than that of kerosene lamps, so there is also a financial saving in the long term. Solar-based electricity generation is also very reliable and not subject to the power shortages that often affect utility systems in developing countries. Finally, the project creates job opportunities for technicians and provides electric power and light to small businesses.

8.3 NEXT STEPS

The foregoing analysis and the basic facts summarized in Box 8.6 underline the importance of the climate change issue and provide a basis for a series of promising actions to address it. The precautionary principle indicates that uncertainty about what might actually constitute "dangerous" anthropogenic interference with the climate system does not provide valid grounds for inaction or delay in making decisions to reduce the risks of climate change. Consistent with this approach is the conclusion that while the ultimate targets may be unclear as yet, the directions of change are more obvious.

Broadly speaking, the absolute levels of net GHG emissions by industrialized nations should decrease in the future. At the same time, while corresponding emissions from developing countries will continue to grow, the rate of growth must decline. This is illustrated in Figure 8.2, which shows the level of development (measurable by a variety of indicators, including per capita income) in relation to net GHG emissions per capita for both developing and industrialized nations. If historical trends continue, both groups of countries will shift in the northeasterly direction (as shown by the solid arrows), as development and emissions both increase. The crucial point is that unless industrialized countries break the trend, the developing world is almost certain to follow the same developmental pattern.

By contrast, a systematic and early effort by industrialized economies to reduce their GHG emissions significantly, while continuing to develop (as shown by the upper dashed arrow in Figure 8.2), is far more likely to induce a favorable change in the growth pattern of developing countries – where development will take place, but with reduced GHG emissions (as shown by the

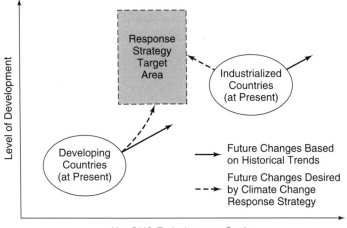

FIGURE 8.2. Desirable directions of change to reduce the risks of global climate change. From Munasinghe (1995).

lower dashed arrow). Thus, an important aim of the climate change response strategy would be to induce all countries to move toward the desired target area (as indicated by the striped rectangle).

BOX 8.6 A Summary of Climate Change

Present State of Knowledge

The balance of evidence suggests a discernible human influence on global climate.

During the past hundred years, the global mean surface air temperature has increased between 0.3 and 0.6°C, while the mean sea level has risen between 10 and 25 cm. Both the horizontal and vertical variations in the patterns of global temperature change, as well as seasonal climate changes, are consistent with the trend of increasing human intervention over time. Nine of the warmest years in the past century occurred after 1980, and glaciers are retreating worldwide. The accumulation of atmospheric GHGs (chiefly CO_2) makes the earth a less efficient emitter of energy back into space. Therefore, the atmosphere tends to warm through the greenhouse effect in order to maintain the essential balance between solar radiation absorbed and radiation reemitted. Aerosols have an opposite (cooling) effect in more localized regions. Energy and land use are the main human activities that contribute to the emission of GHGs.

Potential Future Climate Scenarios

If present trends continue, both GHG concentrations and the impact of climate change will increase significantly in the next century.

If there is no specific human strategy for mitigating climate change, typically the atmospheric concentration of CO_2 will increase from the present 360 ppmv to between 500 and 900 ppmv by 2100 – depending on the rates of global economic and population growth, energy prices, and the application of new technologies. Consequently, the global mean surface temperature will increase between 1 and 3.5°C, and the mean sea level will rise between 15 and 95 cm, during the same period. Because of the thermal inertia of the oceans and the century-long lifetimes of atmospheric GHGs, both the temperature and sea level will continue to rise long after the GHG concentration in the atmosphere has stabilized. Precise predictions are difficult to make because of the complexity of analyzing interlinked geophysical, ecological, and socioeconomic systems on a planetary scale over very long periods of time. Uncertainty, irreversibility, and nonlinear phenomena further complicate matters.

Impact on Ecological and Socioeconomic Systems

Significant widespread effects are likely, with the most severe effects occurring in the tropics.

The impact on ecological systems will be significant, because the projected rate of global mean temperature change will be the highest observed in the past 10,000 years. Typically, forest zones will shift poleward between 150 and 650 km, causing short-term dieback, and coral reefs will suffer severe adverse effects. Flooding of coastal areas due to sea level rise will lead to forced displacement of tens of millions of people, with small island states and river deltas being most severely affected. Increases in vector-borne diseases, particularly malaria, will affect tens of millions of people in the tropics. Changes in global aggregate food production are uncertain, but agriculture in the tropics and subtropics will suffer adverse effects. Changes in the worldwide hydrological cycle will reduce the availability of fresh water in water-scarce regions. Financial, institutional, and human resource shortages will make the developing countries most vulnerable to climate change. Globally, a doubling of atmospheric CO_2 equivalent by 2100 may cause annual damage of between 1 and 2% of GDP, but damage in some developing countries could be as high as 10% of GDP.

Implementing Mitigation Measures

Balanced use of technological options is desirable within an emerging global institutional framework for collective decision making.

The precautionary principle indicates that uncertainty about the future climate does not constitute valid grounds for delay or inaction. "No-regrets" technological and policy measures are presently available worldwide that could significantly reduce GHG emissions in a cost-effective manner, especially in relation to energy and land use. However, achieving even a modest target for stabilization of atmospheric GHG levels within the next 150 years will require significant expenditures of resources and a firm political commitment to accelerate the development and application of new technologies and policies. The situation requires a collective global decision-making process that incorporates both efficiency and equity considerations. At present, the principal institutional framework for addressing climate change issues is the UNFCCC, currently ratified by 164 nations. The Annex I countries may agree to more stringent GHG emission reduction targets at the UNFCCC CoP meeting in Kyoto in December 1997.

Among the specific steps that might constitute such a response strategy are many no-regrets measures, which could be undertaken in the near future. Furthermore, the risks of relatively severe climate change damage in the future, the weight given to the precautionary approach, and the significance of risk aversion in climate change decision making suggest that further actions to reduce net GHG emissions (beyond the no-regrets options) may well be justified, and at least should be considered seriously. A flexible response strategy that incorporates a portfolio of mitigation, adaptation, and other policy measures described in this chapter, which can be updated continuously in the light of new knowledge, appears to be the most prudent approach. Greater reliance on economic incentives and coordination of such policy instruments with noneconomic measures can reduce the costs of both mitigation and adaptation to climate change. In particular, the price system may be used to provide long-run signals to economic agents that will persuade them to adjust cost effectively by reducing net GHG emissions and investing in the research and development of new technologies.

Given the large disparities in the costs of reducing net GHG emissions in different countries, bilateral, regional, and multilateral cooperation among nations could reduce the overall global costs of GHG emission reduction and sink enhancement. In seeking a global consensus, it is especially important to address equity issues, because those who are responsible for the problem and those who will suffer the consequences are different, and are separated both geographically and over time. In particular, both the collective decision-making process and the choices made should incorporate equity considerations – not only for ethical and legal reasons, but also to improve the chances of reaching international agreement and implementing those decisions.

Both research and data gathering appear to have an especially high payoff in terms of reducing uncertainty and improving the effectiveness of response strategies. In the context of the chain of causality associated with climate change (see Figure 1.1), better scientific knowledge about the links between net GHG emissions, atmospheric concentrations, geophysical changes in temperature and sea level, and effects on ecological and social systems is crucial for improving the effectiveness of future decision making. A better understanding of the behavioral responses that determine the reactions of socioeconomic systems to both climate change and policies is also important.

In the context of global optimization of future net emissions and concentration profiles (see Chapter 3), improving the economic valuation of both the costs of mitigation measures and the avoided climate change damage is another worthwhile objective. For example, in order to go beyond an absolute standard for desirable GHG concentrations (based on scientific assessment and a precautionary approach) toward a concentration target based on affordability will require better knowledge of long-range mitigation costs. Such a process might be used to refine a trade-off curve of response costs versus a damage index, along the lines set out in Figure 8.1. In order to proceed even further to a preliminary rough determination of economically justified concentrations – based on comparing economic costs and benefits of a response strategy – will certainly require much better estimates of the monetary costs of climate change effects.

While climate change could have potentially serious effects on physical, ecological, and socioeconomic systems, other phenomena – often not directly related to climate change – are likely to have an equally significant impact. For example, unsustainable agricultural practices will continue to be a major cause of land degradation and desertification. Therefore, it is important to explore the scientific and policy interlinkages among environmental problems (including global ones, like climate change, ozone layer depletion, biodiversity loss, and oceanic pollution, as well as regional and local issues, like desertification and air pollution) that threaten the earth's life support system (i.e., terrestrial and marine ecosystems, atmosphere, lithosphere, and hydrosphere). In brief, climate change analysis should be an important part of an integrated approach aimed at addressing a broad range of sustainable development issues.

In the final analysis, climate change presents humankind with an unprecedented set of problems, which will require significant changes in the way in which the global community makes and implements decisions. Ultimately, an effective response strategy will require not only a portfolio of innovative technological and policy options, but also firm political leadership as well as fundamental changes in human attitudes and behavior. While there is a considerable amount of uncertainty, we believe that a carefully crafted climate change strategy will provide ample opportunities for enhancing the prospects for sustainable development worldwide.

REFERENCES

Barret, S. 1994. Climate change policy and international trade. In *Climate change: Policy instruments and their implications.* Proceedings of the Tsukuba Workshop of IPCC Working Group III, Tsukuba, Japan, January, pp. 17–20.

Bruce, J. P., H. Lee, and E. F. Haites. 1996. Summary for policymakers. In J. P. Bruce, H. Lee, and E. F. Haites (eds.), *Climate change 1995: Economic and social dimensions of climate change.* Contribution of Working Group III to the Second Assessment Report of the Intergovernmental Panel on Climate Change. Cambridge University Press, pp. 1–16.

CCAP (Center for Clean Air Policy) / SEVEn. 1996. *Joint implementation projects in Central and Eastern Europe: Description of ongoing and new projects.* Prague: CCAP/SEVEn, pp. 37–40.

EMF (Energy Modeling Forum). 1993. *Reducing global carbon emissions: Costs and policy options.* EMF-12, Stanford University, Stanford, CA.

Hanslow, K., M. Hinchy, J. Small, and B. Fisher. 1994. International greenhouse economic modeling. Paper presented at the Greenhouse Gas Conference, Wellington, October. ABARE Conference Paper 94.34, Australian Bureau of Agriculture and Resource Economics, Canberra.

Houghton, J. T., et al. 1996. *Climate change 1995: The science of climate change.* Contribution of Working Group I to the Second Assessment Report of the Intergovernmental Panel on Climate Change. Cambridge University Press.

JIN Foundation. 1995. *Joint Implementation Quarterly* (Groningen, the Netherlands), **1**(1), 7–8.

JIN Foundation. 1997. *Joint Implementation Quarterly* (Groningen, the Netherlands), **3**(1), 14.

Jorgenson, D. C., and P. Wilcoxen. 1992. *Reducing U.S. carbon dioxide emissions: An assessment of different instruments.* Discussion Paper no. 1590, April, Harvard Institute of Economic Research, Cambridge, MA.

Larsen, B., and A. Shah. 1992. *World fossil fuel subsidies and global carbon emissions.* Policy Research Working Paper Series no. 1002. Washington, DC: World Bank.

Larsen, B., and A. Shah. 1994. *Global tradable carbon permits, participation incentives and transfers.* Oxford Economic papers, **46**, 841–856, Oxford University.

Munasinghe, M. 1995. *Sustainable energy development.* Environment Department Paper no. 15. Washington, DC: World Bank.

OECD (Organization for Economic Cooperation and Development). 1994. *GREEN: The reference manual.* OECD Department of Economics, Working Paper no. 143. Paris: OECD.

Scheraga, J., and N. Leary. 1992. Improving the efficiency of policies to reduce CO_2 emissions. *Energy Policy,* **20**(5), 394–404.

UNFCCC. 1992. Framework Convention on Climate Change. FCCC Secretariat, Bonn.

Whalley, J., and R. Wigle. 1992. *Results for the OECD comparative modelling exercise from the Whalley–Wigle model.* OECD Economics Department Working Paper no. 121. Paris: OECD.

INDEX

CONCEPTUAL FRAMEWORKS IN GEOGRAPHY
General Editor: W.E.Marsden

Process and Landform:
An Outline of Contemporary Geomorphology

Alan Clowes and Peter Comfort
Both authors teach at King George V College, Southport

Maps and diagrams drawn by Tim Smith

Oliver & Boyd

Acknowledgements

The authors and publishers wish to thank all those who gave their permission for us to reproduce copyright material in this book. Information regarding sources is given in the captions.

Oliver & Boyd
Robert Stevenson House
1–3 Baxter's Place
Leith Walk
Edinburgh EH1 3BB

A Division of Longman Group Limited

ISBN 0 05 003127 9

First published 1982
Third impression 1983

Printed in Singapore by
Selector Printing Co.